BUSINESS/SCIENCE/TECHNOLOGY DIVISION
CHICAGO PUBLIC LIBRARY
400 SOUTH STATE STREET
CHICAGO, IL 60605

CHICAGO PUBLIC LIBRARY

R00582 77520

REF
TS
156.8
.A534
1974

Andrew, W. G.
(William G.)

Applied
instrumentation in
the process
industries

$~~25.~~95

DATE			

D1294195

© THE BAKER & TAYLOR CO.

Applied Instrumentation in the Process Industries

Volume II

Second Edition   Practical Guidelines

Applied Instrumentation in the Process Industries

Volume II
Practical Guidelines

Second Edition

W.G. Andrew
H.B. Williams

Gulf Publishing Company
Book Division
Houston, London, Paris, Tokyo

To my wife, Edna Williams,
and children, Raymond, Rhonda, and Sondra,
for their encouragement and loving kindness

Applied Instrumentation
in the Process Industries
Volume 2/Second Edition

Practical Guidelines

Copyright © 1974, 1980 by Gulf Publishing Company, Houston, Texas. All rights reserved. Printed in the United States of America. This book, or parts thereof, may not be reproduced in any form without permission of the publisher.

Library of Congress Cataloging in Publication Data (Revised)

Andrew, William G
    Applied instrumentation in the process industries.

    Includes index.
    CONTENTS: v. 1. A survey.—v. 2. Practical guidelines.
    1. Process control. 2. Automatic control. 3. Engineering instruments. I. Williams, H. B., 1939–   joint author. II. Title.
TS156.8.A534 1979 660.2′81 79-9418
ISBN 0-87201-382-0 (v. 1)

First Edition
First Printing, *May 1974*
Second Printing, *August 1979*

Second Edition
First Printing, *June 1980*
Second Printing, *July 1986*

BUSINESS/SCIENCE/TECHNOLOGY DIVISION
CHICAGO PUBLIC LIBRARY
400 SOUTH STATE STREET
CHICAGO, IL 60605

Preface

Volume 2, *Practical Guidelines,* is unique in the field of instrumentation literature. In a discipline where so little formal education exists, little writing has been done to communicate the knowledge of the art that accumulates through the experience and knowledge of the practioners. *Practical Guidelines,* for the first time, formally sets forth information, suggests methods, makes comparisons, issues principles and provides guidelines for those who want to master the field of instrumentation quickly and thoroughly.

The treatment is organized from a practical viewpoint; first suggesting the scope of work to be considered on any project involving instrumentation. Design concepts are covered from a broad view to the small details that are a necessary part of detailed design. Comparisons are made of many measurement and control devices with detailed listings of their advantages and disadvantages. The comparisons include flow, level, pressure and temperature measuring methods and control devices such as control valves and relief valves. One chapter is devoted to the design of relief systems. Another chapter is devoted to comparisons of several control valve sizing formulas. Suggestions are made concerning their use.

Other chapters are devoted to topics that seldom receive broad coverage in trade and professional magazines. Topics discussed are: the design of sample systems for analytical devices, control panel designs, instrument air systems, including compressors, driers and distribution systems.

One chapter lists the variations of accuracies obtainable from most of the measurement and control instruments. It also discusses in considerable detail other sources of error that are associated with control systems.

Chapter 10 places special emphasis on an area that is particularly troublesome for instrument people—the selection of measurement and control devices for slurry service applications. Advantages and disadvantages of different types are compared and helpful suggestions are made for slurry and heavily viscous fluid applications.

Chapter 12 of this unique book discusses construction, calibration and startup activities. This facet of instrument work is part of every processing plant project. Little is written, however, about the organization and execution of this important phase of activity. This chapter offers many ideas and suggestions that will help the initiated as well as the uninitiated to successfully execute this work phase.

The final chapter provides an in-depth study of many of the factors which surround the selection and installation of a process oriented computer system. Details of customer/vendor bargaining sessions are treated as a matter of fact.

Acknowledgments

Any technical book draws material from a large number of sources. Although many of these are referenced in the text, it is not feasible to include all the contributors to whom the authors are indebted. Data and information were furnished by many industrial companies.

The authors were encouraged to undertake the work by A.C. Lederer, former president of S.I.P., Inc. The cooperation of W.L. Hampton, Manager of Engineering, is gratefully acknowledged in producing the work on schedule.

Appreciation is extended particularly to M.J. Sandefur for an incisive treatise on computer system projects, and to B.J. Normand and K.G. Rhea for time spent in reviewing and criticizing many chapters and sections of the manuscript. Others who contributed in this area include L. Ashley, W.E. DeLong, D.M. Dudney, L.C. Hoffman, T.E. Lasseter, and J.G. Royle.

In addition, the authors are deeply appreciative of S.I.P., Inc. and its staff for providing the environment and materials for producing this work.

On the second edition: Mr. H.B. Williams gives his profound appreciation to all who have assisted in preparing this revised text—those who have contributed technical information; those who have typed and proofread; those who willingly provided illustrations, suggestions, and encouragement; and to all who have, by their confidence, inspired this effort.

Contents

Safety Considerations 106, Special Purpose Valves 107, Low Flow Applications 107, High Pressure Drop Applications 109

5
Control Valve Sizing 111

6
Pressure Relief Systems. 130

7
Application Guidelines
for Analytical Systems. 179

8
Control Panels 197

9
Instrument Air Systems 220

10
Slurry Service Applications 236

11
Accuracies and Errors. 258

1 Instrument Project Control

William G. Andrew

A guide to the overall duties and responsibilities of the instrument group is needed by those who do the work as well as those who must work with them in any way. Probably, the nature of work done by instrument people is more misunderstood than any other discipline in the engineering and building of process units. The misunderstanding comes generally because few people understand the amount of detailed information that must be assembled, digested and used to apply instrumentation properly to a process.

Guidelines and suggestions are given here which outline the organization and execution of an engineering project as it relates to instrumentation. It lists documents which are necessary for communication between owner and builder, communication within the contractor organization during the engineering phase and the necessary information for construction. Coordination requirements with other groups are discussed, planning hints are given, and a check list for project control is suggested to ensure the successful completion of the project. The principles outlined apply to projects whose capital investments range from a few hundred thousand dollars to the largest of projects.

The viewpoint presented is most applicable to the person responsible for the application, selection, purchasing and installation of instruments and control systems. The documentation necessary for a complete record of a project is outlined. Even though the viewpoint is slanted toward the responsible project instrument engineer, other owner and contractor personnel—project managers, process people and leaders of other disciplines—will profit by an understanding of how the job is organized, developed and executed.

Many jobs appear complex until they are organized and broken down into components or units easily understood and accomplished. This is certainly true of instrument work.

Specific documents are required for the work—for installation and later for maintenance. Scheduling, purchasing, installation and calibration must be done. The following paragraphs discuss what these functions are and how they are carried out.

Documents to be Produced

Most projects require the following engineering documents for a complete job. When the job is small, some of the functions may be combined in the interest of space, time and economy. On large jobs, additional documents may be needed. Generally, however, requirements adhere quite closely to those mentioned below.

Process Flow Sheets

Process flow sheets consist of a pictorial representation of the major pieces of equipment required with major lines of flow to and from each piece (Figure 1.1). Material balances generally are shown. Additional information often given includes operating conditions at various stages of the process (flows, pressures, temperatures, viscosities, etc.), equipment size and configuration and, in some cases, utility requirements. Instrumentation on process flow sheets may or may not be essentially complete. In some instances, practically all of the instrumentation is included; in others, only the major control systems are shown.

In most processes the primary control variables have been determined and verified through laboratory and/or pilot-plant operations if the process is new. On old processes, previous commercial operations have verified proper or improper control techniques. The responsible instrument people may be consulted or may offer suggestions to improve

1

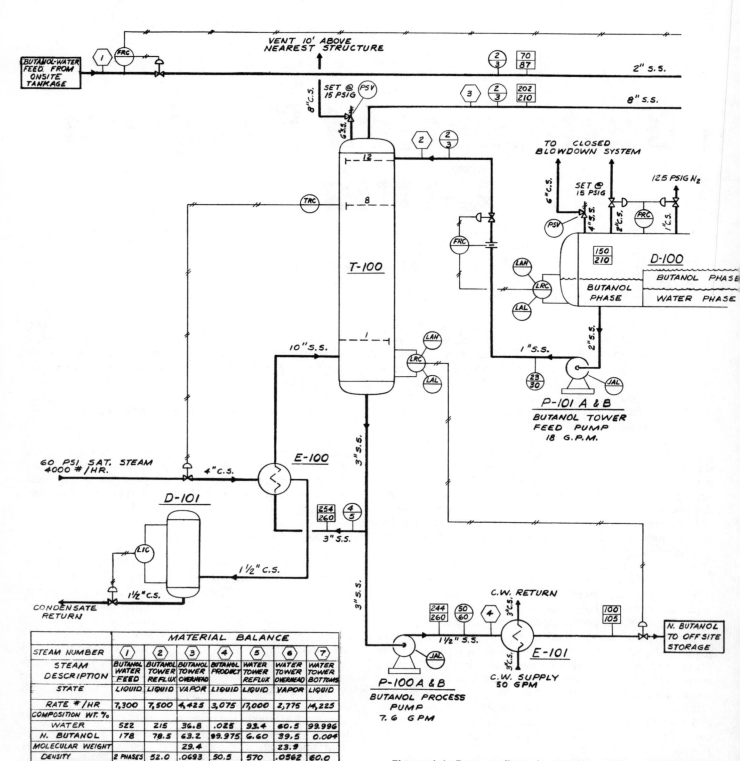

Figure 1.1. Process flow sheets show major process equipment, major connecting lines, material balances and other process information. (Courtesy of S.I.P., Inc.)

MATERIAL BALANCE							
STEAM NUMBER	①	②	③	④	⑤	⑥	⑦
STEAM DESCRIPTION	BUTANOL WATER FEED	BUTANOL TOWER REFLUX	BUTANOL TOWER OVERHEAD	BUTANOL PRODUCT	WATER TOWER REFLUX	WATER TOWER OVERHEAD	WATER TOWER BOTTOMS
STATE	LIQUID	LIQUID	VAPOR	LIQUID	LIQUID	VAPOR	LIQUID
RATE #/HR	7,300	7,500	4,425	3,075	17,000	2,775	14,225
COMPOSITION WT. %							
WATER	522	215	36.8	.025	93.4	60.5	99.996
N. BUTANOL	178	78.5	63.2	99.975	6.60	39.5	0.004
MOLECULAR WEIGHT			29.4			23.9	
DENSITY	2 PHASES	52.0	.0693	50.5	570	.0562	60.0
TEMPERATURE °F	70	116	202	244	116	204	220
PRESSURE PSIG	2.0	2	2	50	2	2	0
VISCOSITY	.0	0.7	—	0.4	0.8		

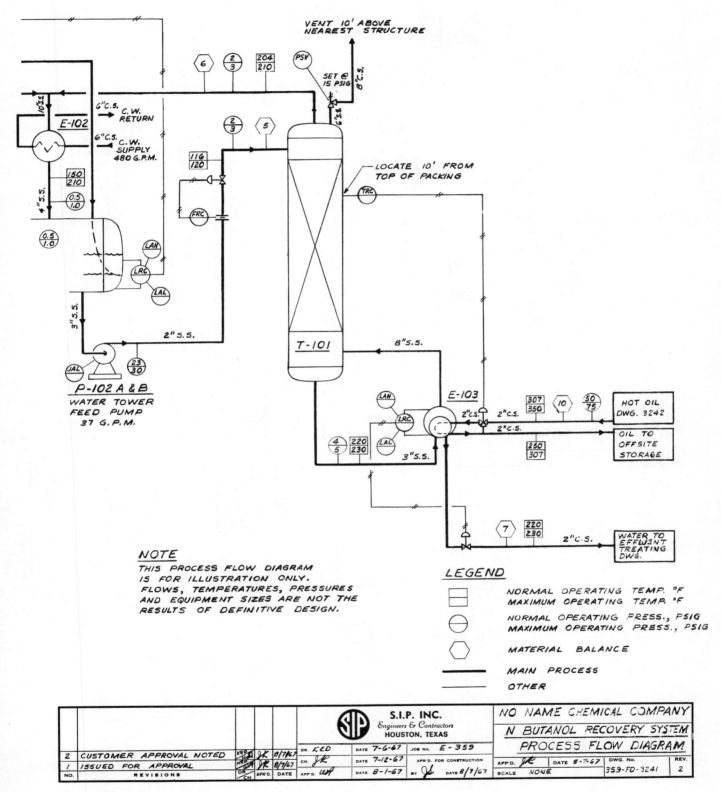

Figure 1.1 continued.

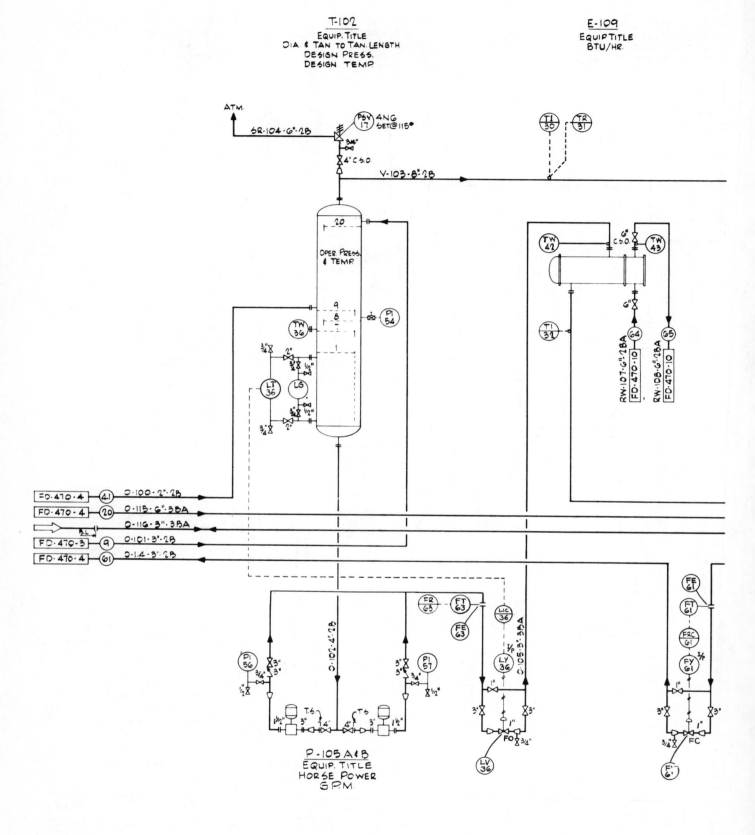

Figure 1.2. Mechanical flow sheets (or P&I diagrams) show detailed mechanical information while omitting much of the process information. (Courtesy of S.I.P., Inc.)

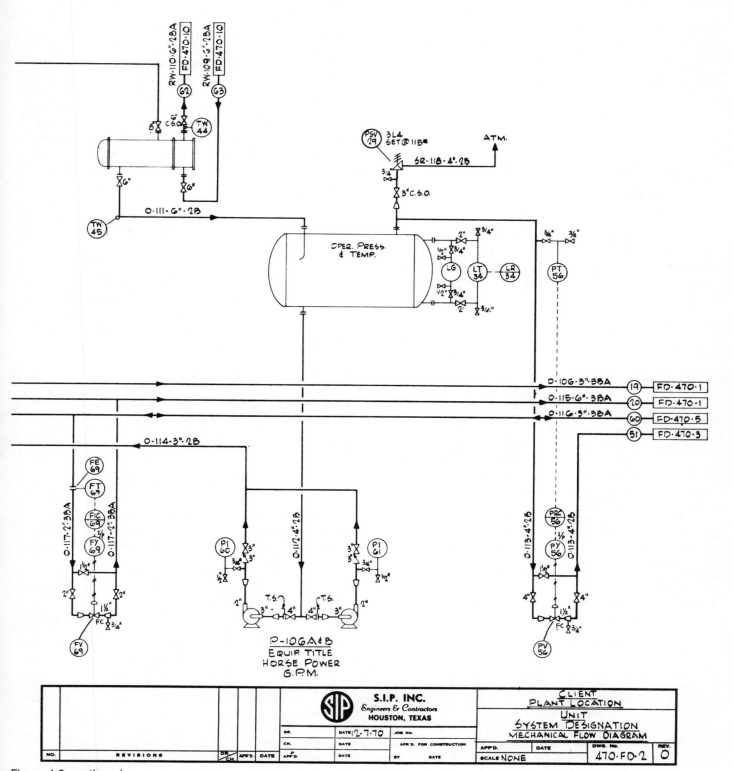

Figure 1.2 continued.

control. The extent of involvement depends primarily on the experience backgrounds of the process and instrument people assigned to the job.

Mechanical Flow Sheets

Mechanical flow sheets or P&I (Process and Instrument) diagrams (Figure 1.2) provide much detailed, mechanical information not shown on process flow sheets, while omitting much of the process information shown on the process flow sheets. They repeat all the major equipment and piping lines as well as show:

1. All other equipment items with design temperatures and pressures
2. All interconnecting piping with size, material and fabrication specifications indicated
3. Utility requirements including pipe sizes, materials and fabrication methods
4. All major instrument devices

In many situations, mechanical flow sheets show schematically every process and utility line that appears on detailed drawings. They provide a valuable reference for proper project installation. The instrument engineer uses it as a source for many documents which must be prepared.

Complete process and mechanical flow sheets are needed prior to the bulk of the engineering effort on a job. Changes are often made as jobs progress, however, particularly on mechanical flow sheets.

Instrument Index Sheets

The instrument index sheets provide a summary of all instruments required for the job, listing each number-identified item of each loop (see Figure 1.3). The list should be made near the start of a job and used to check progress in specification writing, purchasing, expediting, delivery and installation. As items are added, the list increases; deletions should be lined out (not erased), thus serving as a record of changes. Even though information is not complete, the list should be issued early in the job so that project engineers and others concerned with the job may use it to gauge the job requirements.

Instrument index sheets may serve strictly as an index of items required and provide minimum information, or they may be expanded to provide a summary of information about the loop itself, including service conditions of the line or vessel with which it is associated.

Instrument Specification Sheets

To facilitate and speed up the specification and purchase of instrument items, forms have been developed which list the main features available and desirable in various categories of instruments. The Instrument Society of America has been instrumental in this effort and has developed standard forms for 27 categories of instrument items. Figure 1.4 is typical—a specification sheet for pressure instruments. A blank form is included in these standards which can be used to hand-write specifications not already covered on the other 26 forms.

The ISA forms were revised in 1975 and republished in 1976 (reference "Standard ISA-S20", *Standards and Practices for Instrumentation*, 5th Edition, Instrument Society of America, 400 Stanwix Street, Pittsburgh, PA 15222). This upgrading of content and format eliminated many of the deficiencies of earlier forms, although the forms in their present state must still be modified to include information such as electrical area classification and flowsheet numbers. Use of the forms is encouraged by the ISA, and copies are available at nominal cost through their headquarters. The forms may be modified (company headings, new columns, etc.) and reproduced for internal use (not for commercial sale), so long as permission is obtained from the ISA.

Specification sheets serve a fourfold purpose.

1. They contain information relating to the process and/or other instruments which is necessary for complete systems engineering.
2. They provide the purchasing department and other interested people information necessary for fulfilling their jobs efficiently—a communications channel.
3. They serve as permanent records for plant use—for installation, production and maintenance groups.
4. They provide specifications to bidders, using a format familiar to both parties.

Loop Wiring Diagrams

Electronic loop wiring diagrams are electrical schematic drawings which are prepared for individual (or typical) electronic loops. The simplest loop is one that contains only a transmitter and a receiver. Other loops may contain many items—transmitter, recorder, controller, alarm units, control valve, transducer, integrator and perhaps other items.

The amount of documentation on the schematic varies. Some are relatively simple, showing only the locations of the instruments, their identification numbers and termination of the interconnecting wiring. Cable routing, wire size, intermediate terminal points and other pertinent information are necessarily shown on other drawings.

Other loop wiring diagrams are more comprehensive (Figure 1.5), providing not only the information previously described but also showing intermediate junction box terminals properly identified, wire and cable information, complete terminal layout of individual instruments and other useful information. Included may be the transmitter range and calibration and pneumatic hookup information on the transducer and the control valve.

Some loop diagrams are much more complex, especially if the loop contains several components. Complete documentation, however, makes prestartup checkout and maintenance much easier. Since the loop information is well documented, additional drawings are seldom needed for maintenance of the instruments.

INSTRUMENT INDEX

S.I.P. INC.
Engineers & Contractors
HOUSTON, TEXAS

CLIENT A.B.C. CO.
PROJECT Project "B"
LOCATION United States
CLIENT JOB NO. CAR-0A102

S.I.P. JOB NO. E-783

REV.	BY	DATE	APP'D	REV.	BY	DATE	APP'D

SHT. ___ OF ___

TAG NO.	SERVICE DESCRIPTION	LINE OR EQUIP. NO.	MFG.	MOD. NO.	SPEC. SHT.	P.O. NO.	FLOW SHT.	PIPING DWG.	INSTRUMENT DWG. PLAN	INSTRUMENT DWG. INST'L.	INSTRUMENT DWG. LOOP	L.O.C.	SIZE	CALB. OR ACTION	REMARKS
FIC-101	WATER TO STG. TANK T-12	---	FOX	52A	I-12	SIP-1321	D-I-1023	D-P-2041	D-I-1332	SKH-102	SKH-114	P		0-102	
FE-101	"	2"-CS-1201	DANIEL	--	I-12	SIP-1419	D-I-1023	D-P-2041	D-I-1332	SKH-2102	SKH-114	F	2"	0-100" H20	VERTICAL RUN
FT-101	"	2"-CS-1201	FOX	13A	I-15	SIP-1101	D-I-1023	D-P-2041	D-I-1332	SKH-2102	SKH-114	F		100" H20	W/AIRSET PRE-PIPED
FV-101	"	2"-CS-1402	F.G.	657 ES	I-72	SIP-1202	D-I-1023	D-P-2041	D-I-1332	SKH-1135	SKH-114	F	1"	A/O	

Figure 1.3. Instrument index sheets comprise a summary list of all the instruments required for a job and serve as a check list for job progress. (Courtesy of S.I.P., Inc.)

S. I. P., INC. Engineers — Contractors Box 34451 Phone 946-9040 HOUSTON, TEXAS 77034	S.I.P. NO. *E-783* PROJECT *"B"* CLIENT *A.B.C. Co.*		**PRESSURE INSTRUMENTS**			
			SPEC. NO. *II. P-1*			
	LOCATION *UNITED STATES* CLIENT NO. *CAR-0A102*		BY	DATE	ITEM	SHT. *1 of 1*

	1	Tag No. *PT-122*	Service *C-32 PRESSURE*
GENERAL	2	Function	Record ☐ Indicate ☐ Control ☐ Blind ☐ Trans ☒ Other _____
	3	Case	MFR STD ☒ Nom Size _____ Color: MFR STD ☒ Other _____
	4	Mounting	Flush ☐ Surface ☐ Yoke ☒ Other _____
	5	Enclosure Class	General Purpose ☐ Weather proof ☐ Explosion proof ☒ Class *CL.1, GR.D, DIV.II* For Use In Intrin. Safe System ☐ Other _____
	6	Power Supply	117V 60Hz ☐ Other ac ☐ _____ dc ☒ *24* Volts Pneumatic ☐
	7	Chart	_____ Strip ☐ _____ Roll ☐ _____ Fold ☐ _____ Circular _____ Time Marks _____ Range _____ Number _____
	8	Chart Drive	Speed _____ Power _____
	9	Scales	Type _____ Range 1 _____ 2 _____ 3 _____ 4 _____
XMTR	10	Transmitter Output	4-20 mA ☒ 10-50 mA ☐ 21-103 kPa (3-15 psig) ☐ Other _____ For Receiver See Spec Sheet *II. R-3*
CONTROLLER	11	Control Modes	P=Prop (Gain) I=Integral (Auto-Reset) D=Derivative (Rate) Sub: s=Slow f=Fast P ☐ PI ☐ PD ☐ PID ☐ If ☐ Df ☐ Is ☐ Ds ☐ Other _____
	12	Action	On Meas. Increase Output: Increases ☐ Decreases ☐
	13	Auto-Man Switch	None ☐ MFR STD ☐ Other _____
	14	Set Point Adj.	Manual ☐ External ☐ Remote ☐ Other _____
	15	Manual Reg.	None ☐ MFR STD ☐ Other _____
	16	Output	4-20mA ☐ 10-50mA ☐ 21-103 kPa (3-15 psig) ☐ Other _____
ELEMENT	17	Service	Gage Press. ☒ Vacuum ☐ Absolute ☐ Compound ☐
	18	Element Type	Diaphragm ☐ Helix ☐ Bourdon ☒ Bellows ☐ Other _____
	19	Material	316 SS ☐ Ber. Copper ☐ Other *NI-SPAN C W/ 316SS CONN'S*
	20	Range	Fixed ☐ Adj. Range *1000-6000* Set at *0-5000 PSIG* Overrange protection to *9000 PSIG*
	21	Process Data	Press: Normal *4000* Max *4500* Element Range *6000*
	22	Process Conn.	¼ in. NPT ☐ ½ in. NPT ☒ Other _____ Location: Bottom ☐ Back ☐ Other *SIDE*
OPTIONS	23	Alarm Switches	Quantity _____ Form _____ Rating _____
	24	Function	Press ☐ Deviation ☐ Contacts To _____ on Inc Press.
	25	Options	Filt-Reg. ☐ Sup Gage ☐ Output Gage ☒ Charts ☐ Diaph Seal ☐ Type _____ Diaph _____ Bot Bowl _____ Conn _____ Capillary: Length _____ Mtl. _____ Other _____
	26	MFR & Model No.	*FOXBORO E-11GH-11NM-2*

Notes: Flow Sheet *D-I-1023*

NO	REVISIONS	BY	DATE	APPVD	DATE
△					

REPRINTED AND MODIFIED WITH PERMISSION OF THE COPYRIGHT HOLDER: © INSTRUMENT SOCIETY OF AMERICA

Figure 1.4. This pressure instrument specification sheet is typical of 27 instrument specification sheets available from the ISA.

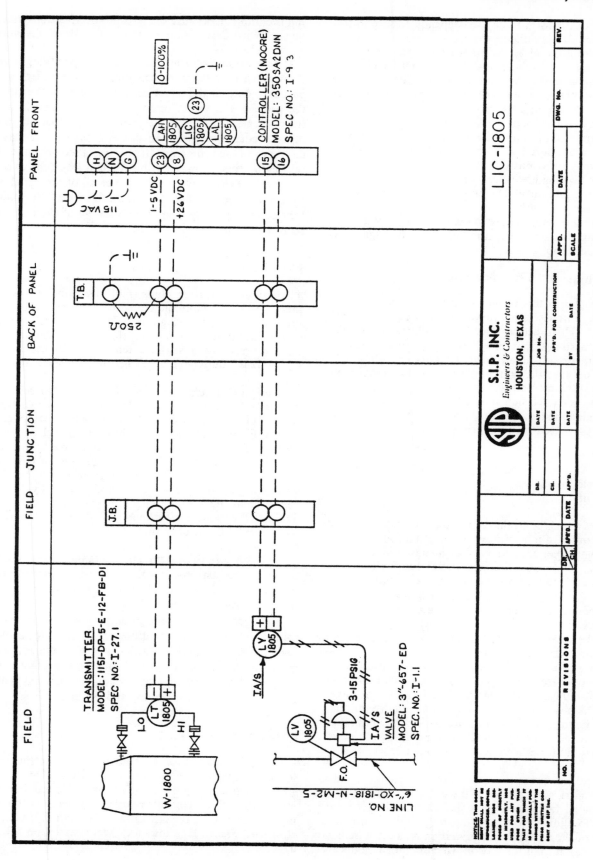

Figure 1.5. Electronic loop wiring diagrams provide detailed hookup information for all loop components—in the computer room, at the control panel and in the field. (Courtesy of S.I.P., Inc.)

The choice of the form to be used depends on several factors—the needs of the builder, the records required and the extent of information given on other drawings. Loops may be drawn on 8½ x 11-inch sheets, 11 x 17-inch or larger if preferred. Loops similar to that shown in Figure 1.5 are often drawn on "D" size sheets (24 x 32 inches). One to four loops may be shown on a "D" size drawing, depending on the complexity of the loop.

The preparation of loop wiring diagrams normally is needed prior to the purchase of the central control panels to furnish wiring information to the panel fabricator. Complete field wiring identification may not be available at this stage of the job, but the essential facts pertinent to the panel fabricator will be known.

On jobs using pneumatic instruments, loops may also be drawn showing all instruments and their hookup. Figure 1.6 shows a typical loop.

Panel Drawings and Specifications

Another chapter of this volume discusses in detail various panel arrangements and layouts as well as panel specifications. This section does not duplicate these discussions but assumes that panel arrangement and layout philosophy are already determined and work execution only is needed. Panel specifications and instrument specifications for all panel instruments must be prepared.

Early in the job a decision should be made concerning the extent of engineering to be done by the panel fabricator. This decision is based on four primary factors.

1. Cost
2. Time
3. Available manpower
4. Capability

The factors are so interrelated that they need not be listed in a definite order of importance. Fabrication cost is always important, but other factors such as installation schedule, startup time, etc., can minimize initial fabrication cost. Timing and cost, then, must be considered together.

If the timing or schedule is critical, panel fabrication may be started before engineering is complete with a gain of several weeks in panel delivery. The added cost of such an action may be small compared to the advantage of earlier onstream commitments or other factors.

Capability and availability of manpower are also closely related. In most cases the buyer knows the panel fabricator well enough to judge his engineering capabilities. The buyer also knows his own department's workload and capabilities. From these factors, he then must determine the amount of engineering to be done by each. In some cases the panel fabricator will do little or none of it; in other cases, he may do all of it.

Having decided which route to take, the engineer must schedule the work accordingly. If the detailed engineering is left to the panel fabricator, the contract must be let at an earlier date to allow time for engineering by the fabricator. The quality of the written panel specifications must be better for this option than for one on which detailed engineering is complete, because the written specifications must convey in word description what detailed drawings normally show pictorially.

If detailed engineering is done by the specifying engineer, the written specifications can often be simplified, and the purchase date of the panel can occur later in the course of the project. For either option, scheduling and planning of the work associated with the panel must be coordinated with other events of the job. In either case, the purchase and delivery of instruments to be installed by the panel fabricator must be made with sufficient lead time to prevent delay of panel completion. Some or all of the panel instruments may be purchased by the fabricator. Normally the fabricator buys only a small percentage of the installed instrument items.

Plot Plans

Instrument location plans (plot plans) include all instrument items that interconnect with other instruments. They usually do not show single items (such as level or pressure gauges and in-line rotameters) that connect only to process lines or vessels because those items normally appear on piping and/or equipment drawings and on mechanical flow sheets.

Many instrument items are shown on two sets of plan drawings because their installation involves two different crafts—pipefitters and electricians. Electrical plot plans normally show all items that have electrical connections. Instrument plot plans show items that have both electrical and piping connections. The drawing scale normally used is ¼ or ⅜ inch per foot. An advantage of the ⅜-inch scale is that it duplicates the scale normally used by the piping department. Background layouts may then be traced or duplicated with little effort and cost. The location of instrument items also is easy since many of them are shown on piping drawings.

Location of all instrument junction boxes should be shown on the instrument plans. Instrument cable and tubing trays, if not shown on electrical drawings, may also be shown on the instrument plans.

Underground cable and tubing runs must be carefully routed. Preliminary locations and routing should be reviewed by piping, electrical and civil engineering design disciplines to avoid conflicts during construction with new design and also existing pipes, conduits, electrical grounding systems, foundations, drains, and sewers.

Installation Details

Installation details or sketches are needed to show mounting and piping methods and preferences. They show the size, pressure rating and type of materials required for the installation. Sufficient information is given to allow correct installation even if the craftsman is not familiar with the instrument. Figure 1.7 shows a typical installation detail for a d/p cell.

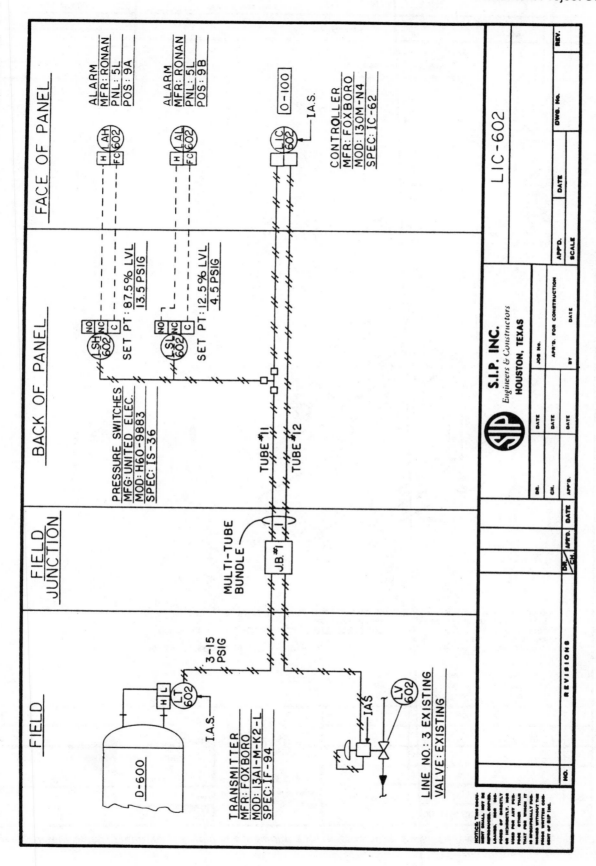

Figure 1.6. Pneumatic loop diagrams provide detailed connection information for all components of pneumatic loops.

NOTE: THIS DATA IS OF A CONFIDENTIAL NATURE AND IS THE PROPERTY OF S.I.P., INC. HOUSTON, TEXAS, U. S. A. AND SHALL NOT BE TRACED, PHOTOGRAPHED, PHOTOSTATED OR REPRODUCED IN ANY MANNER, NOR USED FOR ANY PURPOSE WHATSOEVER EXCEPT BY WRITTEN PERMISSION OF S.I.P., INC.

BILL OF MATERIAL

S.I.P. INSTRUMENT PIPING SPECIFICATION _____

INSTALLATION TYPICAL FOR

MARK NO.	QTY.	PART NO.	DESCRIPTION
1			1/2" O.D. TUBING
2	4		1/2"T X 1/2" NPT MALE CONN.
3	1		1/2" NPT 3-VALVE MANIFOLD

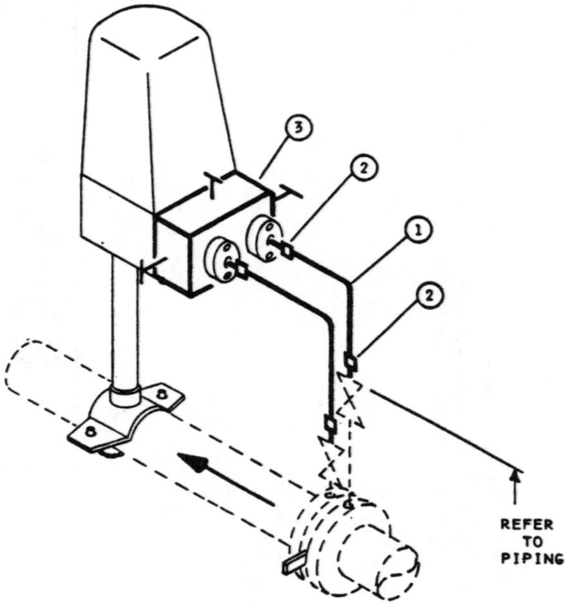

REFER TO PIPING

NOTES:
1.) SLOPE 1" PER FOOT MINIMUM "DOWN" TO INSTRUMENT CONNECTIONS. DO NOT EXCEED 25' LENGTH NOR ALLOW POCKETS OR TRAPPED SECTIONS.
2.) REFER TO SIP INSTRUMENT PIPING SPECIFICATION (PER ABOVE) FOR MATERIALS AND RATINGS FOR THIS DRAWING.
3) CLOSE- COUPLED IS DEFINED AS THE DEVICE BEING MOUNTED WITHIN 4 FEET OF PROCESS CONNECTION.

STANDARD APPROVAL: _____ ;DATE _____ ;BY _____ ;REV. _____ ;DATE _____

REV.	DESCRIPTION	APP.	APP.	APP.	DATE
	PROJECT REVISIONS				

S.I.P. INC.
Engineers & Contractors
HOUSTON, TEXAS

CLIENT _____
PROJECT _____

STANDARD INSTRUMENT INSTALLATION
D/P CELL
GAS SERVICE - CLOSE COUPLED

S.I.P. JOB #	DWG. NO.	REV.
INSTR. STD. 02.14.103		

Figure 1.7. Installation details show the physical arrangement for equipment and piping of instrument items. Materials are also identified. (Courtesy of S.I.P., Inc.)

Typical connection details are also made for instrument electrical connections. Normally these show the proper entry to the instrument, the size and number of wires needed, the size and type of conduit used and the electrical fittings required. Figure 1.8 is typical of this type detail.

These sketches, properly drawn, save many hours of installation labor and ensure proper installation and operation even when installed by inexperienced craftsmen. They also provide a good basis for material take-off and purchase and material inventory control during the building phase.

Special Drawings

In addition to those previously listed, some or all of the following drawings are needed for many jobs.

Instrument Tubing Support Layout

Multiconductor tubing bundles of copper, aluminum or plastic normally are used for interconnecting pneumatic instruments. The tubing bundles are installed between a central control room and field junction boxes which serve as distribution points for instruments in their area. Typical supports for tubing bundles are shown in Figure 1.9. There are several support forms available, including prefabricated metal trays of various sizes and configurations. If desired, field fabricated supports may be used, but the high cost of field labor usually dictates the use of prefabricated forms. On the support detail of Figure 1.9 or on other drawings, junction box details show the incoming tubing bundles and show how they connect to the single tubing runs to individual instruments. Supports for single conductor tubing runs may consist of a smaller size tray of the same type used for multiconductor bundles, or angle iron or small channel iron may be used.

Schematic Control Diagrams

Schematic control diagrams are required for both electrical and pneumatic circuits when the complexity of the system requires additional emphasis to explain and illustrate the functions of the various components. Electrical schematics are made to show all the electrical devices used and are helpful for a quick understanding of the system in calibration, troubleshooting and checkout.

Schematics of pneumatic systems are also helpful, especially when a number of devices such as relays, solenoid valves, integrators and other similar devices are used along with the usual recorder-controller or indicator-controller system.

Emergency Shutdown Systems

The hookup and operation of an emergency shutdown system is a typical example of drawings needed for a special purpose. Because of the nature of the system, many people are required to know how the system operates, including those who check it out periodically and maintain it. This type drawing gives some of the hookup details, schematically shows the entire system and provides additional operating information for a complete understanding of the system. It provides information not usually shown on the other type

drawings, including equipment information, sequence of operation and other operating instructions.

Instrument Wiring Details

Point-to-point wiring details such as cable terminations at control panels or terminal boxes, thermocouple connections at junction boxes or interconnection of instruments behind control panels are needed so that inexperienced craftsmen can wire complicated systems without difficulty. Detailed wiring information allows an entire job to be shown completely and wired easily. The primary requirement for understanding is the ability of the craftsmen (or others) to read and interpret drawings.

This type drawing differs from schematics in that it indicates physical arrangement without showing functions of control devices. Schematics reveal an understanding of the function of control systems but do not reveal the physical layouts of the wiring. Each drawing serves its distinct purpose.

Alarm and Shutdown List

This document is probably less likely to be made than any of the others listed. An alarm and shutdown list, however, is convenient and time saving, especially if the process has a larger number of alarms and interlocks.

Figure 1.10 shows a list made in alphabetical and numerical sequence which gives item numbers, service, switch action, location, setting, function and other useful information. Proper notations can be made to show installation, calibration and checkout so that the document provides a convenient record of job progress related to these functions. A properly completed list also serves as a record for use by the production and maintenance group.

Purchase Requisitions

The purchase requisition is the line of communication between the one who specifies and the one who buys instruments. Delivery requirements and other pertinent information are passed on to the purchasing agent through the requisition forms.

If the instrument specification sheet accompanies the requisition, the requisition need not duplicate the information on the specification sheet. If the requisition is used alone, then it must necessarily contain all the information needed for purchasing. It should also contain tagging information for instrument identification, pricing information (when available), delivery requirements (time and place) and pertinent clauses relating to correspondence, operating instructions, spare parts list and drawings (if required).

Other Documents

The documents listed will suffice for most jobs. Occasionally more may be required. For example, general construction specifications may be needed to describe in written form how the work is to be performed. Complicated control schemes may need written descriptions to accompany drawings for clarification. Common sense and logic will determine whether additional documentation is needed.

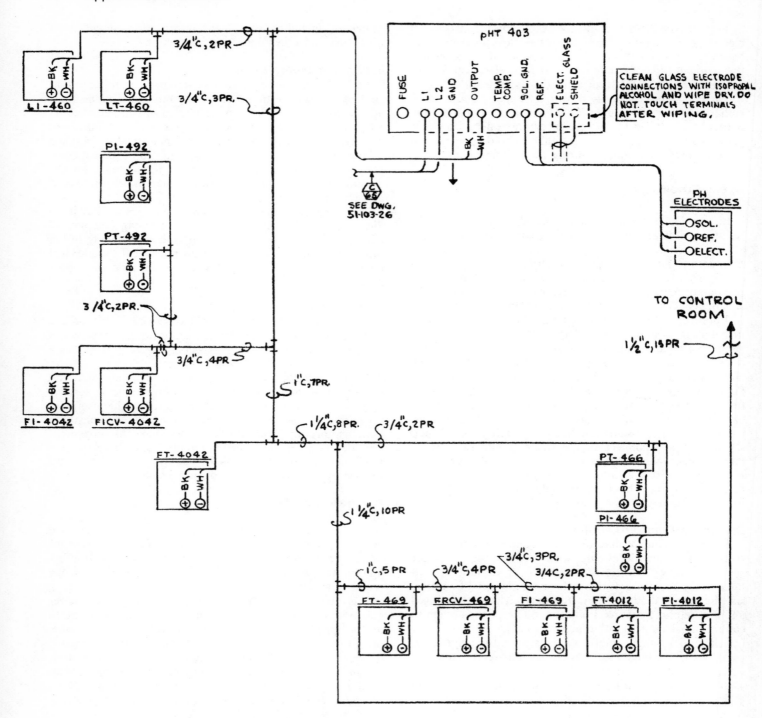

Figure 1.8. Detailed electrical hookup wiring diagrams are also required to ensure proper wiring of control systems.

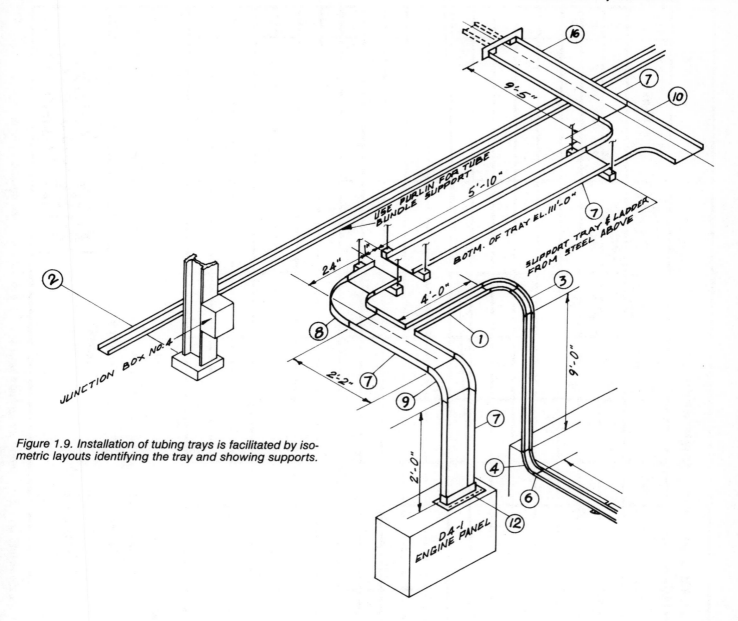

Figure 1.9. Installation of tubing trays is facilitated by isometric layouts identifying the tray and showing supports.

Information Required

Basic guidelines in some form are always available for job execution. The amount of such information varies a great deal. On some jobs extensive and comprehensive standards have been developed and are given to ensure quality and economy and to prevent misunderstandings of the nature of the job requirements. This does not imply that quality and economy are not likely when guidelines are minimal, but assurance of satisfaction is provided when established guidelines are followed.

In new organizations and in many small companies, the above mentioned guidelines often do not exist. In these instances, reliance is made on verbal communication and understanding to achieve the desired objective. This approach sometimes requires changes and modifications during progress of the project.

The information needed to ensure efficient job progress is listed in the following paragraphs.

Process Information

Process flow sheets are prepared by others, but they seldom contain all the information needed for specifying and designing control systems. The needed information includes the desired plant capacity, raw feed requirements, safety valve relief rates, auxiliary stream information, fluid

ALARM AND SHUT-DOWN DEVICES

ITEM NO.	REF. SPEC.	SERVICE	TYPE ALARM	H/L	FUNCTION	SET POINT PROCESS	SET POINT SIGNAL VALUE	CONTACTS OPEN WITH MEASURE	REMARKS	REV. NO.
SLSD 654		DRYER ROTATING BLADES SPEED	S	L	A-SD	10 RPM	10 RPM	DEC.		0
THSD 641		RECTIFIER	T	H	SD	150 °F	150 °F	INC.	REL. CAB. NO.1	0
THA 655		EXTRUDER BARREL COOLING WATER	T	H	A	150 °F	12 PSI	INC.		0

LEGEND:

P	PRESSURE	d	DIFFERENTIAL	H	HIGH
L	LEVEL	AN	ANALYZER	L	LOW
T	TEMPERATURE	A	ALARM	INC	INCREASE
F	FLOW	SD	SHUT DOWN	DEC	DECREASE
S	SPEED	SU	START UP		

REV _____ BY _____ DATE _____

JOB _____ BY _____ DATE _____

S.I.P. INC.
Engineers & Contractors
HOUSTON, TEXAS

Figure 1.10. The alarm and shutdown list provides needed information on switch action, location, setting and function. (Courtesy of S.I.P., Inc.)

properties of all streams, knowledge of equipment and knowledge of all operating conditions. Much of this information can be gleaned from process and mechanical flow sheets, and the remainder must come from other people associated with the project.

Instrument Specifications and Standards

Most long-established companies, often confronted with expansions, have developed, in the form of instrument specifications and standards, written guidelines for instrumenting their processes. Examples of this type information include (a) control valve selection guides relating to body and trim types, (b) percentage of C_v rating for design capacity, (c) valve actuator type, (d) valve accessibility, (e) controller selection guides that specify which loops should have three-mode controllers, which may use two-mode and those that may utilize proportional band only.

These standards or guidelines usually are organized so that groups or categories of instruments are covered in particular sections. Categories covered include temperature, pressure, flow, level and analytical instruments, control panels, control valves, relief valves, panel mounted controllers, local controllers, annunciators and local indicators.

The specifications may include industry codes and practices which must be followed and most certainly would list company preferences and practices which deviate from the commonly accepted codes.

Standards often included for guidance include mounting preferences for instruments (Figure 1.11 is a typical example), dimension requirements for locating meter runs, valving requirements for pressure gauges and mounting and piping methods for level gauges and level controllers. A few such standards along with the written specifications referred to in the preceding paragraphs combine to give the designer sufficient information to follow some previously established design methods.

Piping Specifications

Process piping specifications must be known so that instrument piping will be compatible with them and with the vessels and equipment used. In many instances, sensing devices (d/p cells, pressure transmitters and thermowells, for example) must be of the same material (or better) as the pipe or vessel to which they connect. The piping specifications list material pressure ratings, valve types, branch connection sizes and other similar bits of information necessary to connect the instruments properly. Since costs for special alloy materials and for high pressure flanges and fittings are so much greater than for standard materials and low pressure fittings, the necessity for knowing and following the piping specifications is readily understood.

Electrical Specifications

Specifications governing the design and installation of instruments and instrument systems are found in the electrical specification section as well as the piping section since both crafts are involved in instrument installation work. In many company standards and specifications, a great deal of electrical information relative to instrument design, materials and installation methods is given in the instrument section. Usually, however, reference must also be made to the electrical section to obtain such information as conduit and conduit fitting types, conduit supports, wire and cable specifications, segregation of power and instrument wiring, wire termination methods and test procedures.

Bid Documents

These documents are involved only on design and construction projects on which bids have been made and accepted. They are important to both the owner and the contractor. The responsible instrument engineers for both parties should be familiar with the basis on which the work is proposed to be done.

In some instances specifications and standards may be clear enough that no clarifications or exceptions have been necessary and where no alternatives have been offered. In such cases the specifications and standards along with the flow sheets and other drawings are the bid basis. These, however, are often subject to interpretation. They must be studied and understood by the contractor's engineer; presumably they are well understood by the owner's representative.

If exceptions (to the specifications) or clarifications (where specifications are not clear or may be misinterpreted) have been made or if alternatives have been offered, both parties should be aware of them to make certain that they do not affect the total instrument budget or the engineering manhour estimate.

Prior to beginning work, whether from the owner or contractor viewpoint, the involved instrument engineers should be familiar with the bid documents and with the final contract which reflects the bid proposal. Apparent discrepancies in the two documents should be brought to the attention of the project managers involved.

Project Procedures

The lack of communication between individuals and groups is a major obstacle in most endeavors. Engineering projects are not exceptions. Project procedures are set up to assure good communication among all involved parties.

They should include the following information in some form:

1. A project organization chart
2. The function of key personnel assigned to the project, their responsibilities and duties
3. Accounting procedures—a code of accounts for proper allocation of material and labor costs
4. Distribution of all types of communication forms, including periodic progress reports, special reports, job schedules, drawings, purchase requisitions and orders and day-to-day memoranda (these are examples, not a complete list)

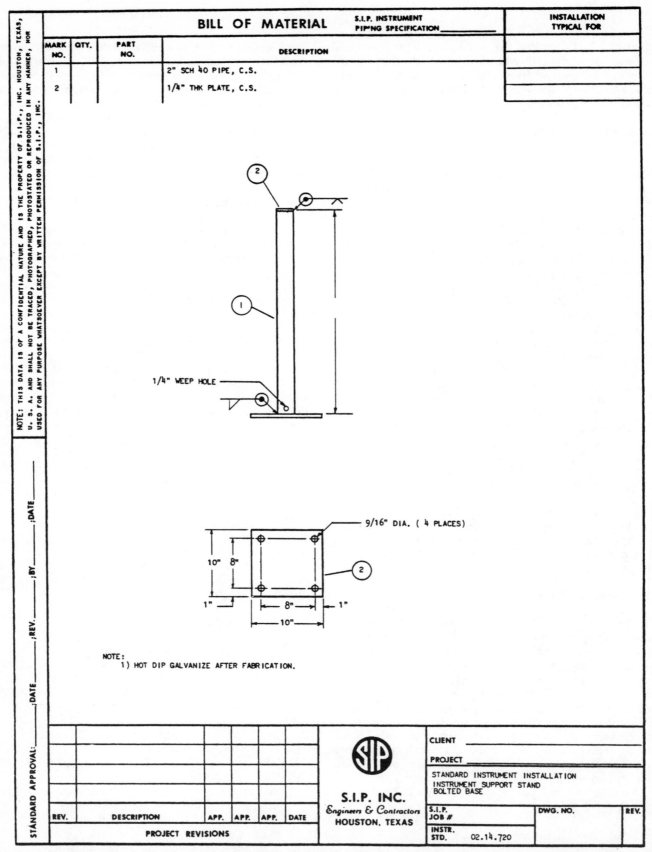

NOTE: THIS DATA IS OF A CONFIDENTIAL NATURE AND IS THE PROPERTY OF S.I.P., INC. HOUSTON, TEXAS, U. S. A. AND SHALL NOT BE TRACED, PHOTOGRAPHED, PHOTOSTATED OR REPRODUCED IN ANY MANNER, NOR USED FOR ANY PURPOSE WHATSOEVER EXCEPT BY WRITTEN PERMISSION OF S.I.P., INC.

BILL OF MATERIAL   S.I.P. INSTRUMENT PIPING SPECIFICATION _____

INSTALLATION TYPICAL FOR

MARK NO.	QTY.	PART NO.	DESCRIPTION
1			2" SCH 40 PIPE, C.S.
2			1/4" THK PLATE, C.S.

1/4" WEEP HOLE

9/16" DIA. (4 PLACES)

10" 8"

1"   8"   1"

10"

NOTE:
1) HOT DIP GALVANIZE AFTER FABRICATION.

STANDARD APPROVAL: _____ ;DATE _____ ;REV. _____ ;BY _____ ;DATE _____ ;DATE

REV.	DESCRIPTION	APP.	APP.	APP.	DATE
	PROJECT REVISIONS				

S.I.P. INC.
Engineers & Contractors
HOUSTON, TEXAS

CLIENT _____

PROJECT _____

STANDARD INSTRUMENT INSTALLATION
INSTRUMENT SUPPORT STAND
BOLTED BASE

S.I.P. JOB #	DWG. NO.	REV.
INSTR. STD. 02.14.720		

Figure 1.11. Showing mounting methods and dimensions simplifies installation of instruments and reduces field labor costs. (Courtesy of S.I.P., Inc.)

5. Approval procedures for the release of engineering drawings, specifications, requisitions, purchase orders and other documents
6. The extent, frequency and complexity of planning, expediting and scheduling activities
7. Equipment and material procurement methods and procedures
8. Engineering drawing sizes and numbering system
9. Identification systems for equipment and instrument items

This list is not intended to be complete, but it indicates the necessity for job organization to assure quick and efficient flow of information. It assures orderly progress and prevents costly mistakes due to lack of information. Primarily, it saves the valuable time of many people. It includes the flow lines of communication among the contractor, owner, vendors and subcontractors, as well as within the contracting organization.

Project Schedule

At the beginning of a project, the responsible instrument engineer determines from the project schedule the dates for purchasing equipment, for starting field installation and for completion of various phases of the project. These factors are interrelated, of course. The various facets of instrumentation (purchase, delivery, installation, calibration and checkout) are seldom critical in the completion of a job, except the last phase—calibration and checkout. By its very nature, instrument checkout is one of the last work items required prior to plant startup.

Other facets (such as delivery) may be critical if instrument design is delayed or if special equipment is needed. On small "crash" projects, standard instrument items may have longer delivery times than other equipment needed for the job. In any case, the instrument engineer should determine the amount of work to be done and correlate it with the project schedule.

Instrument items that need to be specified and purchased early are those which are mounted in-line and whose dimensions need to be known by other disciplines. For example, the piping people need the dimensions of control valves, meter runs and other in-line devices to complete their work.

Some other examples of scheduling problems include:

1. Purchase of panel instruments early enough to meet panel fabricator's requirements.
2. Purchase of relief valves to meet piping erection schedules.
3. The requirement of starting up one or more phases of a project ahead of others. All items required for that phase must be purchased early.

One good approach in scheduling a job is to work from the completion date backwards to determine logical dates for the completion of various stages of the job. Allow plenty of time for custom fabricated items, delivery of equipment

after shipment and communication delays that normally affect a job.

After the instrument schedule is temporarily worked out, review the items again, checking with equipment vendors to verify their probable delivery commitments.

Equipment Information

Equipment size, nozzle size and orientation (on vessels and other equipment), materials of construction and pressure ratings of equipment and flanges must be known prior to the purchase of instrument items related to that equipment. Most of this information is usually available prior to equipment purchase, and it must be assembled for reference when purchasing related instrument items.

Even after equipment specification and purchase is complete, a final check of vendor's drawings of the purchased items should be made to verify compliance with purchase specifications. A final check of this nature often avoids costly field changes.

Vendor Drawings

Vendor drawings of purchased instruments must be checked thoroughly to determine that they comply with purchase orders and their specifications and drawings. The assumption should not be that no mistakes will be made by the many people who handle purchase orders from their origination to the delivery of the finished items. It can be assumed that some will be made. A thorough check of vendor drawings will likely catch some errors in time to prevent costly delays and exchanges.

Work Coordination

Coordination among those responsible for the various phases of engineering is necessary for an efficient, well run job. The project instrument engineer must cooperate and work with the following people or groups. He must be aware of their duties and functions in order to coordinate the work effectively.

Project Manager

Among the many functions of the project manager, those listed below have the greatest effect on the work in the instrument department.

1. Project procedures
2. Schedules
3. Manpower allocation
4. Job progress
5. Cost control

Project Procedures
Project procedures is the term applied to an organized method for getting a job done. The project manager is responsible for determining that the procedures are followed. He directs the flow of information to and from the

operating company (owner), contractor, vendors, subcontractors and between departments of the contracting firm. Drawings and purchase requisitions usually are funneled through and initialed by him.

Schedules

The schedule for a job is determined by the project manager or is worked out with him. The project instrument engineer should be aware of the overall project schedule and must plan to meet its requirements—he schedules his part of the work to that end. Figure 1.12 is a typical schedule which the instrument project leader might make. Normally, instrument item deliveries are short enough that neither deliveries nor instrument installation falls into the critical path schedule of a job. Complete checkout of instruments and control systems is usually among the last work items to be completed on a job, however.

Significant deviations from the instrument schedule should be brought to the attention of the project manager and to anyone else to whom the instrument project leader is responsible.

If instrument deliveries fall behind schedule or if it becomes apparent that they might fall behind, expediting is needed to determine the reason and to take corrective actions. Someone is usually designated for that purpose.

Manpower Allocation

Manpower requirements to meet job schedules are usually discussed among the project manager, the instrument project leader and other interested people. Job requirements and personnel capabilities are usually known well enough that assignments can be made to satisfy completion schedules with a fair degree of accuracy. When additional manpower is needed to meet a commitment or if too much manpower is being used, the project instrument leader should advise the project manager and other management people who are responsible for reallocating manpower.

Progress Reports

Periodic progress reports, usually on a monthly basis, are made to the project manager. It is the most reliable source that he or other interested parties have for gauging the project status. Proper communication through this channel of information highlights difficulties, achievements and schedule changes. Reports from each discipline are completed and issued to designated personnel. The report is used to avoid misunderstandings between owner and builder and within the contractor organization by keeping everyone posted on the job completion picture. The owner needs this information for meshing new plant activities with those existing to reduce overall operating expenses.

The positive psychological effects of just having a monthly status report are enormous. It stands to reason that the absence of such an indicator will only cause doubts in the minds of the project manager and the customer. Conversely, the mere presence of a status report leads interested parties to the conclusion that the report writer understands the project situation. An overall confidence develops in the managerial skills of the engineer, a confidence which builds first in the engineer, and which often is noticed by the person who can recommend a salary increase for that engineer.

Cost Control

Factors that cause significant changes in the cost of instrumentation for a project must be brought to the attention of the project manager. Many different situations arise to alter the original estimate. Escalation of material and labor prices, greater-than-anticipated complexity of control systems, oversight of alloy material requirements, premium payments to meet schedules for fabricated items, loss of key manpower, misinterpretation of quotations—any of these things may occur to increase the cost. Seldom is one confronted with cost decreases.

Since the project manager is the one ultimately responsible for project cost, the final decision usually rests with him. The project instrument leader should make his views known and should advise to the best of his knowledge the proper action.

Process Engineer

During the course of a project, many questions arise relative to process conditions that require close cooperation between process and instrument people. For example, the range of flow for a flowmeter may be higher than that obtainable by a single orifice meter. A discussion of the problem with responsible process people will determine whether to use one or two meters or an entirely different type flow device.

There are many instances in which range requirements, material requirements, etc., have been given but which need not be considered inflexible. They need to be discussed with the proper people before changes are made, however. The process engineer is the logical person to consult for such problems.

Equipment Engineer

Major pieces of equipment ordinarily are purchased during the initial phase of the job because delivery times are usually long. This is particularly true of large or complex items, ones requiring a large amount of engineering, those made of exotic materials or nonstandard items. Most of this type equipment will have instrument related items—level gauges, thermowells, pressure connections—that must attach to it. The instrument project leader must work with the equipment engineers to assure proper connection sizes and locations for the instruments.

When complete packages (or systems) are purchased, such as compressors, extruders, furnaces and refrigeration systems, along with their associated auxiliary units, the responsible instrument leader should check equipment specifications, purchase orders and vendor prints to assure the use of proper instruments and to document information necessary to tie them into the rest of the plant instrument system.

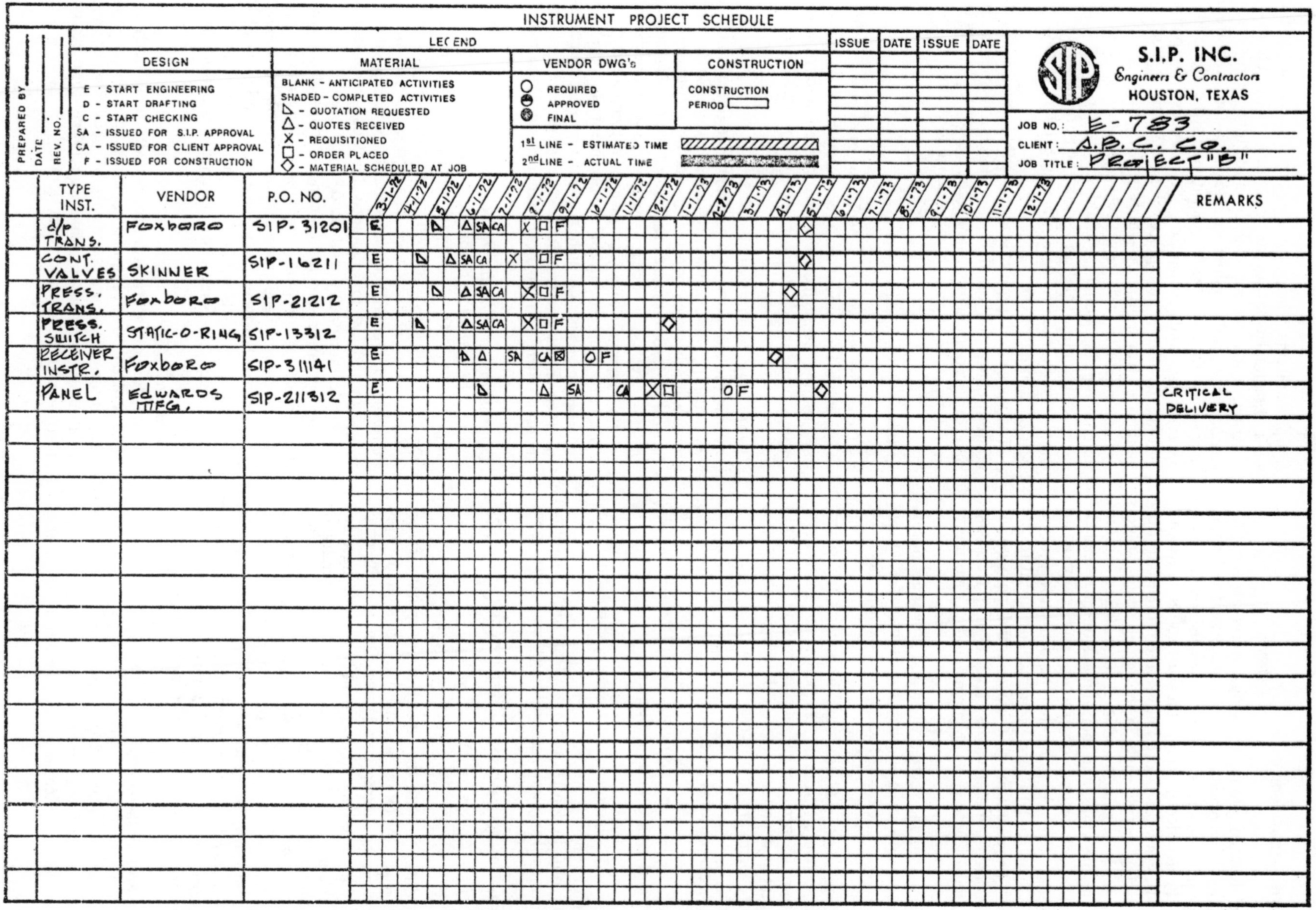

INSTRUMENT PROJECT SCHEDULE

S.I.P. INC.
Engineers & Contractors
HOUSTON, TEXAS

JOB NO.: E-783
CLIENT: A.P.C. Co.
JOB TITLE: PROJECT "B"

LEGEND

DESIGN
E - START ENGINEERING
D - START DRAFTING
C - START CHECKING
SA - ISSUED FOR S.I.P. APPROVAL
CA - ISSUED FOR CLIENT APPROVAL
F - ISSUED FOR CONSTRUCTION

MATERIAL
BLANK - ANTICIPATED ACTIVITIES
SHADED - COMPLETED ACTIVITIES
△ - QUOTATION REQUESTED
△ - QUOTES RECEIVED
X - REQUISITIONED
□ - ORDER PLACED
◇ - MATERIAL SCHEDULED AT JOB

VENDOR DWG's
○ REQUIRED
◐ APPROVED
● FINAL

CONSTRUCTION
CONSTRUCTION PERIOD

1st LINE - ESTIMATED TIME
2nd LINE - ACTUAL TIME

TYPE INST.	VENDOR	P.O. NO.																		REMARKS
d/p TRANS.	Foxboro	SIP-31201	E	△	△	SA	CA	X	□	F								◇	
CONT. VALVES	Skinner	SIP-16211	E	△	△	SA	CA	X	□	F							◇		
PRESS. TRANS.	Foxboro	SIP-21212	E	△	△	SA	CA	X	□	F						◇			
PRESS. SWITCH	Static-O-Ring	SIP-13312	E	△	△	SA	CA	X	□	F			◇						
RECEIVER INSTR.	Foxboro	SIP-31141	E	△	△	SA	CA	SA	CA	OF					◇				
PANEL	Edwards MFG.	SIP-21812	E	△	△	SA	CA	X	□	OF		OF						◇	CRITICAL DELIVERY

Figure 1.12. The instrument schedule shows instrument categories, vendor selections, purchase order numbers and dates pertinent to selection, ordering and delivery. (Courtesy of S.I.P., Inc.)

Piping Design Supervisor

The preparation of piping drawings starts as soon as equipment layouts and orientation are determined and other pertinent information is available. Costly revisions to piping drawings can be avoided if meter run sizes and lengths and control and relief valve sizes are known and dimensions are given to the piping section early in the piping design stage. Piping details for level gauges and level controllers are often shown on piping drawings, and their dimensions are needed. Connection sizes are needed for pressure and temperature devices. Forms that contain the needed information (Figures 1.13 through 1.17) are prepared and given to the piping section. They contain all the size and dimension information needed for relief valves, level transmitters and controllers, control valves, level gauges and orifice meter runs.

They can be dated and revised when and if changes occur. The preparation and issue of such forms provides an efficient and clear means of communication between departments.

Before piping plans, elevations and isometrics are issued, they should receive a "squad" check by the instrument department to make certain that all process connections are shown and that instrument location drawings (plot plans) and piping drawings are in agreement.

Structural, Architectural and Civil

Areas that require coordination with the structural, architectural and civil groups include:

1. Tubing and conduit run supports, including underground runs
2. Panel layouts in control rooms and for local locations
3. Foundations for analyzer buildings and instrument racks
4. Platforms for accessibility of instruments

Electrical

Close coordination with the electrical group is necessary for efficient work by either group. The division of work between them varies among companies. One area in which both often are involved is that pertinent to the purchase and installation of control panels. Agreement must be made on terminal layouts, identification of instrument wiring, the preparation of schematics and pertinent information on electrical items required in or on the panel. Interlock, emergency shutdown, sequence control and other similar systems that involve both instrument and electrical groups need to be coordinated. The responsibility for emergency or standby power systems may lie with either group.

Purchasing and Expediting

The purchasing and expediting functions may be performed by a combined department or by separate departments. Their concern will be material as well as equipment items. The purchasing department should furnish a list of standard clauses that become a part of the purchase contract. Such clauses include directions relative to correspondence and equipment identification; requirements concerning approval prints, operating instructions and spare parts lists; and inspection procedures. Delivery requirements—time and place and in some instances, the method of transportation—need to be given.

If competitive bids have been obtained by others, the purchasing department should be given copies of the successful quotes. In all instances the purchase order prices should be checked to determine agreement with verbal or written quoted prices.

The instrument project leader should be aware of delivery commitments and requirements. If commitments are not met, or if requirements change, the expediting group should be informed so it can take appropriate action.

Others

There are several other people or groups with whom the responsible instrument engineer may have to work from time to time. These include owner representatives, operating groups, vendors, contractors and one's own design group.

Owner Representative

From the viewpoint of a contractor engineer, this is a vital relationship. In addition to the written forms of communication between contractor and owner that convey the control philosophies and practices of the owner, a great deal of this exchange occurs through the owner's project people—in many cases, an instrument engineer.

Usually the extent of involvement by the owner's representatives is established in early project meetings. The involvement ranges from a few periodic meetings to a continuous on-the-job liason throughout the project life.

Operations and Maintenance

From the owner's viewpoint, it is often the practice (and a desirable one) to check with operating and maintenance people. These groups provide a valuable source of information on the performance characteristics of instrument hardware and control systems that have been used previously. High and low maintenance items can be discussed and their effects properly evaluated.

The operating and maintenance groups also have information on spare parts inventories for various types of hardware items, and price evaluations can be weighed correctly when hardware bids are evaluated. The need for additional spare parts often makes an apparent low bid more costly when all aspects of total cost are considered.

Vendors

The flow of information from instrument hardware vendors is often slow. Frequent contact is often necessary to obtain information on hardware dimensions, connections, modifications and sometimes functional characteristics of instruments or instrument systems.

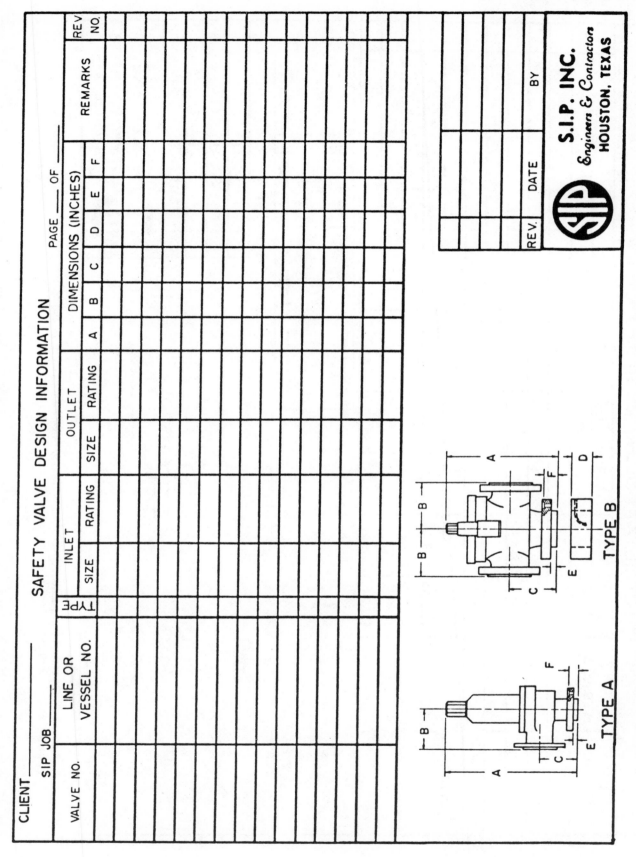

Figure 1.13. Pertinent information on safety relief valve size, rating and dimensions is listed for use by the piping department. (Courtesy of S.I.P., Inc.)

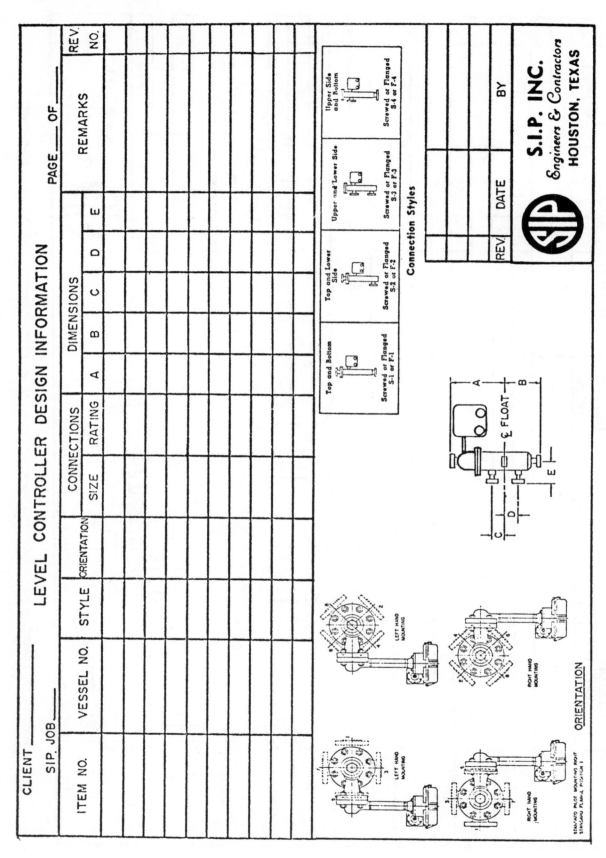

Figure 1.14. Connection orientation, dimensions, sizes and ratings are given to the piping department through this level controller design information sheet. (Courtesy of S.I.P., Inc.)

CONTROL VALVE DESIGN INFORMATION

CLIENT _____

SIP JOB _____

PAGE ____ OF ____

VALVE NO.	LINE NO.	END CONNS.		DIMENSIONS (INCHES)									DET.	REMARKS	REV. NO.
		SIZE	RATING	A	B	C	D	E	F	G	H	J			

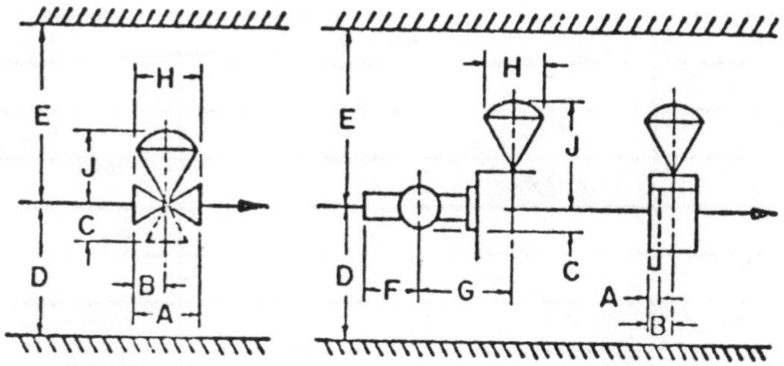

REV.	DATE	BY

S.I.P. INC.
Engineers & Contractors
HOUSTON, TEXAS

DETAIL A
GLOBE OR CAGE VALVE

DETAIL B
BUTTERFLY VALVE

Figure 1.15. Control valve information is itemized on a control valve design information form. (Courtesy of S.I.P., Inc.)

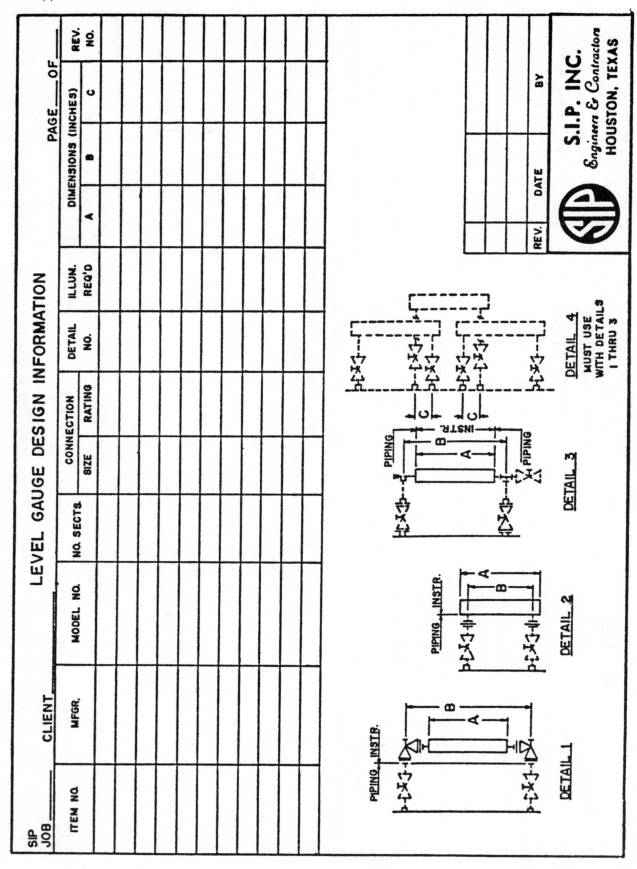

Figure 1.16. Design information is contained on this form for level gauges. (Courtesy of S.I.P., Inc.)

METER RUN DESIGN INFORMATION

CLIENT _____

SIP JOB _____

PAGE ____ OF ____

ITEM NO.	LINE SPEC.	SIZE	PIPE SCHD.	END CONN INLET	END CONN OUTLET	DIMENSIONS A	B	C	D	E	REMARKS	REV. NO.

NOTES:

"A" = UPSTREAM LENGTH

"B" = DOWNSTREAM LENGTH

"C" = MIN. CLEARANCE

"D" = F to F OF FLANGE

"E" = STRAIGHT RUN OF PIPE UPSTREAM BY PIPING

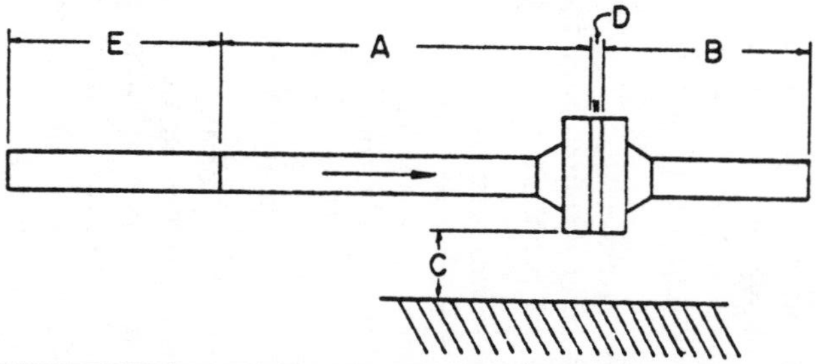

REV	DATE	BY

S.I.P. INC.
Engineers & Contractors
HOUSTON, TEXAS

Figure 1.17. Meter run information is given to the piping department as indicated by this form. (Courtesy of S.I.P., Inc.)

Contractor

From the owner's viewpoint, there is a need to know the ability of the contractor's instrument group. Control philosophy and practice are given to the contractor through standards and specifications, but personal contact is needed for assurance that these will be followed capably and with diligence.

Design Group

The responsible instrument engineer should work closely with the design group that produces the necessary drawings for a job. The relationship between the engineer and the design group varies. In some instances, the design group works apart from the engineer or engineers. Others work closely together, and the engineer is always available for consultation.

Either method is satisfactory if communication is established and maintained. Engineering changes and additions that are made should be passed on immediately to minimize the amount of double effort sometimes caused by such changes.

Department Supervisor

The instrument project leader must keep his supervisor informed about the status of the project. The relationship varies from company to company. As a minimum, the department supervisor should be advised on:

1. Job progress
2. Manpower requirements and changes
3. Unusual problems

In some cases the department supervisor may take a much more active role in the project: working on particular problem areas, advising on measurement methods, helping with schedules, evaluating bids, etc.

Job Execution

The approach to job organization may vary, depending on job size, personnel and many other factors. The methods suggested in this section embody the primary elements of good organization that apply to any job.

Listed below are some helpful hints that should be followed during the job. Most of these need to be made in the earlier phase of the job; some are continuing checks to be made throughout the job.

Planning Hints

1. Review the job (flow sheets, specifications, standards, procedures and other available information) to ascertain its extent and complexity. Determine the quality of help needed to do the work. Ask for the help only when needed but make the request as far in advance as is practicable.
2. Prepare an index or summary of instrument items required. At this stage, the list probably is incomplete. Keep the list up-to-date and use it as an index of job progress by noting work that has been completed in terms of specification sheets, purchase, installation details, etc.
3. Make a list of major work categories (the items can be checked off as work is completed). Suggested categories include:

   a. Control valves
   b. Relief valves (including rupture discs)
   c. Meter runs
   d. Control panel instruments
   e. Transmitters (flow, pressure, temperature, level)
   f. Level gauges
   g. Local controllers
   h. Analytical instruments
   i. Thermometers and thermocouples
   j. Switches and solenoid valves
   k. Miscellaneous instruments (ones not covered in the above categories)
   l. Control panel specifications and drawings
   m. Loop wiring diagrams
   n. Installation details
   o. Plot plans

   Assign these various categories to members of the project group for work execution—the preparation of specification sheets, purchase requisitions, drawings and sketches.

4. Check the project schedule and then make a schedule for the instrument group that conforms to the project schedule. Draw a "manhour versus estimated time" curve to fit the projected instrument schedule (see Figure 1.18). This plot may be used for another curve of actual manpower versus time. It can be used as a check on job progress, assuming that the manpower estimate is relatively close.
5. Set priorities on work and make job assignments to meet the schedule. Job assignments are those mentioned in Step 3. The time requirements of several categories will be such that most men can cover more than one major work category.
6. Set up a filing system for easy reference. Filing may be done by instrument categories, by vendors, by function or by combinations of these. All reference documents and working documents must be filed. There should be an orderly record of job changes, correspondence with owner, contractor, vendors and other departments. Copies of requisitions and purchase orders are filed for easy reference.
7. Use a notebook or calendar as a reminder of actions and decisions that need to be made to maintain the schedule. When decision time comes, press for the necessary action.
8. Report job progress to the proper people at the scheduled time. Project managers, owner, contractor, department manager and others who are involved or have job management responsibilities need to know if the schedule is maintained.

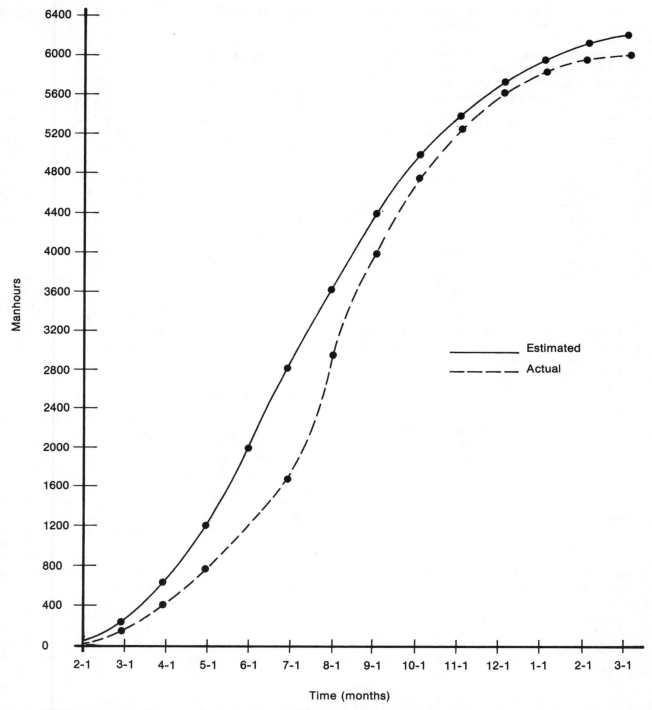

Figure 1.18. Engineering progress is charted against estimated progress as illustrated by the time versus manhour curves.

9. Make sure that additions and deletions to the job are kept up-to-date. Records of changes should appear on flow sheets and the instrument index. Some deleted items may have already been purchased requiring cancellations of purchase orders. Increases to purchase orders should be made as items are added. Records must be dutifully kept to avoid confusion. Drawings and specifications which have become obsolete should be marked and then filed, *not thrown away*. Drawings that were considered complete may require changing.

10. Check occasionally with the project manager to determine that no changes have been made in the job schedule.

11. Make periodic checks on instrument delivery status. It can be anticipated that some items will not meet the promised delivery commitment. The following are some typical difficulties that may occur:

   a. The entire plant (personnel) takes a two-week summer vacation. It may delay anticipated delivery of an item or group of items for 3 to 4 weeks.

   b. A subcontractor to a vendor may fail to deliver a component part.

   c. Alloy or exotic material may be in short supply.

   d. Nonstandard items may have unusually long delivery times.

   e. Vendors may suddenly be awarded more jobs causing a delay in deliveries.

   f. Even standard items of a large manufacturer have scheduled production times with long waits between runs.

   g. Strikes affecting production, craftsmen or transportation may delay equipment deliveries.

   h. Sensitive electronic computer hardware must often be shipped in a special van and receive extra care in handling, requiring individual attention in the preparation for shipping.

   i. Humidity- and temperature-controlled storage for computer or microprocessor equipment is required in the event that such systems are not yet operable at the final destination. Temporary power requirements must also be researched and provided.

Scheduling

Seldom does the schedule of instrument work coincide with any part of the critical path schedule for a project. Except in rare cases, the deliveries of instrument items are much shorter than for major pieces of equipment. Occasionally the deliveries of large size control or relief valves, special alloy devices or some nonstandard or custom-made equipment may run several months, but these instances are rather infrequent.

If the engineering time and deliveries are known to be relatively tight, the normal sequence of work can be adjusted to allow more time for the long-delivery items. This can be worked out rather effectively by working backward from the required installation date. Allowing the necessary time for installation, for shipment and for fabrication time, a date can easily be set for equipment purchase. Engineering effort can be concentrated on specifying and purchasing to meet that date.

The same sort of planning can be done for other instrument items and for other phases of the project—in engineering and/or construction. Thorough planning and wise use of manpower usually provide the answer to difficult project schedules.

Specifying Instruments

Normally the most efficient procedure in writing instrument specifications is to group them into major categories and prepare those that are alike or similar at the same time. For example, flow, pressure, level and temperature transmitters can be specified and purchased in groups. Variations from this procedure may be worthwhile if different types of devices are purchased in the same category—magnetic meters instead of orifice meters, for example. In such a case, the mag meters would be treated as another category.

The same type of grouping applies to control valve sizing and specifying, the calculation of relief valve sizes, the preparation of panel specifications, etc.

It is also desirable in most cases to prepare all the instrument specifications and requisitions that go to a single vendor so that all items from that vendor may be purchased at the same time. An exception to this practice often occurs, however, when panel instruments and field instruments are ordered from the same vendor, and it becomes necessary to purchase panel instruments early to have them ready for the panel fabricator.

Vendor Selection

Quite often there are some overriding reasons for the selection of a certain vendor, particularly for panel mounted instruments. The reasons vary; it may be a desire to match existing equipment, the avoidance of additional spare parts, computer compatibility, a belief in the superiority of a certain brand or some other reason.

From the contractor point of view, such preferences must be noted. Unless there is some other restraint, the client's preference will likely be followed. Often an approved bidders list is provided in which equipment of equal quality is listed for various equipment and material categories. If no preference is expressed, the vendor equipment which meets the required specifications and is most economical should be used.

In evaluating bids, equipment costs alone must not be the only consideration. Other material costs and labor costs must also be considered. For example, two electronic temperature transmitters may be equal in cost and quality. If one were a two-wire system and the other a four-wire system, the two-wire should be chosen to reduce installation costs, unless some other factor offsets that difference.

If two control valve vendors quote essentially the same price for the project valve requirements, the valve manufacturer whose valves are already in use would probably be chosen to reduce valve spare parts inventory.

It almost goes without saying that delivery requirements have been responsible for choice of vendors on many occasions. For most projects, instruments suspected of having long deliveries should be specified early, on a priority basis.

Shipping, Receiving and Storing Instruments

The shipping, receiving and storage of instruments often receive too little attention with consequent loss of money and time. Panel mounted instruments, for example, may sometimes be shipped directly to the project site rather than to the panel fabricator. Unless some nonstandard devices need to be checked along with their associated wiring, a few standard units can be used to check panel wiring, and most instruments can be shipped directly to their point of use, thus avoiding extra shipping costs and handling. The instrument mounting cases must, of course, go to the fabricator in either case.

The receiving of instruments at the plant site should be handled by people who can verify that the equipment complies with the purchase order specifications. It is often necessary to find out the receiving inspector's name and to make sure that the inspector has received final specifications. Often, a copy of the purchase order is sent without the specification sheets. When orders are incomplete or when the wrong equipment is delivered near the project completion date, delays and extra labor costs are likely to occur.

Instrument equipment should be stored in clean, dry atmospheres. Storage buildings should be provided in time to prevent early instrument shipments from setting outside in dusty or wet atmospheres. Control panels, analytical equipment and all other electronic equipment should be especially noted and cared for.

Installation and Checkout

This topic is covered in greater detail in Chapter 12. It is well to note at this time, however, that from the early part of the project, efforts should be made to ensure that all instruments are available when needed. Valves, meter runs and other process-connected equipment (especially the in-line variety) should be ready when their associated lines are installed.

When underground runs of conduit and/or tubing are used, they must be installed in the early stages of construction. The responsible instrument engineer must determine the requirements of this nature and make sure the work is done at the proper time to avoid extra labor costs.

The checkout of the entire project should start well ahead of the expected startup schedule, allowing time to work out unforeseen difficulties. In many instances, some systems of a project may be required for an early startup. The checkout of that portion must be scheduled early. This may require power sources from other portions of the plant or activation of systems in an unfinished area. It may also necessitate an early completion of a section of control panels involving other plant areas. This type of problem can present some unusual difficulties.

Project Check List

As soon as familiarity with the job is gained, consideration should be given to the usual as well as special re-quirements of the job. Lists given below are not conclusive but are among the items which need to be considered.

Design Considerations

Check and plan for:

1. Underground runs that may be required
2. Sufficient instrument air supply
3. Emergency air supply requirements
4. Supports that may be included with other structural requirements
5. Electrical area classifications
6. Proper connections (sizes and locations) for all process-connected instruments
7. Sufficiently long, straight run clearances for flow meters
8. Proper instrument specifications that are furnished with equipment packages, such as compressors, furnaces, extruders, dryers, storage tanks, etc.

Miscellaneous Questions

Pose questions about the job to determine the known and unknown quantities, improvements that might be made and whether oversights have occurred. Typical questions may include the following.

Does the design basis follow conditions stated in the bid proposal? (Contractor oriented question.)

Is there a more economical way to design it?

Have competitive bids been requested for equipment and material?

Is efficient and economical use of manpower being made in the engineering effort? In the construction effort?

Are doorways large enough to admit panel sections?

Have winterizing requirements been met?

Have extended delivery times jeopardized the schedule?

Have late changes been incorporated into drawings, and have all purchases been made? Do they affect the schedule?

Other questions that are pertinent to the job can be added to this list.

Equipment Delivery

Normal delivery times will usually be known fairly well. Give special consideration to

1. Instruments made of exotic materials
2. Pipe, tubing, valves and fittings of exotic materials
3. Unusual or nonstandard items
4. Panel fabrication time
5. Early startup for a portion of the job
6. Possible strikes
7. Vacation shutdowns
8. Writing of last-minute algorithms for computers and microprocessors

Conclusion

Many factors determine the methods for handling an engineering and/or construction contract—the size of the job, whether its location is new or an addition to an existing facility, whether it is a new or existing process, how well defined it is, the personnel assigned to it and the time or schedule for the job. All these factors must be evaluated.

There is a tendency by many people to proceed too quickly and with too little coordination with other groups and individuals. There is an almost equal tendency by others to wait until all facets of a problem are clear, leading to delays affecting others. This can be as costly as proceeding too quickly. The solution to these tendencies is proper communication with other members of the project team to chart progress and schedule action time for the various project phases.

Evaluation, planning, scheduling, coordination, and execution are keys to a successful project. Evaluate all facets of the job, plan and schedule the work for economic execution, communicate and coordinate the work with team members, and proper execution usually follows.

2 Engineering Design Criteria

William G. Andrew

Regardless of the ability or the competence of an engineer to perform a given task, there is a need for guidelines or criteria that should be followed in its pursuit. The design of instrument control systems is no exception. Whether the task is examined from an owner or from a contractor viewpoint, there needs to be a standard of acceptable practices to which the task or job can be compared. The standards used for comparison usually have been adopted after proven performances in many similar applications.

Many operating companies have developed design criteria that must be followed for expansions or for changes and alterations to existing processes. Standards are given for economic reasons to assure the continuance of good control philosophies, to ease the amount of maintenance, to reduce spare parts requirements, but primarily to assure an acceptable performance level.

When standards are not available in written form, reliance is placed on oral communication between those responsible for project management and those to whom the job is assigned. In some instances, complete responsibility may be placed on a person or group because of their knowledge of the process.

This chapter lists many areas where questions arise in plant control systems design. It discusses various design concepts, answers questions relating to the selection of instruments, control center design, future capacity, transmission systems, identification methods and many other design considerations. In many instances, suggestions are given relative to design concepts. These may be adopted or revised for specific job requirements.

Pneumatics Versus Electronics

Since the introduction of electronic instrumentation in the early 1950s, there has been a steady increase in its use in the process industries. There continues to be a great deal of competition between the two methods of control. Improvements and new design concepts of each of these methods add to the difficulty of their comparison. Looking at the long-range trend, however, it seems evident that electronic instrumentation will continue to gain acceptance. There is no intent to predict an end to pneumatics or to say how swiftly their use will decline. There are several factors that favor their continued use. These are discussed as the facets of cost, dependability, safety, maintainability and adaptability to the process.

Cost

Initial hardware cost of electronic instrumentation is now greater than comparable pneumatic hardware. However, a comparison of installed costs is more difficult to obtain accurately. Reports over a period of several years generally concede that installed costs of pneumatics are less for small plants (with short transmission distances), while electronics have the advantage, cost-wise, for large installations.

Comparisons are difficult to make because there are so many variations in transmitting methods. Copper tubing is more expensive than plastic tubing. The type of protective covering used on either affects the price. The labor required to connect copper tubing is greater than for plastic. These factors along with many others cause pneumatic installation costs to vary. Similar variations affect the cost of installing electronic systems. One of the greatest factors is the electrical area classification which determines conduit and wiring methods that must be used. This cost difference is minimized, however, when intrinsically safe systems are used. The recognition and adoption of intrinsically safe instruments contribute significantly to the reduced installed cost of electronic instruments.

Dependability

There is little question concerning the dependability of either pneumatic or electronic instruments. Because of the earlier use of pneumatic instruments, some of the difficulties encountered by electronic instruments had already been overcome for pneumatics. For a while, at least, pneumatics appeared to be (and probably were) more dependable. With experience, though, and with advances in technology and art, little argument can be made for one case over another. As an example, back-up or emergency power systems for electronic instruments presented some problems for several years. These problems generally have been overcome.

Safety

In chemical, refining and petrochemical processes, most plants have areas with hazardous atmospheres. This is not a problem for pneumatic instruments. Charts for recorders can be powered with pneumatic drives so that the use of electrical power can be avoided altogether. This is seldom done, however. The use of intrinsically safe electronic systems which are now available also tends to offset this apparent advantage. The testing and acceptance of intrinsically safe systems is slow, however, and it is still difficult to obtain all the devices and components of complete plant systems without additional engineering time and attention.

Electrical equipment is available for most electrical area classifications, although some types of instruments are not available for Groups A and B classifications. This problem is usually overcome by using pneumatic devices within those areas and by converting to electronics immediately outside for the long runs to the control room or other areas.

The safety aspect may be summarized by saying that pneumatics are better for troublesome areas, but the trend toward use of intrinsically safe instruments is overcoming that handicap.

Maintenance

When electronic devices were first introduced, maintenance was a problem because many instrument mechanics had little or no experience with electrical devices. That problem, too, has largely been overcome. Now many instrument maintenance trainees have had some previous experience or background training for electrical systems and have had little or no experience with pneumatic devices. They usually have little difficulty in understanding and working with pneumatics.

Maintenance and troubleshooting of recent electronic instruments are reduced by being able to repair sections on a card replacement basis.

Process Control Requirements

The response of pneumatic instruments is sufficiently fast for the majority of process control loops. Some processes (or some loops within many processes) need the fast response which only electronic devices offer. One such process is the high pressure polyethylene process where changes occur within a few milliseconds. For this and similar processes, electronic systems are superior to pneumatic.

Summary

The case of pneumatics versus electronics can be summarized briefly as follows.

Electronic
1. Electronics respond faster and have a decided advantage if fast response time is needed.
2. The installed costs are probably less for large plants and where transmission distances are long (excess of 250 to 300 feet).
3. They provide convenient, economic interfacing with supervisory digital computers and data processing or data acquisition systems.
4. Reliability is good and is becoming increasingly better.
5. There is a gradual lowering of electronic component and subsystem cost.
6. Safety presents few problems.
7. Accuracy is about ¼ of 1% compared to ½ of 1% for pneumatics.
8. Complex control functions require less hardware.

Pneumatic
1. The installed cost of pneumatics is lower for small plants with short transmission distances (up to 250 to 300 feet).
2. The speed of response is adequate for most applications.
3. Final control elements continue to be pneumatically powered.
4. Reliability is good.
5. Maintenance presents few problems.
6. Safety is excellent.
7. Improvement in hardware continues.
8. Local control systems are very economical.
9. Plant standards tend to maintain a technical status quo.

Pneumatic instruments will remain the choice of many people for several years. Reasons include resistance to change, familiarity with pneumatics, less training required, cost, compatibility with existing instruments and proven reliability.

Electronic instruments will find increasing use because of their compatibility with computer systems (present or future), reduced cost for large installations, speed of response, improved reliability and because future potential cost reductions and improvements rest in the electronics field.

Control Centers

The entire project staff is concerned with some aspects of control center design, while other aspects are left entirely to the discretion of the responsible instrument people. Considerations included in the following discussion concern its location and layout, its electrical classification, the utility needs and the requirements to consider in instrument panel and console design.

Location

The location for a control center depends on many factors. It should be near the center of the area it serves, if possible. This depends, however, on whether the location is a safe one, what its electrical area classification is, whether it is readily accessible to outside traffic and many other considerations.

If pneumatic instruments are used, the transmission distance must be considered because of the longer response time and the greater installation costs of pneumatic systems.

Other factors to be considered include operator convenience, accessibility to utilities and suitability to future expansion capability.

Layout

The layout of the control area is covered in detail in Chapter 8. In addition to the main control room, however, many modern control centers contain office space, kitchen facilities, comfort facilities and sometimes operator locker rooms. Occasionally, an electrical substation will be an integral part of the control center.

The control center should be laid out to eliminate interference of incoming instrument and electrical lines with other equipment in the center. The incoming lines should cross only the control room and no other areas of the center.

Future expansions should be considered, and the design should allow for additions with a minimum of interruption to existing operations at a minimum cost.

The arrangement of panels in the control center may need to follow a physical sequence which relates to the plant physical layout. The point may seem trivial, but tangible benefits may be obtained in operator training and understanding of the process.

The layout should also take into consideration some or all of the following factors:

1. Vibration and noise problems.
2. Location of accessory devices such as transducers, alarm switches and termination panels and junction boxes.
3. Air regulation and distribution systems for pneumatic instruments.
4. Emergency power systems for electronic instruments
5. The possible need of spares for minor additions or modifications to the process, which include spare instrument spaces, spare disconnects to power supplies, spare terminal points, spare air supplies, spare alarm points, etc.
6. Interconnections between board-mounted instruments and their respective input/output modules located in racks, to minimize wiring, particularly beneath computer floors.

The layout of individual instrument panels and consoles should consider:

1. The arrangement of instruments for easy readability and operability.
2. Related instrument groupings for operator convenience—instruments associated with a given section of the process grouped together.
3. Minimum spacing of instruments consistent with operating and maintenance limitation requirements
4. Operating switches and controls that must be within easy reach of the operator.
5. The grouping of key control systems as closely together as practicable to eliminate unnecessary movement by the operator.

Electrical Classification

Modern control centers are usually located so they may be classified electrically as general purpose. With the increased use of electrical instrumentation, costs would be prohibitive otherwise. If the area does not meet the general purpose classification requirements by virtue of location, the control center may be pressurized to meet that classification by using an acceptable source of clean air. Practices governing electrical classifications for refineries are given in the API RP 500A. This recommended practice is accepted as a standard by most refining, chemical and petrochemical industries.

When a control center requires pressurization to meet a classification of lesser hazard, the installation of a gas detector is advisable for alarm and shutdown of hazardous electrical systems should gas be detected.

Where additions are made to existing control rooms in hazardous areas, the use of pneumatic instruments is usually advisable if pressurization is not feasible.

Utilities

The utilities needed in control centers include electrical power (120 volts for instruments and 220 volts and/or 440 volts for lighting and power), water (drinking and cleaning) and steam and/or gas for heating requirements. Air conditioning needs fall into one or more of the requirements listed. Instrument air is required for pneumatic control systems.

Future and Spare Capacity

Questions of future and spare capacity are interesting and debatable. Future capacity, as used here, applies to major plant or process additions. The decisions of this nature are likely to be made by top management and hinge on the type process, market conditions and many other factors. If provisions are made for major equipment additions, control center layouts and transmission facilities must also provide for these future requirements. Space allotment is not difficult to determine for these requirements for it usually relates directly to existing requirements.

When major additions are not anticipated and need not be considered, spare capacity for the system is still a matter of concern. If the process is new, the likelihood of significant changes is relatively certain. Spare capacity should range between 20 and 50%, depending on the item under consideration. Even for proven processes, 20% spare capacity is advisable to allow for improved control techniques that result from process improvements, the possible availability of new and better instrument devices or a desire for further automation. The last facet may result from economic considerations, reduction of manpower, improved quality or better efficiency of the process.

Percentage figures (20%, for example) for spares must not be used too rigidly. For example, there need not necessarily be space for 20% more controllers, 20% more recorders and 20% more indicators on a particular job. Instead, there might be space for an approximate increase of 25% controller addition with little or no space for new recorders or indicators (especially if additional pens may be added to existing recorders).

One category of instruments that consistently runs short of space is that of alarm functions. Almost invariably new processes require the addition of alarm functions as process experience is gained. Twenty or 25% spares should be allotted as a minimum; 50% spare capacity should be allowed in many situations.

Panel space, conduit, wires and terminals, tubing (on pneumatic jobs), rack space for transmission media and alarm space are items that must be remembered when spares are considered. Twenty percent spare capacity for these items is a nominal figure. In many instances, it should be 50% or more.

Specifications for Various Measurement and Control Groups

The purpose of this section is to present some general and some typical guidelines for designing various control systems. The examples given illustrate the need for this type of information. In presenting these specification needs, a choice must be made either to generalize or to be specific. To generalize completely is to ignore the need for specific information. To be specific in all instances is to confine the information to a specific set of circumstances that may be foreign to the reader. The median approach made is general enough to apply to a large number of needs, and when specifics are given, they can be interpreted in the light of needs for a particular job.

General Considerations

In the design of new plants or in making additions to existing installations, project specifications for instrument design are needed to clarify design philosophies, to ensure a minimum level of quality and to provide a degree of conformity to existing systems.

One of the first considerations usually is the type of measurement desired and sometimes even the make and model number of the instrument to be used. If it is not a grass roots plant, a decision may be made to duplicate existing equipment to minimize spare parts inventory. The decision is sometimes arbitrary, and sometimes it is weighed along with cost, delivery and other factors if a thorough evaluation is made. In making such an evaluation, initial cost, spare parts inventory and similar factors provide a firm basis for comparison. Other factors, such as dependability, maintainability, performance and operator familiarity, are, for the most part, intangibles which are more difficult to evaluate.

Project instrument specifications are divided into several sections. The first section is usually categorized as "general" and includes information guidelines concerning industry codes and standards to be followed, signal transmission systems to be used, control modes, selection guidelines, alarm and interlock philosophy, engineering units required, charts, scales and ranges desired and the documentation wanted. Some of these are discussed in greater detail later in the chapter.

Other sections cover specific categories such as flows, pressures, levels, temperatures, control valves, control panels and analytical devices.

Flow Measurement

The flow section typically reveals such information as maximum and minimum beta ratios, types of acceptable flow devices, differential ranges of meters, desired type of primary flow element, minimum line size for meter runs and meter installation practices. Some typical design guidelines that may be given follow.

1. The standard primary element shall be a concentric plate.
2. Flow elements such as Dall tubes, nozzles or Venturis may be used where high capacity or good pressure recovery is required.
3. Positive displacement meters or turbine meters may be used if greater accuracy is required than that obtainable with orifice meters. Displacement meters and turbine meters shall meet the following requirements:

Strainers shall be used to protect from entrained solids. If vapors are present to affect meter accuracy, air vapor eliminators shall be installed. Manufacturer's recommendations for minimum length of straight pipe before and after the meter must be adhered to. Automatic temperature compensation is required where temperatures are expected to fluctuate more than 10°F.

4. Differential pressure instruments shall be force balance or bellows displacement type. The force balance is preferred. Line-mounted meter manifolds shall have two take-off connection valves and a meter balance valve. Pedestal-mounted meter manifolds shall have two take-off connection valves at the line, and two meter valves and one meter balance valve at the meter.

5. Rotameters shall be metal tube type for general services. Glass tube enclosed types may be used if the fluid is air, inert gas or water, up to a maximum pressure of 50 psi.

6. Meter ranges shall be selected in accordance with the following guidelines. For orifice meters, normal flow rate shall be between 40 and 80% of capacity, provided anticipated minimum and maximum flow rates will be between 25 and 95% of capacity. For rotameters, the maximum shall be selected to use manufacturer's standard tube and float, if possible. Normal flow rate shall be between 40 and 80% of capacity, provided anticipated minimum and maximum flow rates will be between 15 and 95% of capacity.

7. The minimum line size for the meter runs in conventional metering services is 2 inches NPS.

8. Orifice meter differential ranges shall be as follows:

   a. Orifice meter differential range and pipe size shall be selected so that the ratio of orifice diameter to actual internal pipe diameter (d/D) does not exceed 0.7.

   b. Orifice meter differential range shall be selected so that the d/D ratio exceeds 0.25. With 2-inch pipe and 20 inches of water meter differential range, d/D ratios smaller than 0.25 may be used. In any case, actual orifice diameter shall not be less than 0.25 inch.

   c. Orifice meter differential range shall be 20, 25, 50, 100 or 200 inches of water calibration.

   d. For compressible fluids, differential range in inches of water shall not exceed the normal upstream static pressure in pounds per square inch absolute, except that for exhaust steam (approximately 15 psig), 50-inch range may be used.

9. Flange taps shall be used for orifice plates.

10. Orifice diameter shall be calculated using the data published in the AGA *Gas Measurement Committee Report No. 3* or *Principles and Practice of Flow Meter Engineering*.

11. Flow controllers shall have fast reset adjustable to .02 min/repeat or less.

Pressure Measurement

The pressure section of specifications provides guidelines on the selection of ranges (for example, the range might be twice the operating pressure), the use of blind or indicating transmitters, the type and use of pressure gauges, connection sizes, overrange protection requirements, measuring element types preferred and minimum material specification requirements.

Typical guideline statements follow.

1. Pressure range shall be selected so that normal pressure will be in the middle third of the span.

2. Pressure controllers shall have an indicator for process pressure.

3. Suppressed ranges shall be used on all controllers to obtain maximum accuracy and control where practical.

4. Pressure gauge measuring element shall generally be the bourdon tube type. For measurement of slurries and viscous and corrosive fluids, diaphragm seals shall be used. The minimum dial size for pressure gauges shall be 4 inches. The case shall be weatherproof and furnished with a blowout back or blowout disc in the back. Local gauges shall have wall mounting cases with bottom process connections.

5. Panel mounted gauges shall have flush mounting cases with rear process connections and shall not have suppressed ranges.

6. Draft gauges shall be slack diaphragm type.

7. Pressure instruments measuring elements may be bourdon tube, spiral, helical, bellows or diaphragm type, depending upon the process service. The measuring element shall be type 316 stainless steel, unless process fluid requires the use of other materials.

8. Instruments shall have overrange protection to the maximum pressure to which they may be exposed.

9. Pressure transmitters may be force balance or motion balance type.

10. Pressure switch measuring element may be bourdon tube, bellows or diaphragm type, depending upon the service and pressure. The measuring element in process service shall be hardened type 316 stainless steel.

11. Connections for process pressure switches shall be ½ inch, and connections for air transmission signals shall be ¼ inch.

Level Measurement

The level specifications section lists the types of devices preferred for various process requirements, ranges preferred, connection sizes to be used and types of level switches that are acceptable. It states when to use reflex or transparent level gauges and when to use frost-proof shields and lighted gauges.

Typical guideline information may include the following.

1. External displacer instruments shall be used for measurement ranges up to 48 inches. For ranges greater than 48 inches, differential instruments shall be used. The application of hydrostatic, capacitance, ultrasonic or internal displacer level instruments and similar devices may be considered if the preferred types do not satisfy job requirements.

2. Connections for external displacer instruments normally shall be 1½-inch NPS minimum size. When flanged units are used, they shall have a minimum rating of 300 pounds. Side-and-side connections are preferred. Rotatable head construction is required for external displacers.

3. Differential pressure instruments for level measurement shall be the force balance type.

4. Proportional control action adjustable to 300% proportional band minimum and reset adjustable down to .05 repeats per minute or less shall be provided except for cases where the outflow liquid goes to the sewer or directly to storage, in which case proportional action adjustable to 100% proportional band will suffice.

5. Reflex level gauges shall be specified for all services, except the following, where transparent types shall be specified:

   a. Interfaces between two liquids
   b. Liquids containing gum, sediment, or other solid materials which may coat the flutes of a reflex glass

6. Protective shields shall be used for level gauges where process fluids attack glass. Frost shields shall be used if specified operating temperature is below 32°F.

7. The visible portion of gauge glass shall cover the range of the associated level instrument.

8. When two or more level gauge columns are required to cover a wide range, the visible portion of the gauges shall overlap at least 1 inch. No more than five gauge units may be used in one column.

9. Gauge glass lighting, when used, shall be provided by the gauge glass manufacturer and be suitable for the area classification.

10. Gauge glass tube take-off connections shall not exceed 15-inch length. If longer take-off connections are required or if they are subject to vibration, they shall be braced.

Temperature Measurement

The temperature section usually states the types of transmitters that are acceptable, gives performance requirements and provides guidelines on measuring circuits, on the use of suppressed ranges and on the selection of temperature ranges to be used. It indicates when thermocouple burnout protection is used, which measurements require duplicate indication on multiple point indicators and where local indication is required. It usually states what types of thermocouples are to be used, thermocouple well sizes and the materials to be used for special applications such as for heating furnaces.

Typical guidelines may include the following points.

1. Electronic potentiometers and transducer systems shall conform to these standards:

   a. The measuring circuit shall be a null balance or feedback balance type.
   b. Standardization of reference voltage shall be continuous and shall be generated by an electronically stabilized source.
   c. Instrument performance shall be such that maximum error in temperature measurement shall not exceed ±0.5% of span, and dead band shall not exceed 0.15% of span.

2. Controllers shall have proportional, plus reset, plus rate action.

3. The range of temperature controllers shall be selected so that normal temperature will be in the middle third of the range when possible. Suppressed ranges shall be used, having the narrowest span practical.

4. Thermocouple burnout feature with bypass provision shall be supplied on instruments used for control. Failure of the thermocouple shall put the control valve in its NO AIR position.

5. Multiple point temperature recorders shall:

   a. Print out in numbers or periodically numbered dots.
   b. Have fixed time cycle printing.
   c. Have time interval between prints not to exceed 15 seconds—maximum cycle time for printout of all points shall not exceed 2 minutes.

6. Multiple point temperature indicators shall be designed to handle 110% of the number of specified input signals.

7. Analog temperature indicators shall meet the following requirements:

   a. Selector switches for multiple point indicators shall have gold or platinum base contacts, high contact pressure and shall be nonlocking spring return to neutral type.
   b. If a dual range instrument is specified, the range selector switch shall be the locking type, operable from the outside front of the case.

8. Digital temperature indicators shall meet the following requirements:

   a. All logic circuits shall be solid-state. Electromechanical devices, such as stepping switches or relays, shall not be used.
   b. All contacts, connectors, plugs and other electronic junctions shall be suitable for use in their specified environment.
   c. Visual digital display shall show point number as well as temperature indications.

9. Filled system primary elements may be used for local controllers, local recorders and alarm indication.

10. Galvanometer temperature instruments may be used only for alarm.

11. Temperature indicating gauges shall be bimetallic with rigid stem. Minimum dial size shall be 3½ inches, and minimum stem length shall be 6 inches.

12. For temperature range 0°-800°F, thermocouples shall be iron constantan. For operating temperatures 800°-2,000°F, thermocouples shall be chromel alumel. For cryogenic service, thermocouples shall be copper constantan.

13. Electric resistance thermometers shall be used only where thermocouples do not meet accuracy requirements.

14. The thermocouple hot junction tip shall be firmly bottomed in the thermowell.
15. Three feet of PVC-covered flexible metallic tubing shall be supplied between the thermocouple head and the conduit.
16. Thermocouple installation shall be protected from heat radiation by extending the thermocouple head. The maximum temperature permitted at the head is about 200°F.
17. Thermocouple heads shall be weatherproof. Body and cover shall be cast iron. Cover shall be threaded and gasketed. Conduit entry shall be ¾ inch. Thermocouple entry shall be ½ inch.
18. Material for thermowells shall be type 316 stainless steel, unless other special materials are required by the process fluid.
19. Thermocouple extension wires shall not be mixed with wiring of other types and other voltage levels.

Control Valves

The control valve specifications section may provide guidelines relative to maximum percent valve opening at design conditions, inner valve characteristics for various services, the types of valves to be used (single or double port globes, v-balls, butterflies, Saunders type, etc.), when to use positioners and hardened trim, the actuator types desired and when split ranges are acceptable. It provides guidelines on packing requirements, connections, the use of radiation fins, minimum body sizes to be used, the use of handwheels, the location of valve accessories and material specifications.

Typical guidelines follow.

1. Flanged valve bodies shall have a minimum ANSI rating of 300 pounds.
2. Control valves used in safety shutoff service shall be single-seated, tight shutoff.
3. Threaded valves shall be limited to body sizes of 1 inch and smaller.
4. Radiating fins or extension piece shall be used in services where the operating temperature is above 450°F or below 32°F.
5. Packing for operating temperatures up to 450°F shall be Teflon asbestos. Packing for operating temperatures above 450°F shall be graphite lubricated asbestos.
6. Pneumatic diaphragm actuators shall normally be used. Piston actuators shall be used where greater strokes or thrusts are required than can be obtained with diaphragm actuators.
7. Valve positioners shall be used for the following applications:

    a. Split-range operation which requires full valve stroke from some fraction of the controller signal (for example, a 3-15 psi positioner output from a 3-9 psi signal input).
    b. When a maximum loading pressure greater than 20 psi is required.

    c. When the best possible control is required—minimum overshoot and fast recovery.
    d. When reversing action is necessary (this may be accomplished with a reversing relay also).
    e. Where the pressure drop across the valve is high (100 psi or more).
    f. In sludge services which may cause sticky stems and guides.

8. Trim type shall be equal percentage inner valve characteristic except for the following, which shall be linear:

    a. For slow processes.
    b. When more than 40% of the system drop occurs across the valve.
    c. When major process changes are a result of load changes.

9. Control valves shall be sized to pass design flow with 70 to 80% of port area open.
10. For control valves in flashing service, both valve body and valve port shall be sized, using the sum of the flow coefficients calculated for liquid and vapor at downstream conditions.
11. Minimum body size shall be 1 inch for plug sizes down to ⅛ inch.
12. Standard diaphragm range shall be 3 to 15 psi.
13. Handwheels shall be provided for all control valves without blocks or bypass valves.
14. Butterfly valves, when used, shall be arranged to operate with the rotating shaft in the horizontal plane.
15. Valves shall have the following identifying information:

    a. The specified equipment identification number.
    b. Pressure rating of pressure holding parts, on the valve body.
    c. The manufacturer's name, model, serial number, operating range, materials of parts exposed to process fluid, size (body and inner valve), type of plug and spring range on the nameplate.

Control Panels

The control panel section usually provides guidelines on panel design such as shape, size, color and general layout. Layout suggestions include relative locations of controllers, indicators, recorders, pushbuttons, alarms and special instrument devices. It gives requirements relative to back of panel clearances, instrument equipment access, wiring and tubing methods, location of accessory equipment, back-up power requirements, identification requirements for instruments, tubing and wiring requirements and panel checkout procedures.

Typical guidelines may include the following points.

1. Panels shall generally consist of freestanding sections 4 feet-0 inches wide. The depth shall meet the re-

quirements of the panel instruments used. The color shall be green.

2. Instruments shall be arranged so that they are readable and operable by operators in a standing position. Alarm units and indicators shall be located near the top of the panel section, controllers located in the most accessible middle area and recorders on lower levels.

3. Minimum spacing shall be used between instruments within the limitations of maintenance requirements.

4. All instruments associated with a given section of the process shall be grouped together.

5. If electronic instruments are used in a high density layout, the maximum operating temperature of the instrument shall not be exceeded. Fans or blowers may be used if necessary.

6. Local instrument panels shall be totally enclosed and provided with hinged doors at the rear. If purging of instruments is required to meet area classification, it shall be accomplished by purging of individual instruments, not by purging the panel.

7. Panel location shall be free of vibration.

8. Panel design shall incorporate the following:
   a. Fifteen percent spare instrument space.
   b. Fifteen percent spare disconnect switches connected to the instrument power supply.
   c. Fifteen percent spare terminal strips.
   d. Fifteen percent spare valved take-off points from the air supply header.

9. Panels shall be purchased completely piped and wired, requiring only connection to the external piping and wiring circuits.

10. Pneumatic pressure switches shall be located so that adjustment of set points can be made without removing instruments.

11. An accessway of at least 3 feet shall be provided between the back of the panel board and the control house wall or wall-mounted assemblies; 5 feet is preferable.

12. The rear of panel wiring shall not interfere with possible future instrument installation on the panels or accessibility for maintenance.

13. When semigraphic panels are used, they shall have a simplified flow plan located above the instruments. It shall include sufficient process equipment to permit a clear understanding of the process and the relationship of the board mounted instruments to the process. Symbols and flow lines shall be color coded. Temperature indicator symbols (TI's) shall have numbers corresponding with those on the TI instruments.

14. All panel instruments, switches and similar devices shall have nameplates on front with item identification and service description. All instruments, switches and similar devices in back of panel shall have permanent labels with item identification.

15. All incoming and outgoing signal or power wiring and air transmission lines shall terminate at terminal strips or bulkheads on panels or consoles. All terminal strips and all disconnect switches shall be individually labeled for instrument identification.

16. Dual air sets with block valves and individual pressure gauges shall be employed to filter and reduce pressure to a common header, supplying instruments in the control house, or field panel if it has more than 12 instruments.

17. Each pneumatic instrument shall have an individual supply line with a block valve.

18. Air piping to panel mounted instruments shall not interfere with possible future instrument installations or accessibility for maintenance.

Analytical Instruments

The specifications section on analytical devices is usually rather specific. The quantity of applications normally is known and applications are few enough that specific guidelines are easy to provide. It gives limitations on lengths of sample lines, location of sample points and accuracies desired. Guidelines on sample conditioning systems, wiring, calibration, documentation and use of enclosures are usually specific. Inspection and testing procedures are usually outlined in detail. Such rigid requirements are set primarily because plant maintenance personnel often lack the specific training required for servicing special instruments. The following may be some of the typical guidelines included.

1. Documentation shall include at least three copies of instruction manuals supplied by the analyzer vendor and shipped with each analyzer. These manuals shall include the following:
   a. Internal wiring diagrams.
   b. Interconnecting diagrams indicating all connections and shielding requirements, wire color coding or numbering and terminal identification.
   c. A drawing of piping or tubing connections for the analyzer and sample conditioning system.
   d. Detailed installation instructions.
   e. Calibration procedure.
   f. Complete parts list showing and corresponding to the model and serial number of the analyzer.
   g. Recommended spare parts list for 2 years' operation.
   h. Trouble-shooting procedure.
   i. A bill of materials for the sample conditioning system showing manufacturer's catalog numbers of all major components.

2. Analyzers shall meet a Class I, Division I electrical classification unless the analyzer is located in a nonhazardous atmosphere and the composition of the sample inside the analyzer is insufficient to produce an explosive or ignitable mixture in the event of component leakage or failure.

3. Reproducibility shall be ±1% of span or better.

4. Sensitivity shall be ±½% of full span.

5. For calibration purposes, the analyzer vendor shall provide a one-year supply of a certified calibration

sample and all equipment necessary for field installation, including valves, pipe fittings, etc. Calibration sample injection shall be by manually operated valves at the analyzer location.

6. A calibration curve for each analyzer shall be furnished by the vendor. Demonstrated deviation from the calibration curve (either linear or nonlinear detector) shall be a maximum of ±1% of span.

7. A completely fabricated sample conditioning system shall be provided by the analyzer vendor. Provisions for measuring flow to the analyzer shall be included. Sample transport time, from entry to sample conditioning module until entry to analyzer, shall be less than 30 seconds. All components in contact with the sample shall be stainless steel.

8. Automatic zeroing shall be provided with all chromatographs.

9. Base line drift during one analysis cycle, without auto zero, shall be less than ±1% of zero adjust span.

10. All analyzers shall be housed in weatherproof shelters, meeting the recommendations of the analyzer vendor.

11. Gas cylinders for calibration or carrier gas shall be rigidly mounted outside the walk-in enclosure.

12. Analyzer enclosures shall be heated if the ambient temperature can drop below 40°F.

13. Auxiliary equipment, such as power supplies, programmers and similar items shall be located in the control house unless otherwise recommended by the vendor. The location shall provide easy access for servicing.

14. To verify proper operation and calibration, provisions shall be made for introducing a standard sample of known composition into the analyzer. Introduction of the sample shall be by manually operated valves at the analyzer location. It shall not be necessary to disconnect any tubing or piping to accomplish this operation.

15. A factory test is required of all analyzers. The factory test shall be an operating test, and it is the vendor's responsibility to demonstrate that the analyzer meets the purchase specifications. An inspector shall witness and approve all factory tests and reserves the right to approve the test procedure. Reproducibility shall be demonstrated continuously for a period of 8 hours minimum, using a standard sample. For chromatographs, a chromatogram of a "standard" calibration sample is required without auto-zero. Attenuation may be automatic or manual during this chromatogram, but the attenuation shall not exceed that used for final calibration. This chromatogram must show attenuator setting, chart speed, sample inject period, autozero period and switching points for all auxiliary valves such as backflush, etc. Each peak on this chromatogram is to be labeled and attenuator setting noted. The composition of the "standard" sample shall appear on the chart with elution times of all components listed.

Miscellaneous Device

Miscellaneous devices such as pressure, flow and temperature switches, annunciators and other similar devices may or may not be covered in the specifications. Where no guidelines exist, past experience or verbally expressed preferences should be followed.

Transmission Systems

A discussion of transmission systems centers primarily around 3-15 psig systems for pneumatics and low-level DC current systems for electronics.

Pneumatic

There are two signal levels generally recognized for pneumatic signal transmission, 3-15 psig and 3-27 psig. The 3-27 psig level is seldom used so the 3-15 psig level is normally assumed.

The most common size of transmission line is the nominal ¼-inch OD tube size. The materials used are plastic, copper, aluminum or stainless steel. Signal lines are installed individually or bundled for convenience and economy.

Although most transmission lines are made of ¼-inch OD tubing, ⅜-inch OD tubing is used occasionally where transmission distances are long enough to present transmission lag problems or where fast dynamic response is required. The question may be asked, "At what distance does a transmission line become long?" This is not an easy question to answer. It depends somewhat on the requirements of the process. Generally, transmission distances exceeding 300 feet are considered long. Most control loops using ¼-inch OD tubing respond quite well at that distance. When better or faster performance is needed, these solutions are suggested:

1. Use ⅜-inch OD tubing for small volume terminations. The response time is about twice as fast.
2. Use a positioner at the final control element (usually a valve) and make sure the transmitter has a high capacity pilot.
3. Install a high capacity 1:1 booster in the line to the final control element (see Figure 2.1).
4. Use a four-pipe transmission system (Figure 2.2).

The first three solutions work well for a two-pipe system; four-pipe systems are rarely used today. At one time the four-pipe systems were quite popular as a solution to dynamic response problems, but the increased use of valve positioners, booster relays and electronic instruments have almost eliminated their use.

Materials

The use of copper for pneumatic transmission lines is traditional. It was used long before plastic tubing was available. Copper withstands fires for a longer period than plastics and does not deteriorate as quickly. When installed by artisans of the trade, copper tubing layouts behind panels

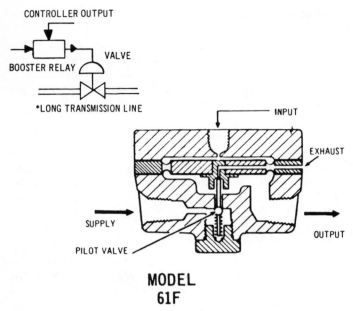

Figure 2.1. High capacity 1:1 booster relay is used to improve response time. (Courtesy of Moore Products Co.)

and in the field are works of beauty. The combination of advantages for copper results in its extensive use despite the cost advantage of plastics.

Copper tubing is installed singly or in bundles, depending on the quantity of runs needed. Individual tubes may be bare or coated with a polyethylene or PVC jacket for mechanical protection. Even when bundled tubing is extensively used, instrument location requirements dictate individual tubing runs from a junction box to particular instruments.

Tubing bundles are available in groups of 2, 3, 4, 5, 7, 8, 10, 12, 14, 19 and 37 tubes in ¼-inch OD tube sizes and in groups of same numbers up to a maximum of 12 in ⅜-inch OD size. Tubes are numbered for ease of identification. Plastic jackets and armored jackets (Figure 2.3) are available for mechanical protection. The tubing may be paralleled or spiraled to make bending easier in the higher quantity tubing bundles.

Armored bundles (Figure 2.3a) provide greater protection from mechanical abuse and are used in installations where corrosion and mechanical abrasion are expected. When direct ground burial is desired, still more protection is available by placing a PVC jacket over the armored cable (Figure 2.4).

Bundled aluminum tubing is furnished in ¼-inch OD size with tube quantities up to 37 in number, but manufacturers do not usually have as many variations in selections as are available in copper. Plastic and metallic armors are available just as they are for copper tubing bundles. Aluminum tubing is not as flexible as copper, and leakages are more likely to occur at fittings. Its use is recommended where atmospheres are particularly suited to the use of aluminum.

Plastic tubing was introduced in the late 1940s and has enjoyed increasing use since that time. It costs less than copper or aluminum, is much lighter and can be installed more easily, reducing labor costs. Some plastic materials become brittle when exposed to outside weather conditions, restricting its use considerably at first. Improvements in its

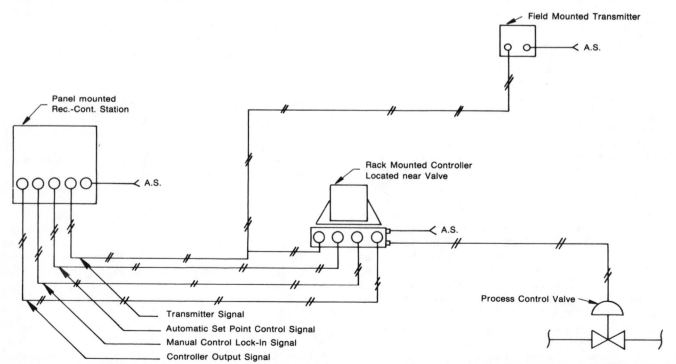

Figure 2.2. A typical four-pipe control system is also used to increase response time, especially where long transmission distances are involved.

a.

b.

c.

Figure 2.3. Plastic sleeves and armored sleeves are used for mechanical protection of tubes. Tubes are numbered for convenience and may be run parallel or spiraled to make bending easier. (Courtesy of TRW Crescent Wire and Cable, A Division of TRW, Inc., and The Okonite Company)

a.

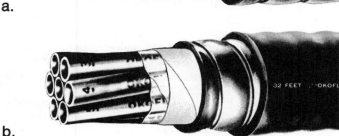

b.

Figure 2.4. Armored and double sheathed tube bundles offer protection from mechanical abuse and corrosion for underground burial. (Courtesy of The Okonite Company)

weather-resisting qualities have largely overcome this objection, however.

Plastic tubing is available in several colors in single tubes so that the ISA color code requirements (ISA RP 7.2) may be followed if desired. Only black tubing is recommended, however, if the tubing is exposed to ultraviolet light or where temperatures are expected to exceed 130°F.

Various grades of plastic are available for varying temperature and pressure requirements. Polyethylene is the most commonly used. Nylon is also used and has several physical property advantages over polyethylene that make its use desirable for rough service. It is more flexible, withstands vibration and impulsing stresses better and is more resistant to pressure and temperature extremes than polyethylene. It is available in black only.

Like copper and aluminum, plastic tubing bundles may contain up to 37 tubes, numbered for ease of identification and with a variety of jackets and armors for various installation conditions. To overcome partially the advantage of copper over plastic, plastic tubing bundles are available with a special thermal barrier (Figure 2.5) that extends its life under flash fire conditions.

Special combinations of 1/4-3/8- and 1/2-inch OD tube sizes can be purchased for special applications. These usually are on special order only, and costs are relatively high.

Distribution and Termination

The introduction of multitube bundles greatly simplified the distribution of pneumatic transmission lines. Prior to their introduction, a common practice was to run single copper tubes in channel iron of varying widths, depending on the quantity of tubes to be run. In many instances layers of tubing were stacked in the channel with iron bars as

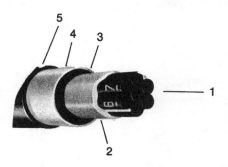

Construction:

1. Multiple number-coded high-density polyethylene tubes
2. Special thermal barrier
3. Barrier envelope of 3/16-inch vinyl
4. Asbestos layer
5. 1/16-inch vinyl jacket

Figure 2.5. Fire-resistant plastic multitube bundle construction extends life under flash fire conditions. (Courtesy of Samuel Moore and Co.)

spacers between layers of ¼-inch OD tubing. The tubing usually was furnished in 50-foot sections, stretched to achieve rigidity and sweated together or joined with compression fittings. The amount of labor required for installation was high, and testing was laborious and costly because of the quantity of joints to be tested. Leaks were common and difficult to locate.

Multiple bundles have largely replaced single tubing runs as indicated in Figure 2.6. Coming from a central control room, bundles of the desired quantity are grouped for ease of support on expanded metal tray or other suitable support systems and run to various areas of the plant. They are terminated in junction boxes and connected to individual tubing runs to transmitters and control valves in the process. Tubing runs from the control panel to the junction boxes are continuous, greatly reducing the chances for leaks and providing more economical installations.

Terminations at control panels are made with bulkhead fittings as shown in Figure 2.7. Individual tubes in the bundle are numbered for identification, and the tubing is also

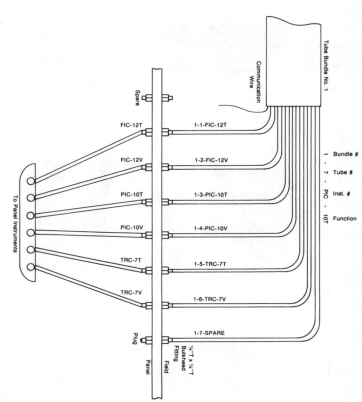

Figure 2.7. Tubing bundles are terminated at bulkheads at rear of panel where connections are made to panel tubing from individual panel instruments.

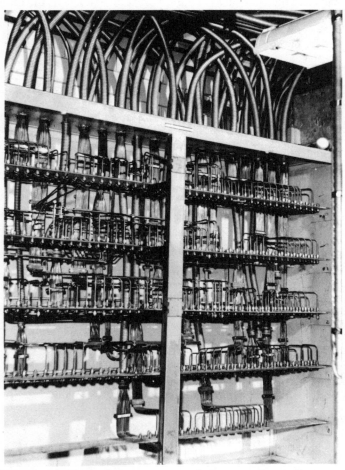

Figure 2.6. Multitube runs can be installed easily and quickly, eliminating much labor and expense of single tubing runs. (Courtesy of TRW Crescent Wire and Cable, A Division of TRW, Inc.)

identified with the proper transmitter or valve to which it runs. At the field junction boxes, similar methods are used to terminate or connect the bundled tubing to the individual tubes. There are two methods of termination, one at bulkhead bars inside the junction box and another that uses the junction box sides as bulkheads (Figure 2.8).

Individual tubing runs are supported by angle iron or other structural forms and, in some cases, are clipped to or supported from large diameter pipe. The latter method is economical, but the tubing is exposed more to mechanical damage than when supported by angle iron, channel or special supports.

To aid in the installation of tubing bundles and support systems, isometric drawings of the entire tubing layout are often made. The tubing tray is usually assigned an area on the pipe support system when tubing runs are made overhead. When the tubing is run underground, it may be encased in concrete, buried directly without additional protection or run in conduit (individual runs or small bundles) or other protective forms.

Interference problems are few for pneumatic transmission lines. Vibration is not a particular problem as long as excessive movements are avoided. Lines should not be run into areas of excessive heat. Exposed plastic tubing should not be run close to hot steam lines or hot process lines.

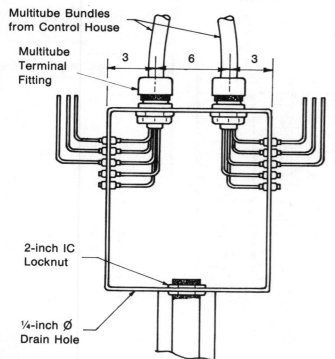

Multitube Bundles from Control House

Multitube Terminal Fitting

3   6   3

2-inch IC Locknut

¼-inch Ø Drain Hole

Figure 2.8. Tubing junction box terminations are made by using its sides as bulkhead bars or by using mounting bars inside the boxes for tubing connections, then running the tubing through grommets.

Electronic

The discussion of electronic transmission lines is limited to applications involving conventional transmitter-controller-valve loops, excluding special devices such as chromatographs, radioactive sensing devices, magnetic flowmeters and similar sensing and transmitting devices which produce nonstandard signal levels. Most of these devices produce low level DC current or voltage signals that are transduced to standard signal levels when used in conventional control loops.

Signal levels used for electronic transmission are not as well standardized as their pneumatic counterparts. Signal level outputs include 1-5, 10-50 and 4-20 milliamp DC and 1-5 volts DC. Signal inputs include the same values as outputs, and some devices accept other levels such as 0.25-1.25, 0-4 and 0-10 volts DC.

Multiconductor cable is normally used for electronic signal transmission. Signal wire sizes range from #16 to #24 AWG. Signal pairs are normally twisted, typically about six times per foot, and multiple pairs are twisted, about twice per foot. The quantity of conductors may range from 2 to 100 pairs; 25 pair cables are used rather extensively. The cables are run from the control center to junction boxes strategically located in the process area where they are terminated and connected to individual pairs going to in-

struments in the area. Figure 2.9 is a typical junction box installation showing cable and wire terminations and their distribution.

Multiconductor cables may be run in cable trays or wireway systems for distribution or may be run in rigid conduit. Plastic coated sheaths are acceptable in many areas; metal armor is used for mechanical protection for rough service. Rigid conduit, though more expensive, provides the safest installation from the viewpoint of safety and insurance from damage.

Cables and conduit may be run underground or overhead. Underground runs should be laid in trenches in horizontal tiers with as little crossing of cables as possible. Cable and wire insulation should be moisture resistant. Burial depth should be at least 18 inches along normal routes and 30 inches under roadways. Underground obstructions and paved areas should be avoided if possible. A minimum clearance of 3 or 4 inches should be maintained between cables and underground piping. They should not be run parallel to underground pipe or to high voltage AC lines.

In new plants where underground runs may be made easily, they can be routed more directly than overhead runs, thus using shorter runs. Cable pulls can be made more easily where short, direct runs are made with consequent savings of materials and labor. Interferences with roadways, vessels and piping may also be avoided.

The disadvantages of underground runs include interference with existing underground piping or electrical systems, inaccessibility for trouble shooting, inability to add new conduits without excessive costs and interference when making plant additions that require excavation—foundations, etc.

Overhead cable and conduit runs are usually made at selected locations on pipe racks. Typical locations are above, below and in a chosen area within the pipe rack.

Interferences present more problems in electrical transmission systems than in pneumatics, primarily from other electrical systems. Routing of signal wires with respect to power or other electrical systems must be carefully considered to prevent spurious noise by:

1. Electrostatic or capacitive coupling
2. Electromagnetic or inductive coupling
3. Leakage current or ground loops

Precautions that need to be taken include the proper separation of cable runs, isolation of low current and low voltage signals, proper grounding techniques and proper shielding when terminals are necessarily close together.

Instrument signal wiring should not be run close to electrical noise-generating equipment such as power transformers, etc. Suggested minimum separation between high voltage sources and thermocouple or low current and low voltage signals levels are given in Table 2.1.

The distribution of conduits (overhead or underground) is similar to pneumatic transmission lines. Cables are run to strategically placed junction boxes, terminated at terminal strips and connected to leads from individual transmitters

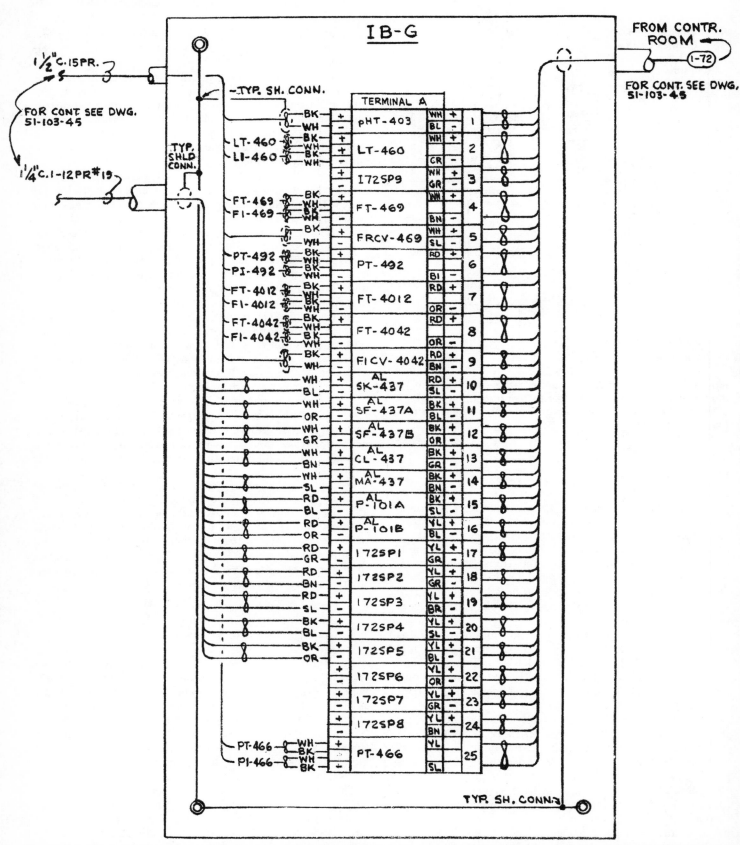

Figure 2.9. These electrical terminations are typical for multiconductor cable that distribute to various instruments within the area.

Table 2.1. Typical Spacings Between Power and Signal Wiring		
Power Wiring Capacity		Min. Spacing (in.)
Max. Voltage less than:	Max. Current less than:	
125v	10A	12
250v	50A	18
440v	200A	24
5kv	800A	48

and valve transducers within the area served by the junction box. From terminals at the transmitter or transducer, the junction box provides the only other set of terminals between the device and the control room. Within the control room, termination is likely to be at a panel terminal strip provided by the panel fabricator. Wires from that terminal strip run to controllers, recorders or indicators mounted in the panel.

Identification

Proper identification of pneumatic and electronic transmission lines is time-consuming and therefore expensive. It is necessary, however, for ease of checkout and testing and is essential for maintenance activities and troubleshooting when difficulties occur.

Pneumatic lines are usually identified at each termination point by a number which probably includes the control loop identification number. The loop number may include a T or a V following the number to identify it as the transmitter line or the valve line. When plastic tubing is used, the color of the tube may identify its function.

Electrical leads to and from junction boxes or other junction points must also be properly labeled or identified. Identification should include the appropriate instrument number such as FRC 302, PRC 713, etc.

Process Connections

The take-offs and connecting piping from pipelines or equipment to the instruments are important to the satisfactory operation of instrument control systems. The methods of take-off, materials, connection sizes and mounting methods are discussed and evaluated in the following paragraphs.

Valves should be installed at each instrument take-off so that instruments may be removed from service for calibration, maintenance or repair. Valves should be located as close to the line or equipment as possible. Gate valves are commonly used for this purpose, but plug valves, needle or other types are used when economics or other considerations (such as slurry service) make them more acceptable. When consideration is being given to the use of ball valves, it should be remembered that ball valves have a quick-opening flow characteristic, giving rise to rapid pressurization and depressurization when putting a transmitter in service or checking its calibration.

Separate take-offs are desirable for each instrument item but are not essential. For example, a pressure connection may be made to one leg of a differential flowmeter installation, or a level gauge and a level transmitter may be mounted from the same vessel connections. If more than one instrument is installed from a common connection, however, each instrument should be isolated by a block valve so that it can be removed from service without interfering with the operation of other instruments.

Screwed connections are more economical than flanged or welded connections and should be used where possible. Special cases may dictate otherwise. For example, some standards require all vessel connections to be flanged. Such a philosophy is based on the possibility of accidental breakage of the more fragile screwed connection. These philosophies usually develop as a result of some costly accident which might have been avoided with heavier connections.

High pressure services sometimes require the use of socket weld or back-welded screwed connections. In most instances, even though the process piping has these requirements, instrument connections may still be screwed for ease of maintenance and repair. Piping specifications usually change to instrument specifications after the first block valve. Exceptions may include hazardous chemicals, hard-to-contain gases, and very high pressure services.

Standard connection sizes for instrument take-offs are ½-, ¾-or 1-inch NPS. The size varies with the type service, but ½-inch NPS is usually specified as a minimum.

One-half inch is adequate for almost all pressure devices, and ½-inch piping or ⅜-inch OD tubing is normally used for connecting the instrument to the process. Connecting piping is usually sloped (about 1 inch per foot) to keep connecting lines drained for gas and vapor services and to keep lines filled on liquid services.

Connection sizes for differential flow elements are normally ½-inch NPS. When pressure ratings exceed 600 pounds, ¾-inch NPS connections are often used. Connecting piping and tubing are about the same as for pressure piping. Slopes are maintained unless the differential pressure units are close-coupled line mounted.

Temperature instrument connections are usually ¾- or 1-inch NPS. Thermowells for thermocouples, resistance bulbs or filled temperature bulbs are normally furnished with either ¾- or 1-inch NPS threads.

Level instruments require a greater variation than any of the other sensing elements. Displacer level gauges vary from 1 to 2 inches in size with 1½ inches quite common. Float devices vary from ¾-inch NPS to 4 inches for flange mounted types.

Table 2.2. Typical Tap Location Schedule for Line and Pedestal Mounted Instruments

Fluid	Takeoff Connection Orientation Horizontal lines	Instrument Location	
		Line Mounted	Pedestal Mounted
Liquids	Horizontal centerline	Below takeoff connections	Below takeoff connections
Gases	Vertical centerline	Above takeoff connections	Above takeoff connections
Steam	Horizontal centerline	Below takeoff connections	Below takeoff connections

Location of Taps

The location of taps for instruments should be carefully considered, particularly for temperature and analytical measurements.

When streams of different temperatures are mixed, temperature measurements should be made 10 to 15 feet downstream to ensure complete mixing. Where possible, thermowells should be located at pipe bends where flow is more turbulent and ensures faster response. Thermowells should be inserted at least to the center of the pipe when installed in pipelines. In equipment of 12-inch diameter or greater, thermowells should extend into the vessel or equipment at least 4 inches, preferably more, unless other reasons forbid. Interference with internals, possible buildup of process material and other considerations sometimes dictate shorter insertion lengths. Other factors that need to be considered relative to location include the amount of agitation, whether the equipment is insulated, etc.

Take-off locations for analytical measurements should be selected to ensure that the sample is representative of the stream. A knowledge of the process is necessary for proper evaluation. Points for consideration include:

1. Insufficient length for mixing where streams of different components are merged
2. Stratification of fluids in lines having laminar flows
3. Condensation or vaporization of certain stream components (this requires a check of the pressure-temperature relationships of the various components)
4. Location with respect to the sensing device to prevent too great a lag in transport time
5. Accessibility for maintenance
6. Entrainment of solids or gases in the sample stream

Locations for level devices are usually dictated by the requirements for level measurement. Displacement sensors have fixed locations while differential devices have some freedom of location provided compensation is made for the difference in liquid heads.

Pressure measurements are less likely to give trouble than most other sensing devices. However, in low pressure measurements, liquid heads may still be an appreciable percentage of the measurement range.

Table 2.2 provides a typical orientation schedule for instrument take-off connections for line and pedestal mounted instruments. It applies particularly to pressure and flow instruments.

Sealing Instruments from the Process

Many process conditions exist which require seals between the instrument device and the process. Following is a list of conditions for which seals should be considered:

1. Where corrosive fluids are present
2. When liquids may vaporize in lead lines
3. When vapors may condense in lead lines
4. Where fluids are extremely viscous or solid at ambient temperatures and where heat tracing may be unsatisfactory
5. Where fluids may polymerize in a line
6. Where slurries or fluidized solids occur

Acceptable methods for sealing include mechanical seals (diaphragms), purging with suitable fluids or the use of immiscible liquids.

Diaphragm seals are effective for pressure measurements but much less so for flow and level measurements because of range and accuracy requirements.

Purging of sensing lines with suitable fluids from reliable sources is often used for flow and pressure measurements, but suitable fluids are not always readily available.

Immiscible fluids for seals on flow and level measurements have been used frequently in the past. Their use has declined with the increased use of suitable diaphragm seals.

Condensing pots are used on steam flows (and other condensible vapors) to seal the instrument from hot steam or vapors.

Manifolds and Gage Valves

Responding to an economic need to reduce construction costs, manufacturers have produced labor saving manifolds and gage valves. Instead of using numerous fittings to connect the process to the instrument, such as is indicated in Figure 3.1, a prefabricated 3-valve manifold reduces the number of fittings dramatically (Figure 12.4) and minimizes time of installation. There also are fewer joints and, therefore, fewer chances for leaks.

Manifold constructions include single-valve arrangements for zeroing differential pressure devices, which only equalize, or which equalize while blocking one process connection so that recalibration can occur at line operating pressure. Three-valve and five-valve manifolds are also available in several styles. Process and instrument connections can be screwed, or flanged for differential pressure transmitter makeup. Some styles are available with OS & Y (outside stem and yoke) construction for use where construction specifications call for it.

Prefabricated manifolds are now available for use between a vessel flange and a differential pressure transmitter used to measure level. Manifolds provide an equalizing leg connection.

Most transmitters in use today are mounted such that sense lines connect to transmitter, transmitter bolts to bracket, and bracket bolts to mounting stand (shown in Figure 2.10a). With this approach, actual mounting sequence is in the reverse order of the steps just listed. Notice that sense lines are not connected until the transmitter has been mounted. If process connections and mounting stands and brackets could be installed without the presence of the transmitter, then transmitter delivery and calibration might not be restraining factors during construction. Also, transmitters could be calibrated and stored in a protected environment until just before loop check and start-up activities. Recent manifold and bracket designs make this judicious timing of transmitter installation possible.

The "AGCO MOUNT" by Anderson, Greenwood & Company uses special brackets and spacers to mount a 3-valve manifold on a 2-inch pipe stand. An optional steam trace block can be factory-attached to the manifold with heat conducting cement. The transmitter is then bolted to the manifold to complete the installation (Figure 2.11). Another optional bracket is used for mounting purge rotameters, air supply regulators, and conduits (Figure 2.12).

The D/A "Unimount" installs directly to a 2-inch pipe stand using U-bolts. It is roddable through the manifold vent valves as well as through optional process connections for close-coupled installations (Figure 2.13). Heat trace connections are also optional.

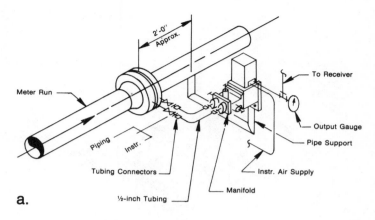

a.

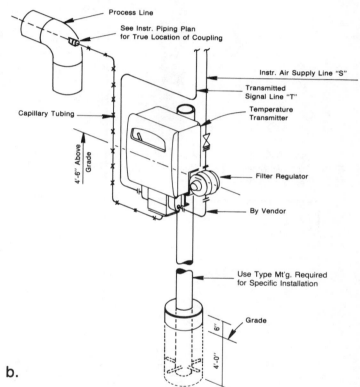

b.

Figure 2.10. Instruments may be line mounted or pedestal mounted, depending on accessibility needs and plant preference.

Instead of using a half-dozen fittings and valves to connect a pressure gage to a process line, many users now permit the use of gage valves. These "valves" are actually prefabricated manifolds, which include connections for process, gage, and bleed (vent). OS & Y construction is available (Figure 2.14), and vent plugs are available as an inexpensive replacement for bleed valves.

Since corrosion is a maintenance problem with most needle valves, stems having packing below the threads are

Figure 2.11. An AGCO MOUNT allows the transmitter to be mounted or removed without disturbing process sensing lines. (Courtesy of Anderson, Greenwood & Co.)

Figure 2.12. Optional brackets can be added to the assembly to mount purge rotameters, air supply regulators, and conduits. (Courtesy of Anderson, Greenwood & Co.)

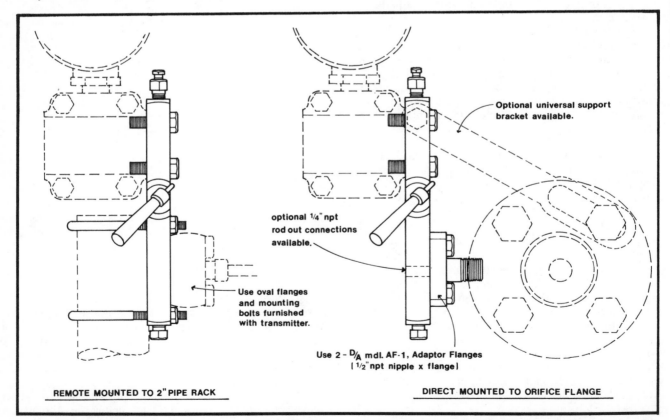

Optional universal support bracket available.

optional ¼" npt rod out connections available.

Use oval flanges and mounting bolts furnished with transmitter.

Use 2 – D/A mdl. AF-1, Adaptor Flanges [½"npt nipple x flange]

REMOTE MOUNTED TO 2" PIPE RACK

DIRECT MOUNTED TO ORIFICE FLANGE

Figure 2.13. The "Unimount" is roddable and installs directly to a 2-inch pipe stand. (Courtesy of D/A Manufacturing Co.)

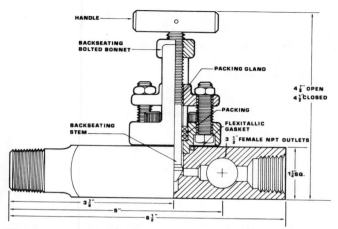

Figure 2.14. OS & Y construction is available in gage valves which are actually prefabricated pressure manifolds. (Courtesy of Hex Engineering Division of Circle Seal Corporation)

Figure 2.15. This 3-valve manifold utilizes needle valves which have packing below the stem threads to reduce corrosion problems. (Courtesy of Hoke Inc.)

desirable (Figure 2.15). Process fluids are prevented from reaching the threads, eliminating stem "freeze-up." Unless care is exercised when attempting to open a stuck valve under line pressure, the plug and packing assembly threads can strip off, causing the assembly to blow out of the valve body. Since a projectile such as this can injure or even kill, OS & Y construction is often required. Most manufacturers now insert a pin adjacent to the flat of the plug and packing assembly to prevent its turning.

Miscellaneous Design Criteria

Other design criteria that need discussion include concepts on mounting instruments, identifying them, selecting ranges and charts, winterizing requirements and meeting the requirements of the various electrical classifications.

Mounting Instruments

Field instruments are either line mounted or pedestal mounted (see Figure 2.10). Line mounting is usually more economical because it reduces material and labor installation costs. This is offset by increased maintenance costs if they are difficult to reach. The usual solution is to line mount them when possible and use pedestal mounting when ease of accessibility is judged to be necessary. Flow d/p cells and other similar units are often line mounted with platforms or ladders used for access. Most locations at grade level or on platforms are considered to be accessible unless the location interferes with normal operating routine. The mounting height above grade or platform ranges normally from 4 to 5 feet.

Areas around heavy equipment such as compressors, generators, etc., are subject to vibration, presenting instrument mounting problems. Pipeline mounting in these areas may need to be avoided. In many cases, walls or columns may be used. Most users will not allow instruments to be mounted on handrails because of occasional erratic readings due to vibration and because of risk of damage.

Selection of Units, Charts, Ranges

The selection of units, charts and ranges may seem trivial, but a surprising amount of time (engineering, construction, startup) can be spent correcting and adding this type of information if it is not settled during the early stages of design.

A common unit for liquid measurement is gpm (gallons per minute), but gph (gallons per hour), bph (barrels per hour) and pph (pounds per hour) are also used. Gas flow measurements may be in pph, cfh (cubic feet per hour), scfh (standard cubic feet per hour) or scfm (standard cubic feet per minute). Pressures are usually given in gauge or absolute units, psig or psia. Temperatures are most often given in Fahrenheit or Centigrade scales.

Range selections for instruments should be made carefully. Pressure ranges are chosen so that the normal operating pressure is about mid-scale. Scale divisions should be easy to read and use.

Flow ranges on differential flow elements are selected so that normal operating flow is about two-thirds of full scale. This causes the indicating pen to be close to mid-scale on a square root chart. On control applications where square root devices are used, control below 25 percent of full scale is not advisable. When linear function elements are used, control can be successful at near-minimum and near-maximum values of full scale.

Charts used for pressure measurement usually correspond directly to the range of the pressure elements. This is not true for flow devices. Scales for differential elements are normally 0 to 10 square root (0-10 $\sqrt{}$), and multipliers are chosen to make scale reading easy. Linear output devices are usually provided with 0-100 linear scales.

Most common for level applications are 0-100% scales, although actual level ranges in distance (inches, feet, etc.) and volume measurements (gallons or barrels) are sometimes used.

Temperature charts most often reflect the actual range selected: 100°-300°F, 0°-75°C, etc. More variations in range exist for temperature measurements than for other variables, except perhaps for analytical devices.

Instrument Identification

Identification practices for instruments are about the same throughout industry. Different methods are used for control room devices, field devices and control valves.

Control room instruments that are displayed on panels or consoles are usually identified by a nameplate with an identification number and service of the instrument such as "FRC-620—Flow to Reactor A." The instrument tag is made of laminated plastic, white letters on a black background, about ¾-inch high by 2 inches wide by 1/16-inch thick, and usually displayed immediately below the instrument. Many recorders and recorder-controllers have provisions for holding nameplates inside the instrument door.

The most common practice is to use a single color for tagging all instrument categories, but a variation quite helpful to operating people is the use of different colored tags for pressure, level, flow and temperature instruments. This helps operators identify instrument functions, particularly during the training phase or during initial operation of a plant or process.

Factory identification nameplates are usually placed on the rear of panel instruments also. This small metal tag contains the serial number of the device (for the manufacturer's identification) as well as the plant instrument identification number.

Field items such as transmitters and local controllers are usually identified by a small metal nameplate or tag attached to the instrument bearing the plant assigned item number (LC-409). The manufacturer stamps his serial number on that or another nameplate. If nameplates are not permanently attached, metal or plastic tags are wired to the device with the appropriate identification number. Items which normally are not permanently marked when received from a vendor include thermocouples, pressure gauges, dial thermometers, solenoid valves and other similar small devices. These are usually field tagged.

Control valves normally are identified by a metal tag permanently attached to the valve yoke on which is stamped the control valve identification number (PRCV-26, for example), the pressure rating of the valve body, the manufacturer's name, model, serial number, operating range, materials of parts exposed to the process fluid, valve size (body and inner valve), type of plug and spring range.

Winterizing

Protection requirements to prevent freezing of process fluids in instruments and instrument lead lines vary a great deal, depending on the climate at the plant location. Most areas within the continental limits of the United States require some protection. Stagnant instrument lead lines and instruments themselves often must be insulated, steam traced and insulated, or, in rare cases, they may use electrical heaters to prevent freezing.

Water lines are nearly always susceptible to freezing, and often other process fluids have low freeze points or may form water or other condensates that can freeze. Fluids that solidify at ambient or temperatures above the freezing point of water must receive special consideration.

Construction Materials

The materials from which instruments are fabricated are of concern from two standpoints:

1. The effect of its contact with the process medium—referred to as "wetted" parts
2. The ability of all the non-wetted parts to withstand ambient conditions and continue to perform adequately

In general, metals used for wetted parts are common— brass, carbon steel, and 316 stainless steel. Alloys, such as Hastelloys and Inconel, and more exotic metals, such as titanium and zirconium, are sometimes recommended to increase corrosion resistance. Because of economic considerations, metallurgical specification for process piping is carried through the take-off valve and is then changed (often to 316 stainless steel) to an instrument piping specification. The rationale is that stainless steel tubing can be replaced several times for the cost of identical exotic materials.

Materials of construction for field instrument housings are not standardized in the industry. The specification of materials should be developed in the light of current practice within the individual plant. It should be obvious that tighter specifications are needed for a humid, fungus-prone, salt water atmosphere than for a cool inland climate.

Packaged Equipment Systems

In recent years a trend has developed to purchase equipment systems that perform a specified purpose. The reason, of course, is to make use of outside expertise in special areas. Extruders, generators, compressors, boilers, refrigeration units, conveying systems, mixers, furnaces, driers, etc., fall into this category. These systems often include an appreciable amount of instrumentation.

It is desirable that the instruments required for such packaged systems be compatible with other instruments in the process to reduce spare parts requirements, training and maintenance efforts. Unfortunately, this often is not the case. Vendors of special systems such as these normally have their own preference of instruments and adhere to them for economic reasons—lower original cost based on quantity discounts and reduced labor costs due to repetitive engineering and installation procedures.

Under such circumstances, a decision must be made whether compatibility problems are offset by other economic advantages. These problems should be anticipated in the early stages of a job to minimize incompatibility as much as possible.

Electrical Safety

Process plants are becoming increasingly dependent on electrical instrumentation and computers. The safe and economical installation of these devices depends on the selection of the appropriate equipment to meet the hazards of the areas in which they are installed. An understanding of these hazards is needed along with a knowledge of how they are classified and the methods used to make them safe for operation.

Electrical equipment can cause fire or explosions in areas which contain flammable liquids, gases or dusts. An ignition (and consequent explosion) takes place when three conditions exist at the same time: (a) the presence of a fuel, (b) the presence of oxygen and (c) a source of heat (sufficiently high).

Flammable gases are most hazardous, and each gas or vapor has a range of concentrations in air under which explosions can occur. When concentrations are outside of this range, the mixture is either too rich or too lean to burn. One method of protecting against explosions is to provide sufficient ventilation to dilute the hazardous vapors or gas below their lower explosive limit (LEL).

Table 2.3 lists some flammable liquids and gases and shows properties of the fluid that are of primary interest from an ignition standpoint. These properties include the upper and lower explosive limits, the vapor density, flash point, ignition temperature and the autoignition temperature.

The enrichment of air with oxygen tends to extend the explosive range of flammable mixtures. In pure oxygen atmospheres, many materials will burn vigorously that are not ordinarily flammable. Electrical systems that might be exposed to enriched oxygen atmospheres must, therefore, receive special attention.

Gases and vapors that are lighter than air diffuse quite readily into the atmosphere and become diluted. In such circumstances they may quickly dilute below their LEL unless they are kept in an enclosed space.

Heavier-than-air gases are less likely to disperse quickly. They have a tendency to creep along floors and low places

maintaining ignitable mixtures. If they reach an ignition source, a fire or explosion results.

Flammable liquids are not as hazardous as flammable gases. Liquids do not ignite until a sufficient amount evaporates for its vapor to reach the lower explosive limit. Gases, however, are flammable immediately upon exposure to an ignition source.

Liquids are classified by the National Fire Protection Association (NFPA) according to their volatility; the more volatile the liquid, the sooner its vapor concentration reaches the LEL. The flash point of a liquid is the temperature at which it gives off enough vapors to form an ignitable mixture (the LEL). Volatile liquids with low flash points approach the danger inherent in flammable gases.

The NFPA has three classifications of flammable liquids. Class I are those very volatile liquids whose flash points are below 100°F. There are three categories in Class I depending on the relationship between flash points and boiling points.

Class II liquids are those between 100° and 200°F, and Class III are those above 200°F. Liquids with flash points above 200°F are not considered very hazardous.

The autoignition temperature is that temperature to which a fuel-air mixture must be heated in order for it to ignite spontaneously (without the introduction of a flame or spark). A hot surface also can serve as the ignition source. The larger the hot surface is, in relation to the volume of fuel-air mixture, the sooner ignition will occur.

Dust explosions occur when flammable dusts are suspended in air and are ignited by flames, sparks or hot surfaces. The properties of flammable dusts that contribute to ignition are their particle size, concentration, minimum explosive concentration, ignition temperature and minimum cloud ignition energy.

The smaller particle sizes are more susceptible to explosions. Most explosive dusts are finer than 200 mesh. Those larger than 100 mesh are usually not hazardous.

Metallic dusts present special problems because their conductivity causes shorting and arcing of ordinary electrical devices and, in effect, creates their own source of ignition. Table 2.4 lists properties of some common explosive dusts.

National Electric Code

The National Electric Code (NEC) is a set of widely accepted code of standards relating to the design of electrical systems—wiring and apparatus. It is recommended by the NFPA and adopted by most fire and insurance underwriters as well as most industrial plants.

Article 500 of the NEC defines three classes of hazardous locations. (These classifications are given also in API RP 500. The API standard refers only to petroleum refineries, although it is widely used in the chemical processing industries too.)

Class I: locations made hazardous by flammable gases or vapors.

Class II: locations made hazardous by combustible dusts.

Table 2.3. Liquid and Gas Properties Can Cause Electrical Hazards

Material	Explosive Range, Vapor			Temperature @F		
	%Vol. in Air		Density	Flash	Boiling	Auto-
	Lower	Upper	(Air = 1)	Point	Point	Ignition
Acetone	2.6	12.8	2.0	0	134	1,000
Acetylene	2.5	81.0	0.9	Gas	−118	571
Ammonia	16.0	25.0	0.6	Gas	−28	1,204
Carbon disulfide	1.3	44.0	2.6	−22	115	212
Ethylene	3.1	32.0	1.0	Gas	−155	914
Ethylene oxide	3.6	100.0	1.5	<0	51	804
Ethyl ether	1.9	48.0	2.6	−49	95	356
Gasoline	1.3	7.0	3.0	−45	100−400	536
Hydrogen	4.0	75.0	0.1	Gas	−22	1,035
Methanol	7.3	36.0	1.1	52	147	725
Propylene oxide	2.8	37.0	2.0	−35	95	
Vinyl chloride	3.6	33.0	2.2	Gas	7	882

Table 2.4. Product Dusts Cause Electrical Hazards Which Vary with the Dust Properties (Properties of Some Explosive Dusts)

Material	Min. Cloud Ignition Energy, Millijoules	Ignition Temperature, °C		Min. Explosive Concentration, Oz/100 ft	Min. Explosion Pressure, psig	Max. Rate of Pressure Rise, psi/sec.
		Cloud	Layer			
Cornstarch	40	400	—	45	106	7,500
Wood flour	40	470	260	35	113	5,500
Pitch (petroleum)	25	630	—	45	82	3,800
Coal (bituminous)	30	610	180	50	101	4,000
Sulfur	15	190	220	35	78	4,700
Benzoic acid	20	620	Melts	30	76	5,500
Aluminum (fines)	50	650	760	45	84	20,000+
Magnesium	40	560	430	30	116	15,000
Titanium	25	330	510	45	70	6,000
Polyethylene (high pressure)	30	450	380	20	81	4,000
Polyurethane (fire-retardant)	15	550	390	25	96	3,700
Rubber (crude, hard)	50	350	—	25	80	3,800
Methyl methacrylate (polymer)	20	480	—	30	84	2,000

Class III: locations made hazardous by ignitable fibers or flyings. This class is not very pertinent to the chemical and petroleum industries.

Each of the three classes of hazards are further divided into Division classifications.

Division I locations are those which may contain hazardous mixtures under normal operating conditions.

Division II locations are those in which the atmosphere is normally nonhazardous but may become hazardous under abnormal circumstances such as equipment failures, failures of ventilating systems, etc.

Locations that are not classified as Division I or Division II are referred to as nonhazardous.

The design of electrical equipment and its associated wiring systems also takes into consideration the characteristics of the atmospheres produced by the specific gases, liquids, vapors and dusts. Accordingly, the classes and divisions listed above are also further divided into groups of varying levels of hazard. The tests required of equipment to meet approval for use in the various hazard levels are more stringent for those mixtures that are more dangerous. There are six groups, A through G, with Group A being the most hazardous. The degree of hazard is in descending order from Group A to Group G. The complete grouping follows.

Group A: atmospheres containing acetylene.

Group B: atmospheres containing hydrogen or gases or vapors of equivalent hazard such as manufactured gas.

Group C: atmospheres containing ethyl/ether vapors, ethylene or cyclopropane.

Group D: atmospheres containing gasoline, hexane, naptha, benzene, butane, propane, alcohol, acetone, benzol, lacquer solvent vapors or natural gas.

Group E: atmospheres containing metal dust, including aluminum, magnesium and other metals of similarly hazardous characteristics.

Group F: atmospheres containing carbon black, coal or coal dust.

Group G: atmospheres containing flour, starch or grain dusts.

Groups A, B, C and D are pertinent to Class I areas, and Groups E, F and G are pertinent to Class II areas.

Selection guidelines for equipment and electrical devices and the installation requirements for electrical systems that must be followed to meet the code are given in the NEC. These requirements are often very costly. There are two primary ways used to reduce the cost of meeting these hazards to provide more economical, but still safe, electrical systems: (a) by purging and pressurizing the equipment and (b) by providing intrinsically safe equipment and systems.

Purging and Pressurizing Enclosures

In both Class I and Class II hazardous areas, the NEC requires that enclosures for electrical devices and for terminal boxes must be constructed in such a manner as to contain an explosion should it occur within the enclosure. The enclosure is not made airtight to prevent the entrance of hazardous vapors but must withstand an explosion that might occur and suppress or extinguish the flame that is produced by the ignited mixture. Such enclosures are referred to as explosionproof and are necessarily heavy for their size, difficult and costly to install and require much space.

The NFPA allows the use of purged and pressurized non-hazardous enclosures in lieu of explosionproof enclosures or fittings provided certain conditions are met in the purging and pressurizing schemes. Complete requirements are given in NFPA Standard 496, *Standard for Purged and Pressurized Equipment in Hazardous Locations* or in the Instrument Society of America Standard, ISA-S12.4. The ISA standard embodies the principle stated in Chapter 5, Article 500, Paragraph 500.1, of the NEC which states that in some cases, hazards may be reduced or eliminated by positive pressure ventilation if effective safeguards are employed against ventilation failures.

ISA-S12.4 is applicable only to enclosures whose volume does not exceed 10 cubic feet. NFPA Standard 496 is applicable to larger enclosures such as electrical equipment rooms or control rooms.

The purge or pressurizing source must be clean air or inert gas.

There are three types of purging classifications, X, Y and Z.

Type Z purging covers requirements adequate to reduce classification within an enclosure from Division II to non-hazardous.

Type Y purging covers requirements adequate to reduce classification within an enclosure from Division I to Division II.

Type X purging covers requirements adequate to reduce classification within an enclosure from Division I to non-hazardous.

When Type X purging is used, an interlock must be used to shut off power to equipment when the purge flow is lost because ordinary arcing or sparking equipment would otherwise be operating in a hazardous area.

When Type Y purging is used, it is not necessary to shut off power to equipment automatically because Division II equipment is in use, and it is unlikely that an electrical failure will occur simultaneously with an explosive concentration. An alarm or indicator is necessary, however, to warn of the loss of the purge source.

When Type Z purging is used, only an alarm or indicator is required to warn of purge failure.

The internal pressure required for enclosures for all three classifications is 0.1 inch of water. For Type X purging a timer must be used to ensure four enclosure volumes of purge fluid before power is applied to the equipment. Four enclosures of air are required for Types Y and Z purging also before power is turned on, but it is not necessary to employ a timer as insurance. Reliance is placed on personnel adherence to this requirement.

Type Y purging also requires that any door opening to the enclosure shall have a door lock to shut off power if the door can be readily opened without keys or tools.

NFPA Standard 496 covers purging and pressurizing requirements for control panels, control rooms and other large enclosures that are in excess of the 10 cubic feet covered by ISA-S12.4. It covers the requirements of the positive pressure air systems and the monitoring and protective devices that may or must be used.

If the air system fails, an alarm is necessary for Types Y and Z purging; for Type X purging, power must be shut off automatically. The air system must be capable of maintaining a positive pressure of at least 0.1 inch of water, and it must come from a source free of hazardous concentrations of flammable gases and vapors.

The air system, in addition to meeting the positive pressure requirements, must also be capable of providing a minimum outward velocity of 60 feet per minute of purge air through all openings, including doors and windows. All doors and windows are considered open when determining velocity requirements. Air for purging must be drawn from a level at least 25 feet above ground. It is assumed that light flammable vapors will disperse at this level and that heavy vapors will stay below it.

Intrinsic Safety

Intrinsic safety involves making electrical equipment safe for hazardous areas by limiting the available energy to a level too low for flammable mixtures to ignite. The concept of designing intrinsically safe systems has been considered for many years, but only in the past few years has the demand been created for its use.

No single design, installation or testing concept is universally acknowledged to govern intrinsic safety. In the United States three groups or agencies recognize its use. The philosophy of the Instrument Society of America is stated in ISA-RP 12.2: "Intrinsically safe equipment and wiring is . . . . incapable of releasing sufficient electrical or thermal energy under normal or abnormal conditions to cause ignition of a specific hazardous atmospheric mixture in its most ignited concentration."

The National Electric Code Article 500 makes a similar statement, adding "abnormal conditions will include accidental damage to any part of the equipment or wiring, insulation or other failure of electrical components, application of overvoltage, adjustment and maintenance operations and other similar conditions."

The National Fire Protection Association NFPA 493 also establishes standards for its use, guidelines for construction and test apparatus and procedures for testing intrinsically safe equipment.

Certifying agencies for the equipment include Factory Mutual and Underwriters' Laboratories. UL Standard Subject 913 combines test procedures and concepts outlined by the three agencies mentioned above in certifying equipment as safe. Certification is granted on a system basis, not to individual devices, because the safety of any one component depends on the characteristics of other elements.

Certification also depends on the specific atmosphere in which the system is to be used. Much research has been done in determining safe energy levels for various hazardous atmospheres. The ISA has published curves (ISA-RP 12.2) showing combinations of voltage, current, inductance and capacitance which do not cause ignition in various atmosphere groups. The most ignitible mixtures of gas in air for each group are used to establish the curves. Each point on the curve represents a combination of circuit parameters which did not cause ignition in 100 or more attempts to ignite. In addition to the conservatism of the data, the ISA and the NFPA recommend that a safety factor of four be applied to the curve data in designing equipment that is not to be subjected to ignition tests. When actual tests are to be made, the four-to-one safety factor is not recommended.

Intrinsically safe systems are almost always installed in combinations of hazardous and nonhazardous areas. Primary sensors and final control elements usually are located in hazardous atmospheres while power supplies, computers, controllers and readout equipment are usually located in control rooms in nonhazardous atmospheres. Interconnecting lines between hazardous and nonhazardous areas pass through barriers which limit energy flow and restrict voltages in the hazardous areas. This is done (a) by limiting both voltage and current with devices such as resistors and zener diodes or (b) by limiting current with resistors and voltage with transformers.

Installation requirements for intrinsically safe systems are simple. The essential requirement is the positive separation of the intrinsic system from power wiring. Wiring should be installed in separate raceways to prevent the possibility of accidental connection or shorting between the two systems, particularly in nonhazardous areas where the two may run close together. In hazardous areas, no real problem is posed because power wiring is contained in conduit or mineral-insulated cable.

Intrinsically safe systems have several advantages.

1. Installation costs are potentially considerably lower than for explosionproof equipment or purged equipment. Explosionproof and purged equipment require wiring to be in rigid conduit, and in Division I areas, the conduit fittings must also be explosionproof and sealed. Purged equipment requires an air supply system and purge failure alarms or shutdowns. Intrinsically safe systems do not require these design features.

2. From a maintenance standpoint, equipment is more accessible for calibration checks, adjustments and repairs. Explosionproof equipment has to be deenergized before being opened so maintenance is often deferred.

3. Safety of the system is unexcelled. Any system is safe if it is properly installed and maintained. However, ordinary carelessness would not render an intrinsically safe system unsafe. The recognized standards are conservative, and the likelihood of failures, faults, etc., causing possible unsafe conditions are remote indeed. On the other hand, explosionproof equipment offers no protection if the cover is left off or is not sufficiently tight. Purging is dependent on the reliability and quality of the purge supply.

Some obvious disadvantages are pointed out below.

1. One disadvantage is the usual necessity for altering the configuration of approved systems. By the very nature of the present approval system, only the devices of a particular manufacturer are used for approval. In practice there is often the need or desire to mix devices of more than one manufacturer. System approval is lost in these instances.

2. A lack of understanding of and confidence in the intrinsic safety concept exists. There is always a resistance to change, and the concept is relatively new in practice.

3. Its low energy limits its use to low-power devices. This can present problems in some instrument systems, and certainly limits its use in power and lighting systems except for remote control circuits.

3 Selecting Measurement Methods

William G. Andrew

The selection of the best type measuring device for a particular process variable is often difficult to make. Even the experienced engineer welcomes a set of guidelines in choosing a method. This chapter lists advantages and disadvantages of many types of measuring devices for flow, level, pressure and temperature, with brief discussions pertinent to each.

Selection is greatly simplified if all the service conditions are known. Some measurements require more information than others. For example, more stream characteristics and conditions must be known for proper application of flow devices than for pressure. It is essential to list all the pertinent information concerning the measurement to be made.

Flow Instruments

The measurement of instantaneous flow rates and the integration of these rates into total quantities are necessary to the operation of modern processes and the economic status of entire plants. While other measurements, such as temperature and pressure, perhaps affect process operation as significantly as flow, they are not as directly involved in the purchase, sale and exchange of raw materials and finished products which determine profit and loss margins.

The importance of flow measurement is generally understood by those who select flow devices. Generally, they also recognize that not all flow measurements have the same importance; each application must be evaluated separately and independently to determine its relative importance and need for accuracy.

Service conditions must be known and listed in detail. They have more effect on the selection of flow measuring

methods than for almost any other measured variable. The following conditions should be known and listed:

1. Line size
2. Range of flow rates—maximum, normal and minimum
3. Fluid characteristics: (a) liquid, gas, slurry, etc.; (b) pressure; (c) temperature; (d) viscosity; (e) specific gravity at standard and flowing conditions; (f) compressibility; (g) molecular weight (for gases and vapors); (h) steam quality (for steam)
4. Corrosive effects (for aid in material selection)
5. Whether flow is steady or pulsating

The flow measurement methods evaluated below include the differential, variable area, magnetic, turbine, target, vortex and positive displacement types.

Differential Meters

The differential or head flow measurement (Figure 3.1) is by far the most popular method in use today for measuring fluid flows in the processing industries. It was one of the earliest methods used, and its popularity has remained almost unchallenged. It is significant, perhaps, that most other methods require in-line mounting of the readout or transmitting device, making it inconvenient and uneconomical to check them or remove them for maintenance purposes.

Although many flow measuring devices operate for long periods of time without difficulty, they must be removed occasionally for checking, for repair, for calibration, etc. In

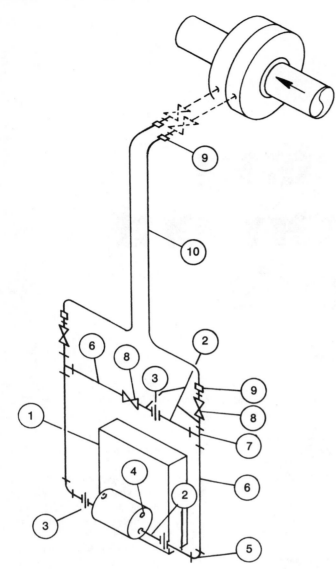

Figure 3.1. The most popular method of measuring fluid flow in the processing industries is the differential pressure or head method.

The accuracy of differential meters is about ±¼ to ±2%. Advantages and disadvantages of the differential method are mentioned below.

Advantages

1. Relatively low cost, especially for large lines
2. Proven accuracy and reliability; well known and predictable flow characteristics
3. Easily removed without shutting down the process
4. The secondary or differential device easily isolated for zero check and/or calibration
5. Adaptable to any pipe size
6. Cost remains almost the same with increasing pipe size—the cost of the primary elements (orifice plates) and flanges are the primary variables

Disadvantages and Limitations

1. Flow rangeability for a given orifice size and differential range limited to a 3:1 to 5:1 range
2. Relatively high permanent pressure loss (considering the thin edge orifice plate)
3. Difficult to use for slurry services
4. Square root rather than linear characteristic
5. Meter run lengths required for accurate measurement sometimes present difficulties
6. Connecting piping sometimes presents problems—freezing, condensation, etc.
7. Low flow rates not easily measured except through the use of integral orifice devices where in-line mounting is required
8. Accuracy dependent on many fixed fluid characteristics such as temperature, pressure, specific gravity, compressibility, etc.
9. Pulsating flow difficult to measure

When differential type meters are used, there still remains a choice for the secondary element. Several types are available.

For many years liquid heads (primarily mercury) were used to measure differential pressures developed by the primary elements. Manometers and mercury float meters were in common use. As mercury increased in cost, however, and because of the higher cost of maintaining mercury meters, dry meters have almost completely replaced them.

The prominent types of dry meters are the bellows and diaphragm units. Their use eliminates some problems inherent in mercury meters, such as overrange, which causes loss of mercury with the resultant mercury contamination of the product.

Bellows meters are often used for local indication and control because they develop forces sufficiently high to operate indicating, recording and control mechanisms.

Diaphragm (force balance) meters are used extensively when signal transmission is necessary. They do not develop the force required for indication and control and thus require the use of other elements for those purposes.

Primary devices for differential measurements are discussed separately later in this chapter.

many instances their operation may be unimpaired; the apparent difficulty may have been an unknown process variation. The uncertainty of the cause, however, dictates a check on the measuring device. When such instances occur, differential meters can easily be isolated, zeroed and checked without shutting down the process, while in-line meters must have bypasses for similar isolation and removal.

Differential meters are likely to retain their popularity until another method is proven that either does not mount in-line or that can be zeroed and checked out easily while mounted in-line. Ultrasonic and radioactive methods offer some hope in this respect.

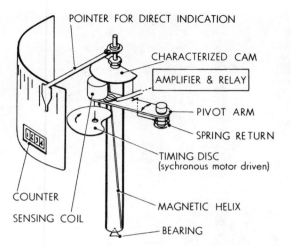

Figure 3.2. Rotameters or variable area meters are used for many fluid flows where rangeability must exceed 4:1 and where small pipe sizes (1½ inches and under) must be used. (Courtesy of Brooks Instruments)

Rotameters

Rotameters or variable area meters (Figure 3.2) have been used rather extensively in the past for measuring flows in line sizes ½ to 1½ inches, particularly when rangeability requirements exceed 4:1. They are available in larger sizes also. Most rotameters have rangeabilities of 10:1.

Glass tube rotameters present considerable hazard for handling toxic, corrosive or otherwise hazardous fluids. This hazard is reduced somewhat by the use of totally enclosed or armored meters, but even the armored designs are avoided by many companies for uses other than air, water or similar services.

Rotameters are widely used for metering purge flows, pump-seal fluids and coolants and lubricants for operating machinery. In these applications, flows are relatively small, and accuracy requirements are not rigid. Accuracy is about ±½ to ±10%.

Advantages

1. Good rangeability
2. Relatively low cost
3. Good for metering small flows
4. Easily equipped with alarm switches
5. No restrictions in regard to inlet and outlet piping requirements (other than a vertical flow requirement)
6. Viscosity-immune designs available
7. Low pressure drop requirement
8. Can be used in some light slurry services

Disadvantages

1. Glass tube types subject to breakage
2. Not good in pulsating services
3. Must be mounted vertically

4. Generally limited to small pipe sizes (unless bypass rotameter is used)
5. Limited to relatively low temperatures
6. Fair accuracy
7. Require in-line mounting (except bypass type)

Magnetic Meters

Electromagnetic meters (Figure 3.3) are used primarily in difficult-to-measure services where other, more economical types do not function well. They are particularly well suited for slurry services but require that the measured fluid be electrically conductive.

Magnetic meters are available in sizes from 1/10 to 96 inches nominal pipe size and are not temperature or pressure limited. A wide variety of construction materials can be used.

Accuracy is from ±½ to ±2%.

Advantages

1. Can handle slurries
2. Can handle corrosive fluids
3. Has very low pressure drop
4. No obstruction in pipe
5. Available in many construction materials
6. Available in large pipe sizes
7. Piping configurations not critical

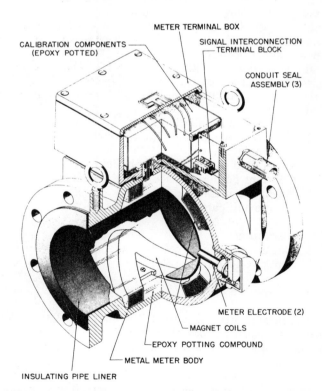

Figure 3.3. Magnetic meters are used primarily in difficult-to-measure services, such as slurries and corrosive materials. (Courtesy of Fischer and Porter Co.)

8. Bidirectional flow measured by reversing connections manually or automatically
9. Measurements unaffected by viscosity, density, temperature or pressure
10. Can measure turbulent or laminar flow
11. Unaffected by conductivity changes as long as the minimum conductivity is maintained

Disadvantages
1. Conductivity must be as high as the minimum required by the particular manufacturer—from 0.1 to 20 micromhos
2. Meters must be full at all times
3. Relatively high in cost
4. Entrained gas bubbles result in measurement errors
5. Electrode fouling occurs in some fluids, causing errors in the generated voltage signals (cleaning methods—electrical and mechanical—are available, however)
6. In-line mounting required
7. Meters must be explosionproof when installed in hazardous electrical areas

Probe-Type Magnetic Meters

A probe-type (or "inside-out") magmeter offers a more economical installation than earlier in-line flanged designs (Figure 3.4). Hot-tap installation methods can be used to mount the sensing probe on virtually any size of pipe 8 inches and up. The device is essentially a velocimeter. An alternate probe design is used with a level sensor to measure open-channel flowrate.

Ranges are 0-1, 0-10, and 0-100 fps. They are switch selectable, and accurate above 0.02 fps to ±2% FS. Sensitivity and repeatability are 0.01 fps. Linearity is ±1% FS.

Advantages
1. Inexpensive installation costs
2. Can be mounted using hot-tap methods
3. Low maintenance
4. Unaffected by solids or air bubbles
5. Constant hardware price for all pipe sizes
6. Negligible pressure drop
7. No moving parts

Disadvantages
1. Not as accurate as other magnetic meters
2. Sensitive to flow profile

Turbine Meters

The turbine meter (Figure 3.5) is another relatively low-use flow measuring device, although its use is increasing because of its adaptability to flow totalizing and to blending operations.

Accuracy and rangeability are also good. Rangeabilities vary from 10:1 to 20:1 generally. Designs to meet military specifications are available with rangeabilities of 100:1.

Accuracy is about ±¼ to ±½%.

Figure 3.4. An "inside-out" magnetic meter can be inserted into the line using conventional hot-tap methods. (Courtesy of Monitek, Inc.)

Figure 3.5. Turbine meters have good rangeability and good accuracies. (Courtesy of Brooks Instruments)

Advantages
1. Good accuracy
2. Excellent repeatability
3. Excellent rangeability
4. Low pressure drop
5. Easy to install and maintain

6. Can be compensated for viscosity variations
7. Adaptable to flow totalizing
8. Adaptable to digital blending systems
9. Good temperature and pressure ratings

Disadvantages

1. In-line mounting required
2. Relatively high cost
3. Limited use for slurry applications
4. Nonlubricating fluids sometimes present problems
5. Straight runs of pipe (15 diameters) required ahead of the meter
6. Strainers recommended, except for the special slurry meters

Target Meters

Target meters are used primarily in slurry and corrosive services where buildup around orifice plates present problems or where lead lines to differential devices are undesirable. Since the meter mounts inline, lead lines that might solidify or otherwise present difficulties are eliminated.

The target meter supplied by the Foxboro Company is available as standard in sizes from ¾- to 4-inch pipe diameters and combines an annular orifice with a force balance transducer.

A similar meter by the Ramapo Instrument Company is called a "Drag Body" flowmeter and detects the impact forces on the probe mounted in-line (see Figure 3.7). The probe is available for pipe sizes from ½ to 60 inches and could be made for any pipe size.

Because of the nature of the device, it cannot be zeroed, except at no-flow condition. If zeroing is desired under operating conditions, a bypass must be provided. Normally this is not practicable, especially in the larger pipe sizes.

Calibration data for these meters is supplied by the manufacturer and varies with pipe and target size (Figure 3.6).

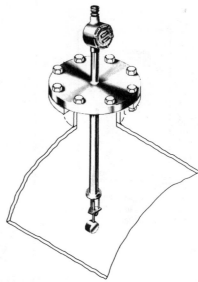

Figure 3.7. Flange mounted target meters are available for large pipe sizes to reduce equipment costs. (Courtesy of Ramapo Instrument Co., Inc.)

The repeatability of target meters is good, and accuracy ranges from ±½ to ±3%.

Advantages

1. Useful for "difficult" measurements, such as slurries, polymer-bearing and sediment-bearing materials, corrosive mixtures, etc.
2. Accuracy good when calibrated for specific streams
3. Good repeatability
4. Good for relatively high temperatures and pressures

Disadvantages

1. In-line mounting required
2. No-flow condition must exist for zeroing
3. Limited calibration data

Vortex Meters

Vortex meters have only recently been introduced; consequently, application data are limited. Two meters that utilize essentially the same operating principle are made by the Fischer Porter Company (Figure 3.8) and Eastech Incorporated (Figure 3.9). They are designed to impart a swirl pattern to the fluid stream by using obstructions which produce pulses proportional to flow. The rangeability of both devices is excellent—about 100:1.

Accuracy of the F&P Swirlmeter is about ±¾%, but repeatability is claimed to be ±¼%. Its projected use is for gases. It is presently available in line sizes from 1 to 6 inches.

The Eastech Vortex Shedding Meter is designed for both liquids and gases. Its repeatability is claimed to be ±0.1% in its linear operating range. It is currently available in pipe sizes from 2 to 6 inches.

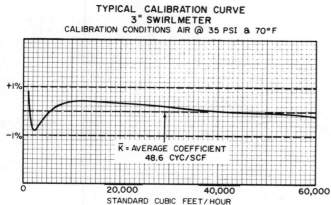

TYPICAL CALIBRATION CURVE
3" SWIRLMETER
CALIBRATION CONDITIONS AIR @ 35 PSI & 70°F

K̄ = AVERAGE COEFFICIENT
48.6 CYC/SCF

STANDARD CUBIC FEET/HOUR

Figure 3.6. Target meters measure flow by the force the flowing fluid imparts to a disc or target located in the stream and are used primarily for difficult-to-measure fluids. (Courtesy of Foxboro Co.)

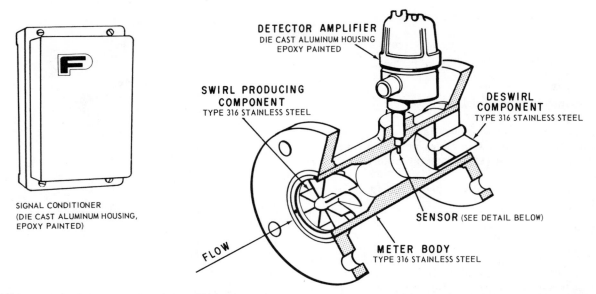

Figure 3.8. The swirlmeter imparts a swirling motion to fluid which causes small temperature variations detected by a sensitive thermistor, the frequency of which is proportional to flow. (Courtesy of Fischer and Porter Co.)

Advantages

1. Excellent rangeability
2. Digital readout lends itself to blending applications and flow totalization
3. No moving parts in the Vortex meter
4. Within the linear range, it is relatively immune to density, temperature, pressure and viscosity variations
5. Very low pressure drop

Disadvantages

1. Limited application data
2. In-line mounting required
3. Limitations imposed on upstream and downstream piping requirements—the Vortex meter has the same constraints as orifice meters and the Swirlmeter must be preceded by 10 pipe diameters on the upstream side
4. Operation is impaired if the temperature elements coat enough to affect adversely the rate of heat exchange between the elements and the flowing liquid
5. Relatively high cost

Positive Displacement Meters

Positive displacement (PD) meters have the nature of separating streams into known volumes or segments, thereby measuring rates and/or total quantities. They are classified as mechanical meters and have moving parts that can be damaged easily and that require some degree of maintenance.

Several types of displacement meters are available, including the nutating disc, the oscillating piston, the reciprocating piston and the rotating vane.

Advantages and disadvantages vary with the type meter but, as a group, may be given.

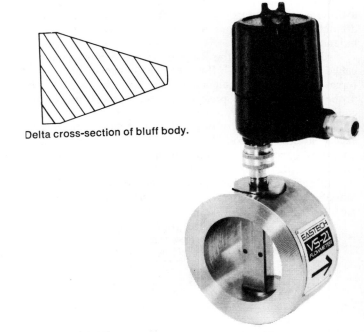

Delta cross-section of bluff body.

Figure 3.9. The Eastech Vortex shedding meter is used for both liquid and gas measurement. (Courtesy of Eastech, Inc.)

Advantages

1. High rangeability—30:1 for some types
2. Ease of calibration
3. Linear readout
4. Flexibility of readout devices
5. Good to excellent accuracy

Disadvantages

1. Relatively high pressure drop
2. Very little overrange protection
3. Susceptible to damages from gas or liquid slugs and from dirty fluids
4. In-line mounting
5. Relatively high cost, especially for high flow rate applications
6. For most meters, problems presented by materials that tend to "plate" or "coat"

When viewed as to particular types, the advantages and limitations of each of four basic types may be listed.

Nutating Disc (Figure 3.10)

Advantages

1. Relatively low cost
2. Moderate pressure loss
3. Applicable to liquid batching systems
4. Several construction materials available

Limitations

1. Limited as to pipe size and capacity
2. Fair accuracy for PD meters—±1%
3. Fluids should be clean

Oscillating Piston (Figure 3.11)

Advantages

1. Good accuracy, especially at low flow rates
2. Easily applied to liquid batching systems

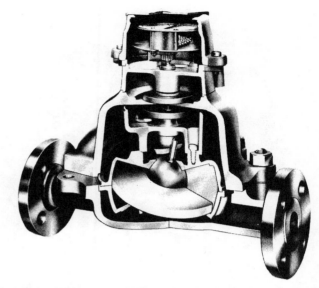

Figure 3.10. The nutating disc, one of several types of positive displacement meters, is available only in small pipe sizes. (Courtesy of Ametrol Division of Hersey Products, Inc.)

3. Good repeatability
4. Moderate cost
5. Easy to install and maintain

Limitations

1. Available only in small sizes, normally 2 inches or less
2. Limited power for driving accessories
3. Fluids must be clean

a.

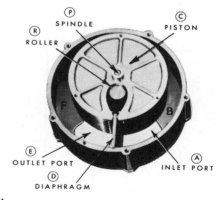

b.

Figure 3.11. The oscillating piston PD meter is also limited normally to small pipe sizes. (Courtesy of Neptune Meter Co.)

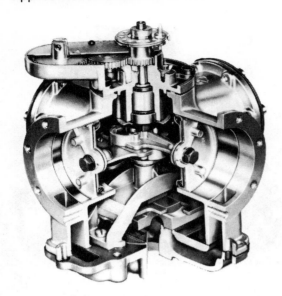

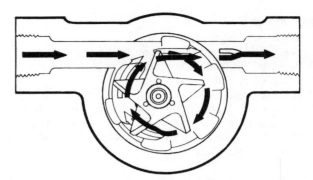

Figure 3.13. The rotating vane PD meter has good accuracy and low pressure drop and is available in larger pipe sizes. (Courtesy of ITT Barton)

Figure 3.12. A reciprocating piston PD meter provides accurate measurement, but its pulsating characteristic is objectionable in liquid service. (Courtesy of A.O. Smith Corp.)

Reciprocating Piston (Figure 3.12)

Advantages

1. Capable of high accuracy
2. Construction materials not limited

Limitations

1. Relatively high cost
2. Subject to leakage
3. Dirty fluids present problems
4. Maintenance costs may be high
5. Restricted to moderate flow rates

Rotating Vane (Figure 3.13)

Advantages

1. Low pressure loss
2. Applicable to a wide variety of gas and liquid fluids, including viscous materials
3. Has relatively high temperature and pressure ratings
4. Available in numerous construction materials
5. Good accuracy

Limitations

1. Tend to be bulky and heavy in larger sizes
2. Relatively high cost
3. Susceptible to damage from entrained vapors and dirty fluids
4. Accuracy decreases at low flow rates; slip factor is high

Figure 3.14. Thin, sharp-edged orifice plates began to be used in the early 1900s. Concentric bores are most common.

Primary Elements for Differential Meters

A wide variety of primary elements are available for differential or head meters. Thin-edged orifice plates are used predominantly, but other elements offer advantages for particular applications. Good and bad features of each type are listed below.

Concentric Orifice (Figure 3.14)

Advantages

1. Low cost
2. Available in many materials
3. Can be used for a wide range of pipe sizes
4. Abundant application data; known characteristics
5. Good accuracy if plates are properly installed

Disadvantages

1. Relatively high permanent pressure loss
2. Tends to clog, reducing its use in slurry services
3. Accuracy dependent on installation care

Eccentric Orifice

Advantages of eccentric orifices (Figure 3.15) are similar to concentric plates except that their bore location makes it useful for measuring fluids containing solids, for water containing oils and for wet steam.

Eccentric plates have the same disadvantages as concentric plates, but, in addition, the probable error can be much higher, and operating data is more limited.

Segmental Orifice

Segmental orifice plates (Figure 3.16) are used essentially for the same services as eccentric orifices, so their advantages and disadvantages are about the same. The segmental opening may be placed at the top or bottom of the line, depending on the service.

Venturi Tube

There are several configurations of Venturi tubes (Figure 3.17) available, and their characteristics vary. Generally the following advantages and disadvantages apply.

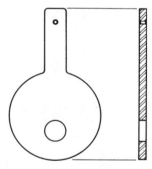

Figure 3.15. Eccentric orifices have the bore offset from center to minimize problems in services of solids-containing materials.

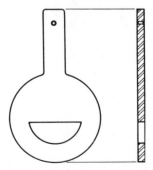

Figure 3.16. Segmental orifices provide another version of plates useful for solids-containing materials.

Figure 3.17. Venturi tubes possess the characteristic of a low permanent pressure loss. (Courtesy of BIF, a unit of General Signal Corp.)

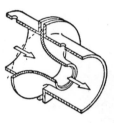

Figure 3.18. Flow nozzles have lower permanent pressure drops than orifice plates but higher than Venturis. (Courtesy of Foxboro Co.)

Advantages

1. Low permanent pressure loss
2. Can handle suspended solids (lead lines may have to be purged)
3. Used for high flow rates
4. Well-known characterstics
5. More accurate over wide flow ranges than orifices or nozzles (can be used at low and high beta ratios)

Disadvantages

1. High cost
2. Not normally available in pipe sizes below 6 inches

Flow Nozzle (Figure 3.18)

Advantages

1. Permanent pressure loss lower than that for an orifice plate
2. Good for solids-containing fluids (lead lines may need to be purged)
3. Available in numerous materials

Disadvantages

1. Higher cost than orifice plates
2. Limited to moderate pipe sizes

Low Loss Flow Tubes

Several low-pressure loss tubes are available, including the Gentile Flow Tube, the Dall Tube and others. They are characterized by their extremely low permanent pressure loss and should be used when only small losses can be tolerated. Several of these designs are well investigated, and operating data can be supplied by the manufacturer.

Laminar Flow Elements

These elements are not widely used but do find application on low flow rate services primarily. They are used occasionally on air services where relatively large flows are required near atmospheric pressure.

Advantages

1. Good rangeability
2. Applicable to low flow rates
3. Used for liquids and gases

Disadvantages

1. High pressure drop
2. Accuracy dependent on calibration
3. Fluids should be clean
4. Subject to viscosity and density errors, hence to temperature and pressure variations

Pitot Tubes

There are several variations in the design of Pitot tubes; Figure 3.19 shows one type. They are rarely used in process streams but are used occasionally in utility streams where high accuracy is not necessary.

Advantages

1. Essentially no pressure loss
2. Economical to install
3. Some types can be removed from the line

Disadvantages

1. Poor accuracy
2. Calibration data needs to be supplied from the manufacturer
3. Not recommended for dirty or sticky fluids
4. Sensitive to upstream disturbances

Annubar Tubes

The Annubar element (Figure 3.20) is a recent addition to the line of primary devices, and it has received quick and relatively wide acceptance in the processing industries. Its characteristics are similar to the Pitot tube, but its multiple sensing arrangement provides better accuracy than the Pitot tube, ranging from $\pm\frac{1}{2}$ to $\pm 1\frac{1}{2}\%$ over a wide range of flows and pipe sizes.

Advantages

1. Available for a wide range of pipe sizes from $\frac{1}{2}$ to 150 inches
2. Simple and economical to install

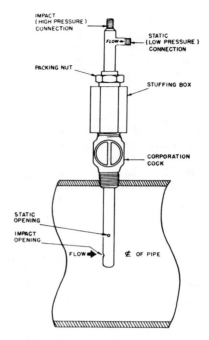

Figure 3.19. Pitot tubes are not used extensively but find some use in large utility streams. (Courtesy of Foxboro Co.)

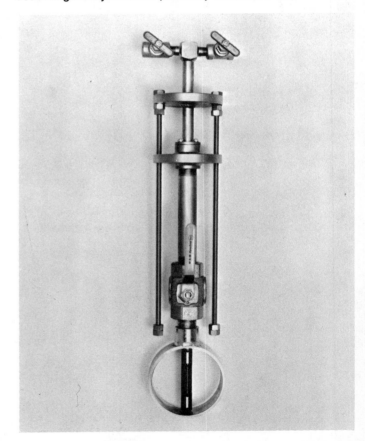

Figure 3.20. The Annubar is an inexpensive, low-loss measuring element which can be installed using conventional hot-tap methods. (Courtesy of Dieterich Standard Corp.)

3. Negligible pressure drop
4. Can be placed in service under pressure
5. Can be rotated in service for cleaning action
6. Long-term measurement stability
7. Cost-effective compared to an orifice plate installation above 6″ diameter pipe

Disadvantages

1. Not applicable to dirty or sticky fluids
2. Operating data still limited

Elbow Taps

Elbow taps (Figure 3.21) are rarely used. One of the occasional useful applications is where flow measurement may be needed in an existing installation at low cost and where good accuracy is not necessary.

Advantages

1. Easy to add to existing installations where elbows exist
2. Low cost
3. No additional pressure loss
4. No obstructions in the line

Disadvantages

1. Poor accuracy—from ±5 to ±10%
2. Comparatively small differential pressure developed
3. Straight pipe runs of approximately 25 pipe diameters needed upstream of measurement

Level Instruments

The selection of the proper level measurement method is probably more difficult than for any of the four major process variables except flow. As in flow measurements, the conditions of the measured media have many adverse effects on the measuring device. The operating conditions which generally must be known include:

1. The level range
2. Fluid characteristics: (a) temperature; (b) pressure; (c) specific gravity; (d) whether the fluid is clean or dirty, contains vapors or solids, etc.
3. Corrosive effects
4. Whether the fluid has a tendency to "coat" vessel walls or the measuring device
5. Whether the fluid is turbulent around the measurement area

Normally there is little difficulty in measuring the level of clean, nonviscous fluids. Slurries, heavily viscous materials and solids, however, present many problems.

Measurement methods to be evaluated include displacement, differential pressure, capacitance, ultrasonic, radiation and several single-point switch types.

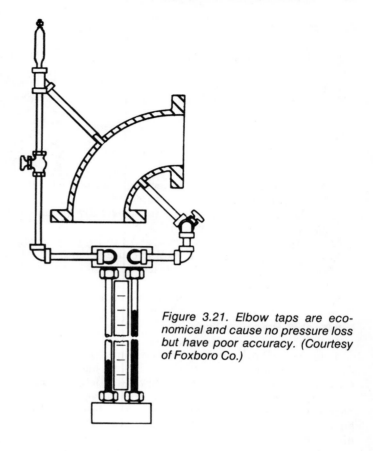

Figure 3.21. Elbow taps are economical and cause no pressure loss but have poor accuracy. (Courtesy of Foxboro Co.)

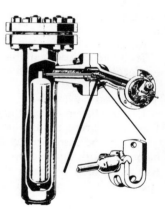

Figure 3.22. The displacement level device is the most common method of measuring clean liquid levels. (Courtesy of Fisher Controls Co.)

Displacement Type

The displacement or buoyancy method (Figure 3.22) was for many years the most widely used technique for measuring liquid levels. It is still popular for clean fluids, but many processes that contain slurries, have tendencies to "coat," have wide ranges, etc., require other, more acceptable methods.

Advantages and disadvantages of the displacement methods follow.

Advantages

1. High accuracy
2. Reliable in clean liquids
3. Proven method
4. Can be mounted internally or externally
5. Externally mounted units can be valved off for maintenance
6. Adaptable to liquid interface measurement

Disadvantages

1. Has limited range—units exceeding 48 inches in length are bulky and difficult to handle
2. Costs increase appreciably for externally mounted units as pressure ratings increase
3. External units may require heating to avoid freezing
4. External units may be in error because of temperature differences between the vessel fluid and the level chamber fluid
5. Internal units may require stilling chambers

Differential Pressure Type

The use of differential pressure measurement, or hydrostatic head, for level determination has increased significantly in the past few years. Several different types of differential devices make it possible to measure wide level ranges for clean or corrosive services, for slurries and for highly viscous materials. Almost any type of differential measuring device can be used for level measurement if it is available in the low ranges normally required for level, usually from about 10 inches of water to approximately 150 inches of water.

Rather than discussing in detail the wide variety of differential devices available, they are classified broadly into two groups, sealed and nonsealed systems.

Nonsealed Systems

Differential units such as the d/p (Figure 3.23) used for flow measurement are used also for measuring levels. They may be used directly in contact with the fluid or may be purged with suitable gas or liquid sources. Other differential units such as those shown in Figure 3.24 are made specifically for level measurement.

Advantages

1. Good accuracy
2. Adaptable to wide level ranges
3. Available in many construction materials and/or can be purged for corrosive services
4. Can be purged for use in slurry service
5. Moderate cost
6. Can be isolated and zeroed in place

Disadvantages

1. Errors caused by density variations
2. Low pressure lead line often undesirable in other than atmospheric applications

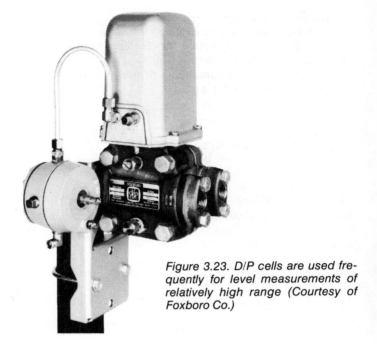

Figure 3.23. D/P cells are used frequently for level measurements of relatively high range (Courtesy of Foxboro Co.)

3. Heating of lead lines sometimes necessary
4. Operating and maintenance problems often presented by purged lines
5. Purging materials often present process difficulties

Sealed Systems

To meet application requirements for some slurries and highly viscous materials, sealed systems often offer suitable solutions for difficult level measurements. Figure 3.25 shows two such varieties where the measuring elements are sealed or isolated from the process fluid.

Advantages

1. Purge not required
2. Good for slurries and corrosive materials
3. Sensing units essentially flush with the operating fluid, eliminating cavities where plugging and freezing are likely to occur
4. Wide measurement range
5. Wetted diaphragms can be furnished in a wide variety of materials
6. Fair accuracy
7. Can be used for open or closed vessels
8. Good for relatively high temperatures
9. Simple and easy to install

Disadvantages

1. Units cannot be removed for checking and/or maintenance without shutting down the equipment
2. Density variations cause errors
3. The mounting location affects calibration since the head of the capillary filled legs must be considered
4. Ambient temperature changes cause errors in the capillary filled system type

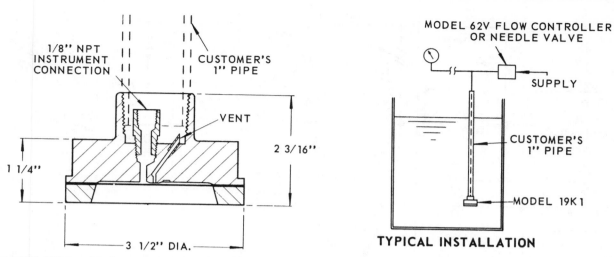

Figure 3.24. This differential device is used only for level measurement as opposed to the d/p cell used for both level and flow measurements. (Courtesy of Moore Products Co.)

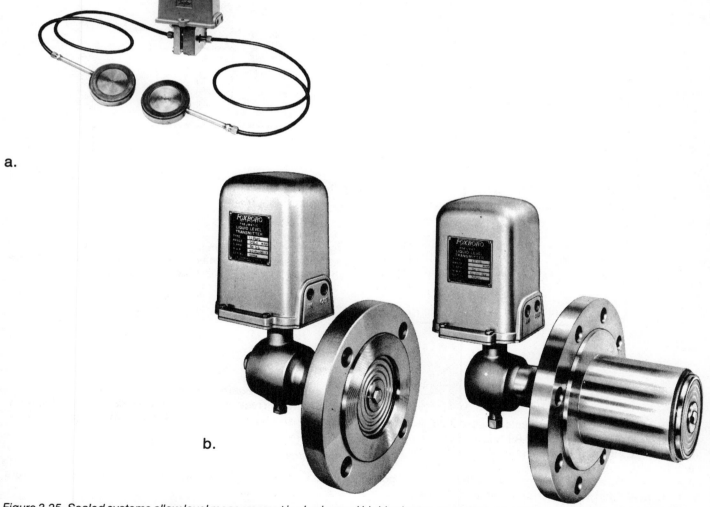

a.

b.

Figure 3.25. Sealed systems allow level measurement in slurries and highly viscous materials where other type measurements will not work. (Courtesy of Taylor Instr. Proc. Cont. Div., Sybron Corp. and Foxboro Co.)

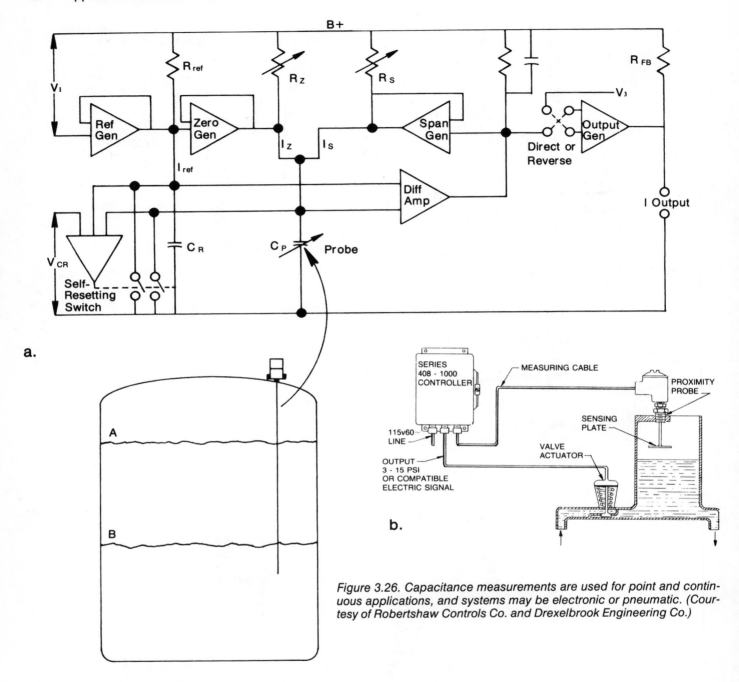

Figure 3.26. Capacitance measurements are used for point and continuous applications, and systems may be electronic or pneumatic. (Courtesy of Robertshaw Controls Co. and Drexelbrook Engineering Co.)

Capacitance Type

Capacitance level measurements (Figure 3.26) need to be separated into two categories for evaluation—continuous and point. Continuous measurements are not made frequently with capacitance probes but work well in some applications. However, they are used quite often for point measurements and are superior for many applications.

Continuous Measurement

For the infrequent use in continuous level applications, the capacitance methods have several advantages and disadvantages.

Advantages

1. Can be used for some applications where other types are not feasible

2. Moderate cost
3. Fair accuracy
4. Can be used for high temperature and high pressure applications
5. Can be used in polymer and slurry services

Disadvantages

1. Special calibration required in many instances
2. Affected by density variations of measured materials
3. Limited application data
4. Erroneous readings occur when coatings form on the probe. (Note: at least one manufacturer surmounts this problem by precoating the probe and impressing a radio frequency signal on the probe.)

Point Measurement

The application of capacitance probes for point measurement has become quite common in recent years. They are particularly good for powders, solids and difficult-to-measure slurries. They cannot be applied indiscriminately, however; there are many installations where their use is questionable. Discussions with vendors and other users are helpful in determining proper applications.

Advantages

1. Reasonable cost
2. Easy to install
3. Useful on applications such as powders, pellets and other solids-containing materials, as well as slurries and corrosive materials where many other level devices will not work
4. Simple in design
5. No moving parts

Disadvantages

1. Accuracy affected by material characteristics
2. Coating of probes troublesome on some designs
3. Lack of data on dielectric constants of some materials

Ultrasonic Types

The ultrasonic level detectors (Figure 3.27) are used primarily for point measurement. They have been in use since about 1960. Somewhat like capacitance probes, they are often used in services which present problems for the more traditional measurement methods.

There are several variations in technique—single sensor systems where the transmitting and receiving elements are in the same enclosure and two sensor systems where they are separately enclosed. Two sensor systems vary in that the two sensors may be built and mounted as a single unit, or they may be built as two separately mounted units.

They are considered below as a general type, and no distinction is made concerning the various measurement techniques.

Figure 3.27. Ultrasonic level devices are used primarily for point level measurement, but models are available from some manufacturers for continuous measurements. (Courtesy of Delavan Mfg. Co.)

Figure 3.28. Fixed distance between transmitting and receiving elements eliminates some application problems for ultrasonic devices. (Courtesy of National Sonics Corp.)

Advantages

1. Essentially no moving parts
2. Utilizes solid-state circuitry requiring little maintenance
3. Accuracy good where application is suitable
4. Applicable to some difficult-to-measure streams such as powders, solids, solids-containing fluids and slurries
5. Easy to install

Disadvantages

1. Insufficient application data
2. Tendency to bridge for some sensor types and for some materials
3. Relatively high cost
4. Difficulty in fixing distance between transmitting and receiving units in two-element systems—this objection is overcome by fixed distances in some models (see Figure 3.28).

Radiation Type

Like many other level measurement methods, the radioactive type (Figure 3.29) is used both for continuous and point level measurement. Generally, it is an expensive method to use and is considered seriously only when it is obvious that other, less expensive, methods are not very likely to be satisfactory.

Continuous Systems

Continuous systems for radiation level gauging are expensive. Not only is the hardware costly, but calibration and testing can also be time consuming, thus adding operating costs. Since they are so often used as a last resort method and are successul in most instances, their overall cost may still be considered economical.

Advantages

1. Sometimes works when no other method is available
2. External mounting often possible
3. Easy zero check
4. Motor-driven models available for high-accuracy applications

Disadvantages

1. Costly to install
2. Requires licensing by regulatory agency
3. Dangerous to handle unless precautions are followed
4. Original calibration and checkout often difficult and costly
5. Errors caused by density variations in measured materials
6. Lack of application data
7. Difficult to obtain linear readout over wide ranges
8. Problems presented by materials that coat walls

Point Measurement

Point measurements are not nearly as difficult to make as continuous measurements. Calibration is also relatively easy.

Advantages

1. Sometimes work when other methods fail
2. Mounted outside the process

Disadvantages

1. Costly to purchase and install
2. Requires licensing by regulatory agency
3. Dangerous to handle unless precautions are followed
4. Lack of application data
5. Problems presented by buildup of material on vessel walls

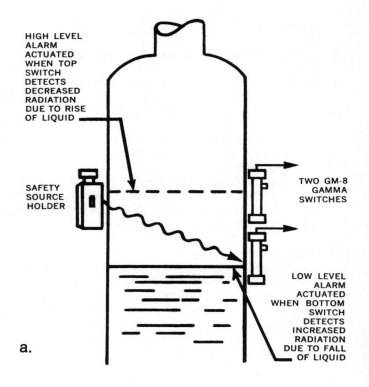

a.

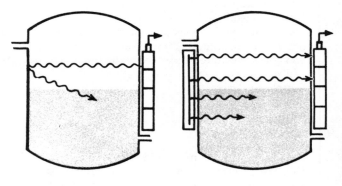

b.

Figure 3.29. Radioactive level measurements are used for applications where few, if any, other methods will work. (Courtesy of The Ohmart Corporation)

Miscellaneous Types

Some general observations are made about various other types of level measurement—principally point measurement devices.

Float Type

Float switches, for example, are simple and comparatively economical. They may be mounted internally or in exter-

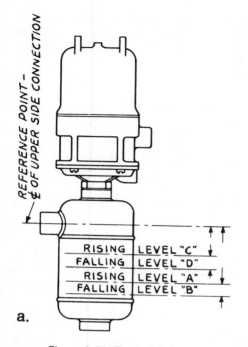

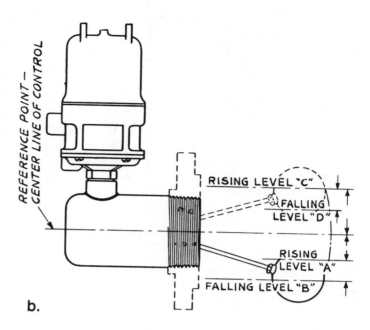

a.   b.

Figure 3.30. Float switches are used for point measurement and are simple and economical. (Courtesy of Magnetrol)

nal chambers (Figure 3.30). Internal units are difficult to work on or check out when operation is questionable. External units cost more because of the additional piping required.

The operation of float units is usually trouble-free if used in clean services. They are not normally recommended for slurries, solids-containing fluids, heavily viscous services or fluids that tend to coat.

Rotating Paddle Type

Paddle switches (Figure 3.31) are used primarily on level applications for solid materials such as powders and pellets. They are relatively inexpensive and easy to install. They need to be placed in a position where they are protected from inflowing streams. When this is not feasible, a protecting baffle should be placed above the unit.

Some switch models have been subject to failures because fine powders enter shaft seals, causing seizure and false alarms. This can be prevented by proper sealing of shafts.

Conduction type

One of the least expensive methods for point measurement or control of level is by conductivity-operated relays. When the conducting liquid rises to connect the probes shown in Figure 3.32a, current flows through the secondary coil, influencing the magnetism to close the armature, which in turn closes both sets of contacts. A current is then established through the liquid from the long electrode to the vessel wall (Figure 3.32b), holding the armature up until liquid level breaks the circuit by falling below the long electrode (Figure 3.32c).

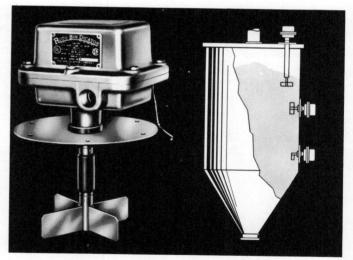

Figure 3.31. Rotating paddle switches work well for many solids materials such as powders and pellets. (Courtesy of Bindicator)

Multiple relays can be used with multiple electrode assemblies to give operation in lined tanks, to start and stop several pumps, and to provide alarm functions. Control package options include relaying controls through remote telephone connections and also alarm bells mounted integral to the relay box. Relay types are electromechanical, solid-state, and intrinsically safe.

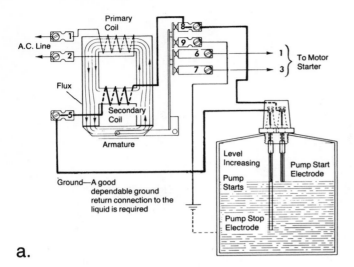

a.

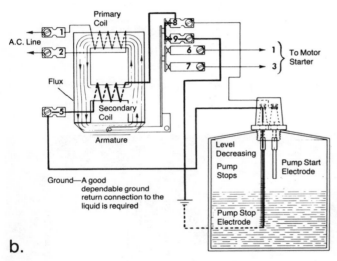

b.

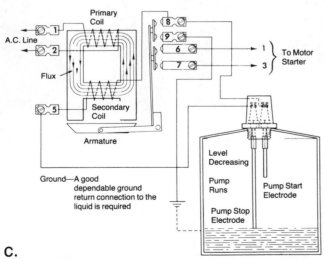

c.

Figure 3.32. When the conducting liquid rises to connect the probes shown in (a), the armature closes, and remains closed (b) until level falls below the bottom probe (c). (Courtesy of B/W Controls Inc.)

Minimum conductivity ranges from 67,000 micromhos/cm to 11 micromhos/cm. Power coils are available for line voltages of 110-120, 208-240, 440-480, and 550-600 VAC, 60 HZ.

Advantages
1. Inexpensive to purchase and install
2. Reliable
3. Low maintenance costs
4. Relay is only moving part
5. Activation from wide range of conductivities

Disadvantages
1. Possibility of malfunction if conductive buildup occurs on electrode holder

Vibration Type
Vibration switches (Figure 3.33) are used in liquid, slurry and powder services. Like several other designs, their use has increased for special applications where early standard float types were undesirable.

Vibration switches should be protected by location or by baffles from incoming streams. They may not work well in fine powder services where materials pack due to paddle vibrations, thus appearing to be uncovered. This shortcoming can be overcome by using differently designed probes.

Pressure Instruments

The selection of pressure devices is not nearly as difficult as selecting flow or level devices. In flow and level measurements the characteristics of the process fluid affect and determine the operability of the particular method as well as the cost of operation and maintenance. In pressure measurement, the emphasis is less on fluid characteristics and more on consideration of accuracy, ranges of measurement and material selection. Even then, there is such an overlap of range capabilities and material availability that element selection is relatively easy. Only in the very low and very high pressure ranges are limitations really felt in the selection process.

For comparison and evaluation, pressure elements are divided into the following categories: manometers, bourdon

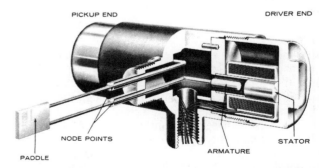

Figure 3.33. Vibration paddle switches are used for liquid, slurry and powder services. (Courtesy of Automation Products, Inc.)

elements, diaphragms and bellows. A discussion is also included on strain gauges and other electronic methods of pressure measurement.

Manometers

Manometers are seldom used for process pressure applications, except occasionally for low pressure services where measurements are in the low range of a few inches of water. In early days of process instrumentation, their use was more frequent, but other, easily maintained, economical, low pressure sensors have been developed which are more suitable.

Advantages

1. Simple and time proven
2. High accuracy and sensitivity
3. Wide range of filling fluids of varying specific gravities
4. Moderate cost
5. Particularly suitable for low pressure and low differential pressure applications

Disadvantages

1. Lack of portability
2. Need of leveling
3. Generally large and bulky
4. Measured fluid must be compatible with the manometer fluid used
5. No overrange protection
6. Condensation may present problems

Bourdon Elements

Bourdon elements include the C-type and the helical and spiral types shown in Figure 3.34. These types comprise a high percentage of all pressure applications. Most local pressure gauges and many receiver gauges utilize the C-type bourdon tube. There is a considerable overlap in the ranges covered by these types.

The C-bourdon is used for ranges from about 15 to 100,000 psig. Accuracies range from ±0.1 to ±5%.

Spiral elements are used generally in the medium pressure ranges, but heavy duty elements are available in ranges up to 100,000 psig. Accuracy is about ±0.5%.

Helical bourdons are used for ranges from 100 to 80,000 psig with an accuracy of about ±½ to ±1% of span.

Advantages

1. Low cost
2. Simple construction
3. Years of experience in application
4. Available in a wide variety of ranges, including very high ranges
5. Adaptable to transducer designs for electronic instruments
6. Good accuracy, especially when considered in relation to cost

Disadvantages

1. Low spring gradient below 50 psig
2. Susceptible to shock and vibration
3. Subject to hysteresis

Bellows Elements

Most bellows elements are used for low pressure and differential pressure services. Figure 3.35 shows units that operate from vacuum service up to 0-400 psig. The majority of applications are in inches-of-water range up to 30 or 40 psig. Units are available, however, that operate up to 0-2,000 psig.

The greatest use for bellows units is as receiving elements for pneumatic recorders, indicators and controllers.

Bellows are also widely used as the differential units for flow measurement, particularly for field mounted recorders or controllers. Accuracy is in the range of ±½%.

Advantages

1. High force delivered
2. Moderate cost
3. Adaptable for absolute and differential pressure use
4. Good in the low-to-moderate pressure range

Disadvantages

1. Need ambient temperature compensation
2. Not suited for high pressures
3. Limited in the availability of metals—some tend to work-harden
4. Require spring for accurate characterization

Diaphragm Elements

Refinements in recent years have greatly increased the use of diaphragm elements. Figure 3.36 shows some typical designs—flat diaphragms, corrugated diaphragms and single and multiple capsule units.

Diaphragm elements are used in vacuum and pressure services, in motion and force balance units and for both pneumatic and electronic instruments.

Slack diaphragm units (Figure 3.37) are available for extremely low spans of 0-0.5 inches H_2O.

The normal application range for diaphragm elements is from vacuum service up to 200 psig. Special button diaphragm units (see Figure 3.36c) are available for pressures to 10,000 psig.

Accuracies range from ±½ to ±1¼% of full span.

Advantages

1. Moderate cost
2. High overrange characteristics
3. Linearity good
4. Adaptability to absolute and differential pressure measurement
5. Available in several materials for good corrosion resistance
6. Small in size
7. Adaptable to slurry services

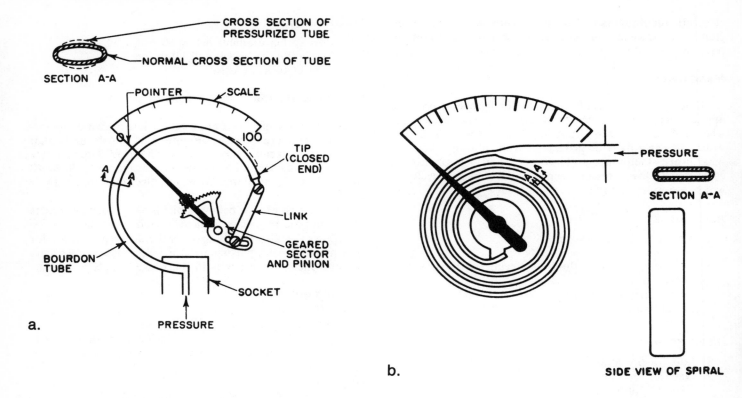

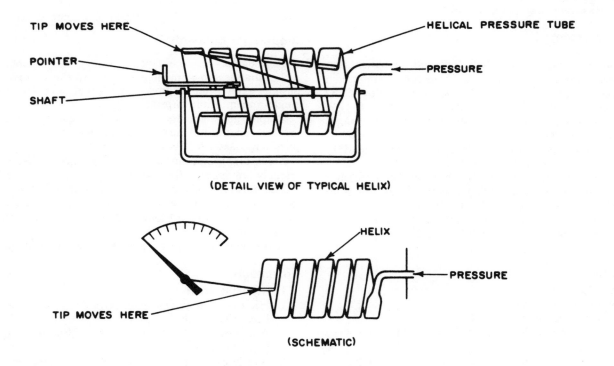

Figure 3.34. Spiral, C-type, and helical bourdon elements are used extensively for pressure measurement. Each element covers a wide range of measurements. (Courtesy of Ametek/U.S. Gauge)

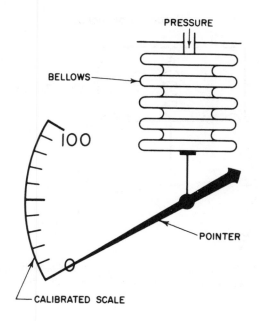

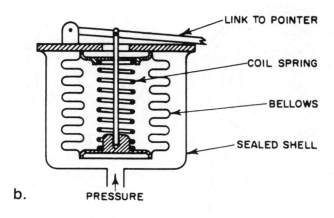

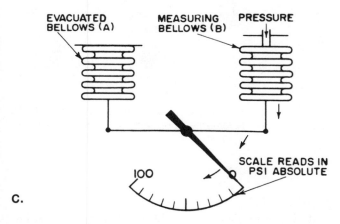

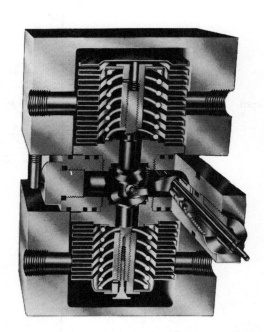

Figure 3.35. Many varieties of bellows elements are used for measuring pressure and are really useful in low pressure ranges. (Courtesy of Ametek/U.S. Gauge and ITT Barton)

Disadvantages

1. Does not have good vibration and shock resistance
2. Difficult to repair
3. Limited to relatively low pressures (except for button diaphragm type)

Strain Gauges

Strain gauges have been used for pressure measurement for several years and have proved to be very satisfactory for special applications. Their acceptance as a standard method for measuring pressure has been rather slow. Because of their continued use in fields other than the processing in-

dustries, much research has been done, and advances in the technology have been notable. It seems likely at this time that the threshold of a new era of wide acceptance has been reached. The use of some newer strain gauge designs could reduce installation time and materials appreciably.

Accuracies range from ±0.2 to ±1.0% of span.

Advantages

1. Small and easy to install
2. Good accuracy
3. Available for wide range of measurement—from vacuum to 200,000 psig
4. Good stability

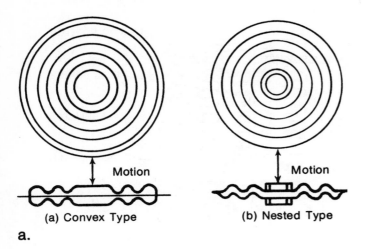

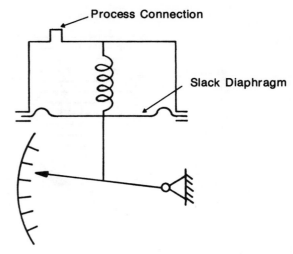

Figure 3.37. Slack diaphragm elements are used to measure low pressure spans in the order of 0-0.5 inch of water.

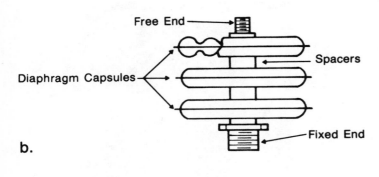

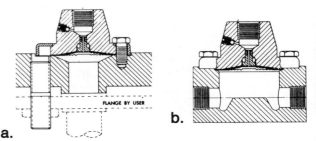

Figure 3.38. Chemical seals are used to isolate process materials from measuring elements. (Courtesy of Dresser Industries, Inc.)

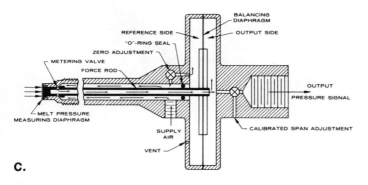

Figure 3.36. The use of diaphragms and capsules has greatly increased in the past few years and now covers a wide range of pressures. (c. Courtesy of Rosemount, Inc.)

5. High output signal strength
6. High overrange capacity
7. Simple to maintain
8. No moving parts
9. Good shock and vibration characteristics
10. Readily adaptable to electronic systems
11. Fast speed of response

Disadvantages

1. Moderate to high cost (could be offset by reduced installation costs)
2. Electrical readout necessary
3. Requires constant voltage supply
4. Some problems presented by temperature variations—temperature compensation usually necessary

Chemical Seals

Chemical seals (Figure 3.38) are used to isolate process materials from measuring elements for three primary reasons:

1. To prevent clogging of the measuring element
2. To prevent corrosive materials from reaching the measuring element because of cost or other factors
3. To prevent process fluids from freezing, vaporizing or setting up

There are some disadvantages to the use of seals.

1. Additional cost (this is not always true since the cost of sealing may be offset by the use of a far less costly measuring element).
2. Possible reduction in accuracy. Theoretically, there is no reduction in accuracy if chemical seal assemblies are properly filled, attached and calibrated. The fill fluid is subject to ambient temperature variations, however, and causes some inaccuracy.

Temperature Instruments

The selection of temperature instruments is somewhat analogous to pressure instrument selection in that process fluid characteristics have less bearing on choosing a method than on flow and level measurements. Unlike flow and level measurements where some methods simply will not work, most of the temperature measuring methods in common use would work in a majority of applications—the choice rests on cost, accuracy, response, maintainability and preference.

Methods which are evaluated include filled systems, thermocouples, resistance bulbs, thermistors, radiation pyrometers, optical pyrometers and bimetal thermometers.

Filled Systems

Filled thermal systems (Figure 3.39) were among the early methods used for process temperature measurement. The method was, and still is, a satisfactory way of measuring temperature for local indication, recording and control.

Its use is not limited to local readout and control but is utilized for pneumatic transmission for remote readout and/or control.

Filled thermal systems are divided into four separate groups or classes. Each class is further divided into groups related to automatic compensation (of the capillary and case) for ambient temperature variations. These classifications are made by SAMA (Scientific Apparatus Manufacturers Association).

Prior to listing the good features and the limitations of filled systems, a brief review of the different types is presented.

Class I systems are liquid filled (mercury excluded) and are divided into three groups: (a) *Class I*— uncompensated; (b) *Class IA*—fully compensated (capillary and case); (c) *Class IB*—case compensated only.

Class I systems are used for measurements from −125° to 600° or 700°F. Minimum range spans are about 25°F, and maximum spans are about 450°F. Scales are uniform except at low temperatures. Response speeds are in the order of 5 to 10 seconds.

Class II systems are vapor filled and are divided into four groups: (a) *Class IIA*—for temperatures above ambient (assuming case and tubing are at ambient temperature); (b) *Class IIB*—for temperatures below ambient; (c) *Class IIC*—for temperatures that may cycle above and below ambient; (d) *Class IID*—for temperatures that may be below, at and above ambient, and where ambient temperature is important.

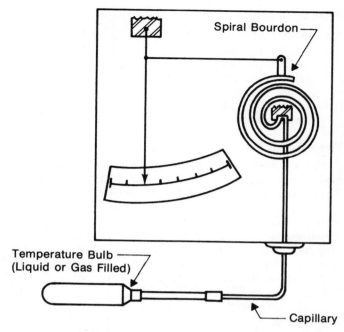

Figure 3.39. Filled thermal systems have provided satisfactory temperature measuring methods for many years.

Class II systems are used for measurements from −430° to 600°F. Minimum range spans are about 40°F and maximum spans about 300°F. Scales are not uniform. Response speeds are in the order of 5 to 10 seconds.

Class III systems are gas filled and are divided into two groups: (a) *Class IIIA*—fully compensated (case and capillary); (b) *Class IIIB*—case compensated only.

Class III systems are used for measurements from −400° to 1,500°F. Minimum range span is about 100°F, and maximum spans are about 1,000°F. Scales are uniform. Response times vary from 1 to 5 seconds.

There is no SAMA Class IV.

Class V systems are mercury filled and are divided into two groups: (a) *Class VA*—fully compensated (case and capillary); (b) *Class VB*—case compensated only.

Class V systems are used for measurements from −40° to 1,000°F. Minimum range span is about 40°F, and maximum span is about 1,000°F. Scales are uniform. Response time is about 4 or 5 seconds.

When viewed as a group, filled thermal systems have the following advantages and disadvantages.

Advantages

1. Simple, time-proven measurement method
2. Relatively low cost
3. No outside source of power required
4. Good selection of calibrated charts available
5. Narrow spans available
6. Ruggedly constructed
7. Presents no electrical hazards in hazardous atmospheres

Disadvantages

1. Limited to measurements below 1,500°F
2. Relatively slow response (fast enough for most applications, however)
3. Bulb failure requires replacement of entire thermal system
4. Transmission distances more limited than for electrical systems

Thermocouples

Thermocouple temperature measurements comprise a high percentage of those made in the processing industries. For remote process temperature checkpoints, they are used as exclusively as bimetal thermometers are for local readout. Their response speed is quite high, depending on thermocouple design and protecting well used. Complete loop hardware for control applications is relatively expensive, but when multiple measurements are made for the same readout device, the average individual cost is relatively low.

Advantages

1. Small units that can be mounted conveniently
2. Wide variety of designs for standard and special applications
3. Electrical output adaptable to a variety of readout and/or control devices
4. Response speed high compared to filled systems
5. Wide measurement range—from near absolute zero to 5,000°F
6. Low cost
7. Good accuracy
8. Calibration checks made easily
9. Transmission distances can be long
10. Good reproducibility

Disadvantages

1. Temperature-voltage relationship not fully linear
2. Accuracy less than that for resistance bulbs
3. Stray voltage pickups must be considered
4. Temperature spans not as narrow as filled systems or resistance bulbs
5. Requires an amplifier for many measurements
6. Hot junctions "age" in same services
7. Requires expensive accessories for control applications

Resistance Bulbs

Resistance bulbs in the past have been used infrequently and only for applications requiring high accuracy and/or narrow spans.

Advantages

1. High accuracy
2. Narrow spans
3. Reproducibility good
4. Remain stable and accurate for many years

5. Temperature compensation not necessary
6. Fast response speed

Disadvantages

1. Relatively expensive compared to thermocouples
2. Bulb sizes larger than thermocouples
3. Self-heating can be a problem
4. Mechanical abuse or vibration can be a problem

Thermistors

Thermistors have been used only on a very limited basis until recently. Improved production methods for semiconductor materials, however, have paved the way for increased use, especially for special applications where other methods are lacking.

Advantages

1. Fast response
2. Good for narrow spans
3. Low cost
4. Cold junction compensation not necessary
5. Negligible leadwire resistance
6. Stability increases with age
7. Available in small sizes
8. Adaptable to various electrical readout devices

Disadvantages

1. Nonlinear temperature versus resistance curve
2. Not suitable for wide temperature spans
3. Interchangeability of individual elements often a problem
4. Experience limited for process application

Radiation Pyrometers

Radiation pyrometers (Figure 3.40) are used primarily for high and very high temperatures above the ranges normally covered by thermocouples. They find limited use in the chemical processing industries.

Advantages

1. Ability to measure high temperatures
2. Does not require contact with target of measurement
3. Fast response speed
4. High output
5. Moderate cost

Disadvantages

1. Nonlinear scale
2. Subject to errors due to presence of intervening gases or vapors that absorb radiating frequencies
3. Measurement affected by emissivity of target material

Optical Pyrometers

Optical pyrometers are classified into two categories, those which operate within the visible spectrum where the

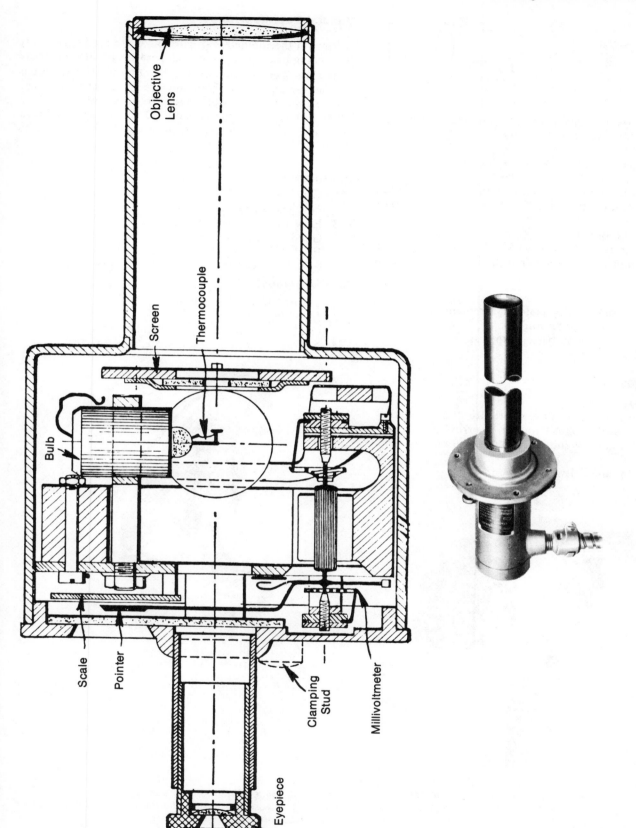

Figure 3.40. Radiation pyrometers are used for high and very high temperature ranges above those normally covered by thermocouples. A sighting tube decreases the effects of intervening gases. (Courtesy of Pyrometer Instrument Co. and Honeywell, Inc.)

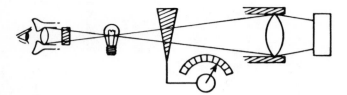

Figure 3.41. The manual pyrometer uses the human eye as a detector. (Courtesy of Pyrometer Inst. Co.)

human eye is used as a detector and those that compare the measured source to a known standard.

The manual pyrometer (using the human eye as a detector) operates within limits of 2,400° to 6,300°F (see Figure 3.41).

The automatic or infrared pyrometer (Figure 3.42) can measure temperatures from 0° to 8,000°F, using a known standard to compare radiation frequencies.

Advantages

1. Need not contact target of measurement
2. Useful for high temperatures
3. Light, portable, with battery operation
4. Good accuracy

Disadvantages

1. Relatively expensive
2. Operator adjustment of temperature dial required of manual model—subject to human errors
3. Manual model not suitable for alarm or control function
4. Subject to emissivity errors

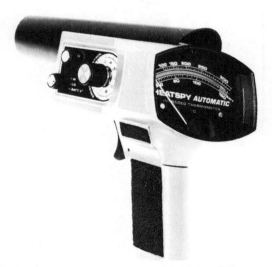

Figure 3.42. Infrared pyrometers are packaged for portability and can sense temperatures of distant or moving objects. They measure temperatures from 0 to 8,000°F. (Courtesy of William Wahl Corp.)

Bimetal Thermometers

Bimetal thermometers (Figure 3.43) are used extensively for local indication of process temperatures. They cost more than glass stem thermometers, but the additional cost is justified because glass thermometers are so easily broken.

Measurements can be made from about −100° to 1,000°F; minimum span is about 30°F. The scale is linear over most of the range. Accuracies range from ±½ to ±2% or higher.

Advantages

1. Low cost
2. Not easily broken
3. Easy to install and maintain
4. Good accuracy relative to cost
5. Fairly wide temperature range

Disadvantages

1. Limited to local mounting
2. Indicating only
3. Calibration may change if handled roughly
4. Not as accurate as glass stem thermometers

Figure 3.43. Bimetal thermometers are used extensively for local indication of process and utility temperatures. (Courtesy of Tel-Tru Manufacturing Co.)

Miscellaneous Types

Other temperature measurement methods comprise a fairly small percentage of process temperature measurements. Some of them, however, are very useful for special applications. Some of these are described and evaluated briefly.

Pyrometric Cones

Pyrometric cones are used primarily in the ceramics industry. They are not used as exact temperature indicators for particular applications. Their cost is low, and they are available in approximately 60 ranges between 1,085° and 3,660°F. They can be used only once and require observation by an experienced person.

Accuracy is limited to about ±15° to ±20°F.

Temperature-Sensitive Materials

Another one-shot method of temperature measurement is the application of temperature-sensitive materials to surfaces for measurement. Generally, the chemicals take the form of crayons, pellets or paint. When the chemicals reach their temperature-sensitive region, they change color (generally black).

Indicators are available in ranges of 10° to 50°F in steps over a range of 100° to 2,500°F.

Quartz Crystal Thermometers

Quartz crystal thermometers find very limited use in the processing industries. They are rugged and accurate with excellent short-term stability. They are larger in size than thermocouples or resistance temperature detectors (RTDs) and comparatively expensive. Their use is normally in laboratory type environments.

Temperature Switch Selection

A variety of temperature switches are available for the many applications where on-off action is needed for alarm, interlock or control. The most common types are (a) bimetallic, (b) filled systems and (c) thermocouples.

Most standard units have adjustable spans so that the buyer may adjust the settings as desired. Some units, however, are furnished with fixed ranges that are factory set—no adjustment can be made.

When selecting temperature switches, it is desirable to select a range so that the set point is near the center of the adjustment range. The greater the variation in ambient temperature, the greater is the need for near midpoint selection, for switches are made with the greatest accuracy at the midpoint. It is also desirable to make the setting when ambient temperature is at the midpoint of the switch's ambient range.

Bimetal

Bimetal switches (Figure 3.44) are available in ranges from —100° to 1,000°F. They are inexpensive and easy to

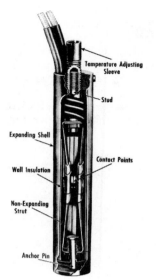

Figure 3.44. Bimetallic temperature switches are inexpensive and easy to adjust. (Courtesy of Fenwal Incorporated)

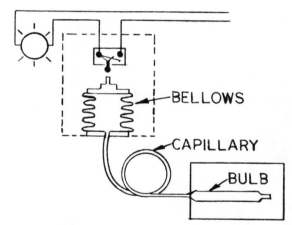

Figure 3.45. Thermal filled systems (liquid, gas or vapor) may also be used as switches in point temperature measurement applications.

adjust in the field. They can be damaged by overheating, and their repeatability characteristic is relatively poor.

Filled Systems

Filled systems consist of closed systems filled with liquid, gas or vapor (see Figure 3.45). The sensing bulbs are connected by capillary to sensing elements such as bourdons, bellows or diaphragms, which move to actuate the switches as temperature changes expand or contract the fill fluid.

Among the filled systems, liquids have the advantages in cost, bulb size and response speed. However, they are more sensitive to ambient temperature variations.

Liquid filled systems are available in ranges from —150° to 1,200°F. They have small bulb sizes, are moderately priced, respond quickly and have low on-off differential values. They may require ambient temperature compensation if the temperature varies appreciably.

Vapor pressure systems are available in ranges from −100° to 550°F. They are relatively insensitive to ambient temperature change, but respond slowly, require large sensing bulbs and are relatively expensive.

Gas-filled systems are available in ranges from −100° to 1,500°F. They are also insensitive to ambient temperature changes and have the same limitations as vapor systems—their response is slow, they have large bulbs, and they are expensive.

Thermocouples

Thermocouple switches are available in ranges from −400° to 4,500°F. Their advantages include wide ranges and fast response characteristics. However, they require expensive accesories.

4 Control Valve Selection

William G. Andrew

Factors affecting the choice of the proper valve in a control loop include a knowledge of the dynamics of the loop, physical properties of the process stream, control requirements of the particular service and the relative economics of available choices. Working directly in the process fluid, the valve may wear due to erosion, corrosion or cavitation. The use of improper valve materials may contaminate the process or cause the valve to wear excessively. The following information and considerations will help in selecting the body designs, materials and mechanical features of control valves for most of the applications encountered.

Function in the System

The function of a control valve in a system is to absorb the proper amount of pressure drop (P_v in Figure 4.1) to maintain system balance under all operating conditions. Other pressure drop values remain fairly constant or have a fixed flow-friction relationship. The control valve is the selected variable in the system.

Any system using a control valve has two types of pressure drops—static and dynamic. Static pressure drop, it is sometimes assumed, does not change in magnitude, regardless of flow conditions in a system. However, this is not always true. A typical system is shown in Figure 4.1. The static pressure drop, P_h, due to the difference in elevation, changes as the specific gravity of the fluid changes. Normally this change is insignificant.

The dynamic pressure drop of a system is related (a) to fluid flow in the system and (b) to dynamic or static forces at each end of the defined piping system. Losses due to fluid flow include P_e, the pressure drop across a flow element; P_v, the drop across the valve; P_f, pipe friction losses; and P_x, the pressure drop across the heat exchanger.

Initial pressure, P_p, varies as process changes occur upstream of the pump. Tower pressure, P_t, varies with process temperature changes and other conditions at the tower and downstream of it. The difference between P_p and P_t is usually considered constant for practical purposes and becomes a part of the static drop, P_h.

A simplified equation for pressure drop, ΔP_s, in the system becomes:

$$\Delta P_s = P_h + P_e + P_v + P_f + P_x \tag{4.1}$$

A knowledge of the hydraulics and the physical layout of the system allows a logical determination of the control valve ΔP and the total pressure drop for the system.

Pressure Drop Requirements for Good Control

The necessary pressure drop across the control valve for good control depends on (a) the total system drop, (b) the installed characteristic of the valve and (c) the variation in flow required. The ΔP required for a valve is usually stated in percent of the total dynamic system drop. Percentages normally considered necessary are 30 to 35% for valves with equal percentage characteristics and up to 50% for linear characteristic valves. There is little doubt that valves will function well when these percentages of the total dynamic system drop are allowed.

Experience has taught, however, that good control often is obtained with as little as 15% of the total system drop across the valve. A careful analysis of the system should be made if the percentage falls below this value. Too often ar-

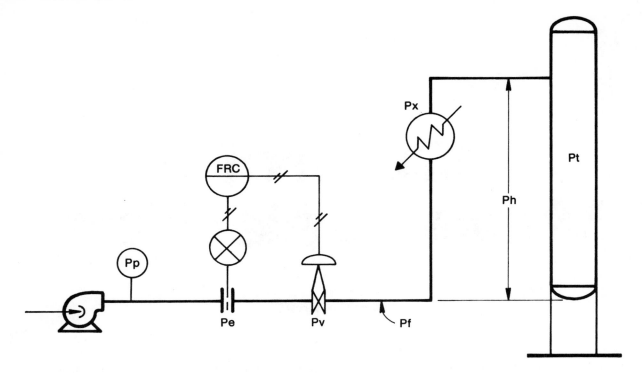

Figure 4.1. Typical hydraulic system showing static conditions and dynamic drops. The control valve drop, Pv, *is the primary variable in the system.*

bitrary values of 10 to 15 psi are allowed for a valve ΔP without analyzing the system. These values are sufficient for some applications but are lacking in many other situations.

The percentages given above apply primarily to globe valves. When the available drop is too small for globe bodies, consideration should be given to high recovery valves such as ball or butterfly.

The economics of control valve design must not be overlooked in the determination of pressure drops and valve sizes. A control valve is recognized as an engineered resistance in the loop. The energy it absorbs is supplied normally by a pump or other prime mover whose costs (installed and operating) may vary with valve size. Lower initial valve cost (small size) may be offset by higher operating costs (high ΔP).

Pressure drop requirements may be summarized as follows:

1. Preferred ΔP—50% for linear valves; 35% for equal percentage
2. Adequate ΔP—15% for many applications
3. Increase the percentage ΔP for wide flow ranges
4. Decrease the percentage ΔP for small flow variations

Control valve sizes usually are smaller than line sizes. If the valve is calculated to be the same as line size, the data and calculations for line and valve sizing should be reviewed for one, or the other may be sized wrong.

Capacity Requirements

Flow sheets for most processes give design rates for the main process and utility streams. Knowledge of normal, maximum and minimum rates, specific gravity, vapor pressure and viscosity are needed for proper valve sizing and valve selection. When these values are known, valve sizing usually presents few problems.

Design Rate

Design rate is the term used to denote the *normal* flow when a plant is operating at its intended capacity. At start-up, during upsets and for other abnormal variations, momentary rates may vary considerably from the normal rate. The control valve usually is selected to control from 25 to 60% opening for this rate. Its opening (at design rate) depends on the valve trim characteristic selected, the ratios of normal/minimum and normal/maximum values and the C_v rating of the valve in the range of C_v values acceptable.

Maximum Rate

Maximum flow rates are listed on mechanical or process flow sheets for many projects. However, sometimes arbitrary rates must necessarily be chosen. It should be kept in mind that in order to control, there must be an ability to deviate up or down from a given flow. If a valve is wide open, it no longer controls because it can go in only one direction from that position. When the maximum rate is not known, 150% of *design* rate should be used to size the valve.

Sources to be checked for maximum rates include pump curves, compressor and other equipment capacities, heating and/or cooling loads and process engineers who have a knowledge of the process requirements.

Minimum Rate

Minimum rates are usually more difficult to determine than maximum rates. When minimum rates are not established, the responsible instrument engineer and the process engineer should discuss the various aspects of the control systems to determine conditions likely to exist during startups and upsets. Most single port valves control quite well down to 3 or 4% of their rated capacity. Control at the low end is sometimes difficult, however, because the ΔP of the valve is highest at these conditions.

When minimum flow rates are low, double port valves may present difficulty for they may have leakage rates as high as the minimum rates desired—1 to 2% (more when worn) of their rated capacity.

Valve Rangeability

Rangeabilities of control valves vary from about 5:1 for diaphragm valves to claimed values of 300:1 for some types of throttling ball valves. Globe bodies have rangeabilities from 30:1 to 50:1. From a practical standpoint, it is better to assume rangeability values of 15:1 to 20:1 for globe valves and 30:1 to 50:1 for ball valves. Controllability at the extreme end of any type valve is questionable. In Figure 4.2 movement toward the extreme ends of an equal percentage characteristic curve, though the change is small in each case, changes the rangeability from 39.2 to 9.5.

Table 4.1 lists typical rangeabilities for the valve body types discussed. Values shown are commonly claimed by valve manufacturers but represent ideal conditions. From a practical standpoint, these values should normally be reduced about one-third.

It has been noted in Chapter 11 of Volume 1 that the *installed* flow characteristic of control valves differs from the *inherent* characteristic because the inherent characteristic is determined at a constant pressure drop whereas operating conditions determine the installed characteristic. Similarly, operating conditions affect the installed rangeability of the valve. For a constant pressure drop, the rangeability as depicted in Figure 4.2 holds true. As the pressure drop changes, however, rangeability changes. Where the pressure drop of the entire loop or system remains constant, the pressure drop across the valve increases with decreasing flow, resulting in an increased valve rangeability under operating conditions.

Choosing the Flow Characteristic

If a control valve were to operate at a constant load without appreciable upsets or changes, the characteristic of

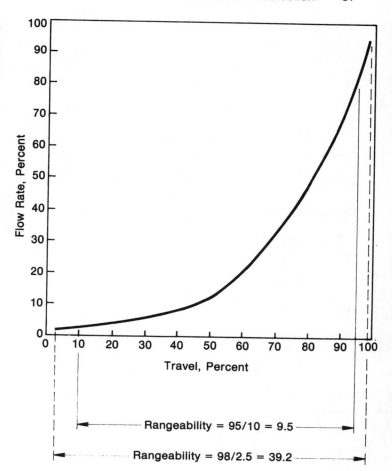

Figure 4.2. Rangeability depends on definition of controllability at extreme ends of valve travel.

Table 4.1. Comparison of Typical Rangeabilities for Various Body Types	
Body Design	**Typical Rangeability***
Double Port Globe	50:1
Single Port Globe	50:1
Split body	30:1
Angle	30:1
Diaphragm	5:1
Butterfly, 60°	15:1
Butterfly, 90°	100:1
Pinch	5:1
Vee-ball	30:1
Full ball	100:1

*Valves with equal percentage characteristics have greater rangeabilities than those of other characteristics.

the valve would be unimportant. Variations from the control point would be small. Rare is the process, however, described by this condition. Instead, load changes and pressure variations occur, requiring quick responses from the valve. The effectiveness of the response is affected by the inherent characteristic of the valve.

Following are some guidelines to use when choosing flow characteristic. For convenience, guidelines are listed both by application requirement and by flow characteristic. Typical curves are shown in Figure 4.3 for the three major types of characteristic plugs—linear, equal percentage and quick opening.

By Application

For specific applications on flow, pressure and level, the following should be considered.

Flow Control
Choose the proper flow characteristic based on the tabulation in Table 4.2.

Pressure Control

1. For liquid applications, use an equal percentage flow characteristic
2. For compressible fluid applications, use an equal percentage flow characteristic if the system has less than 10 feet of pipe downstream of the control valve
3. For compressible fluid applications, use a linear flow characteristic if the downstream system involves a

Table 4.2. Suggested Plug Characteristic for Flow Applications

Flow Controller Measuring Element	Location of Control Valve	Flow Characteristic	
		Wide Flow Range	Small Flow Range†
With square root extractor	In series	Linear	Equal percentage
With square root extractor	In bypass*	Linear	Equal percentage
Without square root extractor	In series	Linear	Equal percentage
Without square root extractor	In bypass*	Equal percentage	Equal percentage

*When control valve closes, flow rate increases in measuring element.
†Large changes in pressure drop occur at valve.

receiver, distribution system or transmission line exceeding 100 feet (if the pressure drop varies more than 5 to 1, an equal percentage characteristic would be the best selection)

Liquid Level Control

1. With constant pressure drop, use a linear characteristic
2. When pressure drop decreases with an increase in load, the linear valve plug is still the best selection, except where the full load pressure drop is less than 20% of the no-load pressure drop (for this exception, use the equal percentage flow characteristic)
3. When pressure drop increases with an increase in load, again use the linear valve plug, except where this increase is greater than 2 to 1 (for this exception, use a quick opening characteristic)

By Flow Characteristic

Equal Percentage
Equal percentage characteristics (Figure 4.4) are applicable:

1. For fast processes
2. When high rangeability is required
3. When system dynamics are not well known
4. At heat exchangers where an increase in product rate requires much greater increase in heating or cooling medium

An equal percentage valve loses its inherent characteristic as it absorbs less and less of the dynamic system drop, tending toward a linear characteristic. It is still the best choice when system dynamics are not well known, for it keeps a desirable control characteristic over a wide range of percent dynamic loss. It also has wide rangeability, which is desirable when system loads are not well known.

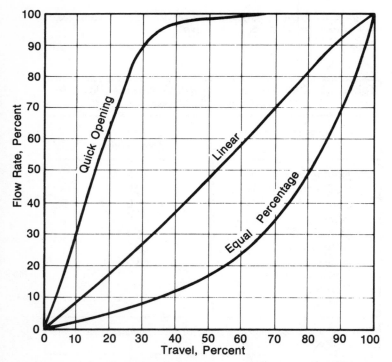

Figure 4.3. Typical characteristic curves for quick-opening, linear and equal percentage plugs.

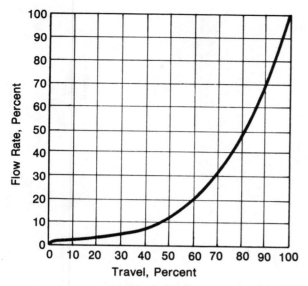

Figure 4.4. Equal percentage characteristic curve.

Quick Opening

Quick opening valves are used:

1. For on-off control
2. When maximum valve capacity must be obtained quickly

A quick opening valve has a linear characteristic for about one-fourth of its travel from shutoff. When sized to operate in this range, it becomes a linear valve. Beyond this point, it has little use, except for on-off control.

Linear

Linear valves are used:

1. For slow processes
2. When more than 40% of the system pressure drop occurs across the valve
3. When major process changes are a result of load changes

Modified

Many variations of the three basic valve trim characteristics are available. Their selection may be based on the system characteristic required, or it may simply be based on the predominant characteristic available from a particular valve manufacturer or with a particular body design.

Satisfactory valve performance is often obtained even with a mismatch between the process and the valve. This is especially true since most controllers now are available with wide proportional bands and reset and rate actions to provide desirable control characteristics.

Figure 4.5 shows plug configurations of various flow characteristics for single port valves, and Figure 4.6 shows plugs for double port valves.

C_v Comparison Table by Characteristics

Table 4.3 lists C_v values for linear, equal percentage and quick opening characteristics in single port, double port and cage bodies for valve sizes ½ inch through 12 inches. These values are from valves of a single manufacturer. The comparison shows that for the same size valve, less flow is obtained from single port valves. Capacities of double port and cage valves are nearly equal in sizes up to 4 inches. Cage valves are not listed as available from this manufacturer in

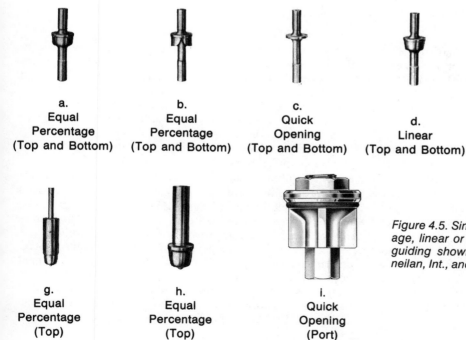

a.
Equal
Percentage
(Top and Bottom)

b.
Equal
Percentage
(Top and Bottom)

c.
Quick
Opening
(Top and Bottom)

d.
Linear
(Top and Bottom)

e.
Equal
Percentage
(Top and Port)

f.
Linear
(Top)

g.
Equal
Percentage
(Top)

h.
Equal
Percentage
(Top)

i.
Quick
Opening
(Port)

Figure 4.5. Single port valve plugs with equal percentage, linear or quick-opening characteristics. Type of guiding shown in parentheses. (Courtesy of Masoneilan, Int., and Fisher Controls Co.)

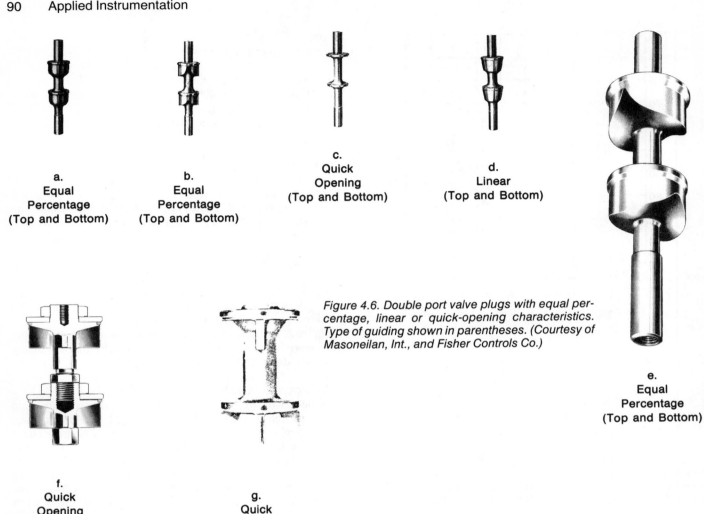

a.
Equal
Percentage
(Top and Bottom)

b.
Equal
Percentage
(Top and Bottom)

c.
Quick
Opening
(Top and Bottom)

d.
Linear
(Top and Bottom)

Figure 4.6. Double port valve plugs with equal percentage, linear or quick-opening characteristics. Type of guiding shown in parentheses. (Courtesy of Masoneilan, Int., and Fisher Controls Co.)

e.
Equal
Percentage
(Top and Bottom)

f.
Quick
Opening
(Port or
Top and Bottom)

g.
Quick
Opening
(Port)

sizes above 6 inches. Table 4.4 makes a comparison of C_v values of single port valves by manufacturer for linear, equal percentage and quick opening characteristics. There is a considerable variation in capacities among manufacturers.

Choosing the Body Design

The choice of body design for a control valve is seldom a difficult one. Reviews of typical jobs reveal that globe valves lead in applications by a large percentage. Standard double port globe valves (Figure 4.7) are most common with single port valves (Figure 4.8) next in popular usage. These have been workhorses of industry as final control elements for many years.

Double seated valves are popular because of their balanced design—the forces tending to close the valve are only slightly different from those tending to open. They are available with reversible plugs—constructed so that increasing loading pressure moves the plug either into or out of the port.

Most single seated valves are of unbalanced design, though balanced designs (Figure 4.9) are made. They are used extensively in small sizes, 2 inches and below, and where tight shutoff is needed. Their use in small body sizes is due largely to their simple design and because, for low flows, the unbalanced forces are still not large enough to require large actuator sizes.

Split body globe valves (Figure 4.10) are offered by several manufacturers. They are simple in construction, are easily maintained and have economical advantages, especially in alloy materials. They are sometimes furnished with separable flanges (Figure 4.11), allowing flanges to be made of a less expensive material than the body. The flanges do not come in contact with the flowing medium.

Another globe valve developed in recent years and increasing in use is the cage valve (Figure 4.12), so named because the plug is guided in a "cage" enclosure in the valve body. It is used in many applications where a double seated valve would be used and has additional advantages of high C_v rating, low noise, good stability, easily changed trim,

Table 4.3. Comparison of Typical *Cv* Values by Trim Characteristics for Single Port, Double Port and Cage Valves

Valve Type	Valve Characteristic	Valve Size, inches										
		1/2	3/4	1	1½	2	3	4	6	8	10	12
Single Port Globe	Linear			10.0	26.0	40	90	152				
	Equal percentage	4.0	7.6	10.0	24.0	43	99	153	348	587		
	Quick-opening	4.8	6.1	10.8	26.0	47	98	179	424	666	1280	1700
Double Port Globe	Linear			12.0	30.0	50	116	189	439	732		
	Equal percentage			12.5	31.0	51	116	196	449	780	1110	1680
	Quick-opening			19.0	45.0	63	154	243	606	1220	1580	2380
Cage Globe	Linear			20.0	34.9	61	135	212	416			
	Equal percentage	4.0	9.0	17.0	33.0	56	115	203	364			
	Quick-opening	6.5	14.0	21.0	38.0	67	150	235	460			

Table 4.4. Comparison of Typical *Cv* Values by Trim Type and by Valve Manufacturer

Valve Size	Valve Type	Manufacturer			
		A	B	C	D
1″	Linear	9.0	10.0	12.0	13.0
	Equal %	9.0	10.1	11.5	13.0
	Quick open	12.0	10.8	17.5	15.0
2″	Linear	36.0	40.0	48.0	52.0
	Equal %	36.0	43.4	44.0	52.0
	Quick open	47.0	47.7	55.0	58.0
3″	Linear	75.0	90.0	95.0	110.0
	Equal %	75.0	99.0	93.0	115.0
	Quick open	100.0	98.0	105.0	125.0
4″	Linear	124.0	152.0	150.0	190.0
	Equal %	124.0	153.0	150.0	170.0
	Quick open	175.0	179.0	185.0	230.0
6″	Linear	270.0		360.0	390.0
	Equal %	270.0	348.0	350.0	340.0
	Quick open	400.0	424.0	400.0	400.0

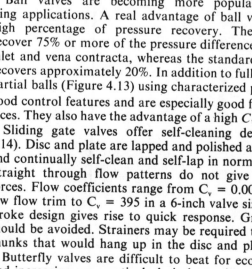

Figure 4.8. Single port globe valves are also widely used. Design is unbalanced. (Courtesy of Fisher Controls Co.)

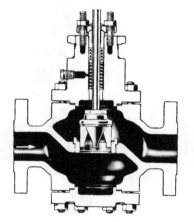

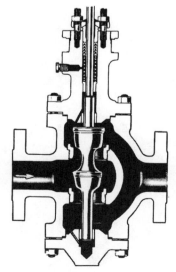

Figure 4.7. Double port globe valves are the most popular body type valves in use. (Courtesy of Fisher Controls Co.)

reduced trim using the same plug and reduced erosion problems.

Ball valves are becoming more popular for throttling applications. A real advantage of ball valves is their high percentage of pressure recovery. They frequently recover 75% or more of the pressure difference between the inlet and vena contracta, whereas the standard globe valve recovers approximately 20%. In addition to full ball designs, partial balls (Figure 4.13) using characterized ports provide good control features and are especially good for slurry services. They also have the advantage of a high C_v rating.

Sliding gate valves offer self-cleaning design (Figure 4.14). Disc and plate are lapped and polished at the factory, and continually self-clean and self-lap in normal operation. Straight through flow patterns do not give unbalancing forces. Flow coefficients range from $C_v = 0.0008$ in special low flow trim to $C_v = 395$ in a 6-inch valve size. The short stroke design gives rise to quick response. Gritty services should be avoided. Strainers may be required to screen out chunks that would hang up in the disc and plate.

Butterfly valves are difficult to beat for economy. They find increasing use, particularly in large sizes. They have a high recovery characteristic, thus reducing the pressure drop requirement. In recent years, rubber lined valves, T-ring types and other designs have overcome one of the main dis-

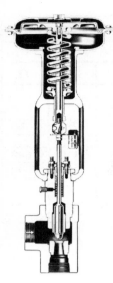

Figure 4.9. Single seat, balanced angle valve. (Courtesy of Fisher Controls Co.)

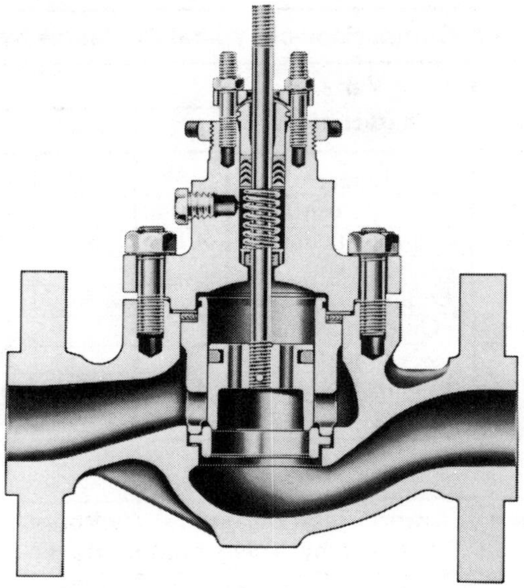

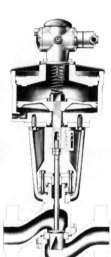

Figure 4.12. Globe valve with cage trim offers good, solid plug guiding. (Courtesy of Fisher Controls Co.)

Figure 4.10. Split body globe valves are economical and easily maintained. (Courtesy of ITT Hammel Dahl/Conoflow)

Figure 4.13. Partial ball valves are especially good for slurry services. (Courtesy of Masoneilan, Int.)

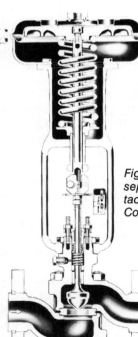

Figure 4.11. Split body globe valve with separable flanges, which do not contact flowing fluid. (Courtesy of Fisher Controls Co.)

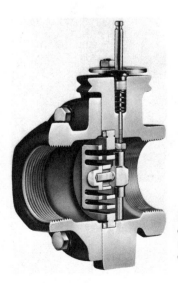

Figure 4.14. Sliding gate valves offer self-cleaning design. (Courtesy of Jordan Valve Div. of Richards Industries Inc.)

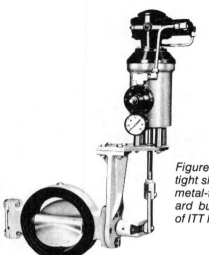

Figure 4.15. Soft seat provides tight shutoff not obtainable with metal-to-metal seats of standard butterfly valves. (Courtesy of ITT Hammel Dahl/Conoflow)

Figure 4.17. The serrated-edge of the Lo-T butterfly reduces torque requirements. (Courtesy of H. D. Baumann Assoc. Inc.)

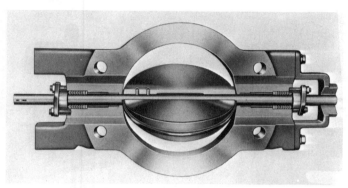

Figure 4.16. Recent disc designs modify the trailing edge. (Courtesy of Fisher Controls Co.)

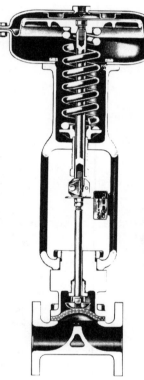

Figure 4.18. Diaphragm valves (Saunders-type shown with spring and diaphragm operator) are also widely used in slurry services. (Courtesy of Fisher Controls Co.)

advantages of butterfly valves—high leakage rates. Tight shutoff (Figure 4.15) is obtained by using the metal disc and a soft seating surface. Butterfly valves do require more power from an actuator, and control characteristics are usually not as good as other body designs.

Recent disc designs of butterfly style valves have sought to improve control stability and minimize operating torque by modifying the trailing edge. The Fisher "fishtail" butterfly valve has an approximate equal percentage characteristic through 90° rotation and a 100:1 rangeability (Figure 14.16).

The serrated edge of the "Lo-T" butterfly (Figure 14.17) not only reduces torque requirements and gives a nearly equal percentage flow characteristic, but also reduces the amount of pressure drop actually recovered across the valve. By forcing valve discharge pressure to remain low, fluids which might have vaporized in the low pressure region (vena contracta) within the valve body remain in a vapor state rather than condensing (bubbles collapse) to a liquid, a condition of cavitation. Noise is less likely in such a valve design.

Diaphragm valves (Figure 4.18) are used primarily for slurry services and corrosive applications. Packless seal con-

struction avoids many difficulties associated with slurry services by isolating working parts of the valve from the process fluid. Generally, they have poor control characteristics and short diaphragm lives.

As an aid for selecting the proper designs, advantages and disadvantages of various body types are listed below.

Sliding Gate (Figure 4.14)

Advantages
1. Reduced turbulence (and noise)
2. Fast response
3. Self-cleaning
4. Tight shut-off

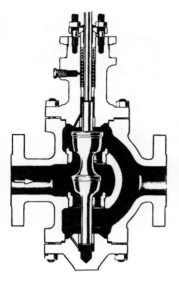

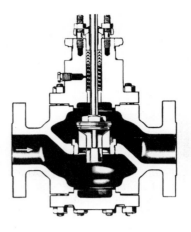

Figure 4.19. Double port globe valve has high capacity but also high leakage rates. (Courtesy of Fisher Controls Co.)

Figure 4.20. Single port globe valves have high rangeabilities but require high operating forces. (Courtesy of Fisher Controls Co.)

5. High rangeability (100 to 1) with a near linear flow characteristic
6. Direct actuators available

Disadvantages
1. Flow fluid must be clean, no grit

Double Port Globe (Figure 4.19)

Advantages
1. High flow capacity compared to single port valves of same size
2. High rangeability
3. Balanced design requiring smaller actuator in comparison to single port design
4. Reversible ports available
5. Frequently used in sizes larger than 2 inches

Disadvantages
1. Relatively high leakage rates at shutoff—1% of rated capacity is a commonly accepted figure (when valve is new or not worn)
2. Low pressure recovery characteristic
3. Erosion may occur on high pressure drop applications due to leakage characteristic
4. Not good for high flow, low pressure drop applications

Single Port Globe (Figure 4.20)

Advantages
1. High rangeability
2. Provides tight shutoff—little or no leakage when new or in good condition
3. Reversible plugs available
4. Frequently used in sizes under 2 inches

Disadvantages
1. Unbalanced design requires relatively large actuator
2. Low pressure recovery characteristic

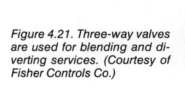

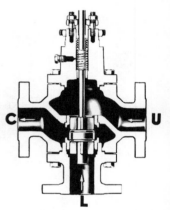

Figure 4.21. Three-way valves are used for blending and diverting services. (Courtesy of Fisher Controls Co.)

Three-Way Globe (Figure 4.21)

Advantages
1. Good for blending and diverting applications
2. Can replace 2 two-way valves in certain applications
3. Frequently used for temperature control systems at heat exchangers

Disadvantages
1. Cannot control total flow
2. May need different size ports which are not available
3. Need to know flow conditions quite precisely

Split Body (Figure 4.22)

Advantages
1. Good control characteristics
2. Tight shutoff
3. Simple, economic construction, especially in alloy materials
4. Easily maintained
5. Relatively free of pockets where sediment or solids may collect
6. Separable flanges available for added economy when alloy materials are required by the flowing medium

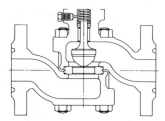

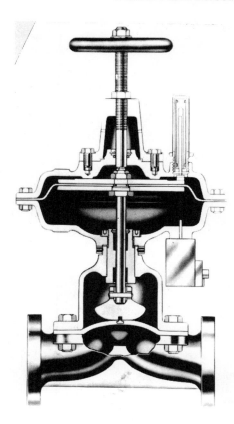

Figure 4.22. Split body valves have good control characteristics and tight shutoff, but pressure drop is limited by top guiding. (Courtesy of Fisher Controls Co.)

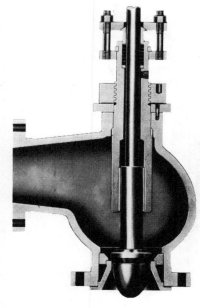

Figure 4.23. Venturi angle valves are good for flashing services and high pressure drops but are not normally available in the small sizes. (Courtesy of ITT Hammel Dahl/Conoflow)

Figure 4.24. Saunders-type diaphragm valves are used for slurry and heavily viscous services but provide relatively poor control. (Courtesy of ITT Hammel Dahl/Conoflow)

Disadvantages
1. Limited pressure drop because of top guided plug

Venturi Angle (Figure 4.23)

Advantages
1. Good control characteristics
2. High capacity, good rangeability
3. Tight shutoff
4. Minimizes erosion problems
5. Conserves space by eliminating 90° ells in piping
6. Can handle sludges and slurries
7. Good for flashing services

Disadvantages
1. Normally made only in sizes 2 inch and above.

Diaphragm (Figure 4.24)

Advantages
1. High capacity
2. Low cost
3. Self-cleaning action, good for slurries
4. Diaphragm seals flowing medium from the working parts of process fluids
5. Provides tight shutoff if pressure is low
6. Can handle corrosive fluids because of variety of materials of construction—good for corrosive chemicals

Disadvantages
1. Poor control characteristics
2. Low rangeability
3. Short diaphragm life
4. Slow response speed
5. Not usually suitable for high pressure drop applications
6. Temperature limited by diaphragm material characteristic

Butterfly (Figure 4.25)

Advantages
1. High capacity
2. Economical, especially in larger sizes
3. High recovery characteristic—low pressure drop through valve
4. Does not permit sediment buildup—good for slurry services
5. Requires minimum space for installation
6. Readily available in large sizes

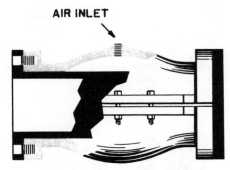

Figure 4.26. Pinch valves are excellent for heavy slurry services, have high capacity and are economical. They exhibit poor control characteristics. (Courtesy of Red Valve Company, Inc.)

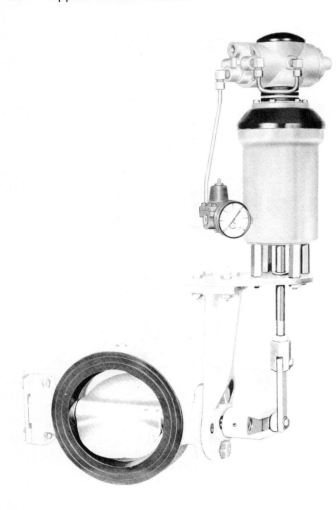

Figure 4.27. Fisher "Vee Ball" valve has high capacity, good rangeability and good control characteristics. (Courtesy of Fisher Controls Co.)

Figure 4.25. Butterfly valves have high capacities and high recovery characteristics and are economical, but they require large operators because of high operating torques. (Courtesy of ITT Hammel Dahl/Conoflow)

Disadvantages
1. Operating torques can be high—large actuator required if valve is large or pressure drop is high (recent designs have reduced to some extent the torque requirements)
2. Tight shutoff depends on use of resilient seats which are temperature limited
3. Throttling control on some designs is limited to 60° travel.

Pinch (Figure 4.26)

Advantages
1. High capacity
2. Economical
3. Self-cleaning action, good for slurries
4. Flowing medium does not contact working parts

Disadvantages
1. Poor control characteristics
2. Relatively low operating temperature and pressure
3. Not good for high pressure drop applications
4. Slow response speed

Vee-Ball (Figure 4.27)

Advantages
1. High capacity
2. Good control characteristics
3. Good rangeability
4. Reasonable cost
5. Good for slurry services

Disadvantages
1. Limited operating pressure
2. Not good for high pressure drops

Table 4.5. Capacity Comparison of Valve Body Types											
Valve Type	C_v Factor Comparison for Various Valve Types by Size, Inches										
	½	¾	1	1½	2	3	4	6	8	10	12
Single port globe	5.0	6.4	11.0	25.0	48.0	95	147	324	436	987	1180
Double port globe			13.0	32.0	53.0	121	190	438	765	1320	1760
Split body		6.4	11.0	25.0	48.0	93	140	307			
Angle			16.5	38.2	68.7	152	293	492	766	1030	
Saunders	4.0	8.5	8.5	31.0	55.0	180	305	515	1130	1450	2350
Cage	6.5	14.0	21.4	38.0	67.2	150	235	460			
Butterfly 60°			15.9	38.8	68.0	152	280	638	1163	1843	2679
Vee-ball					99.0	254	536	1040	1700	2690	3930

C_v Comparison of Body Designs

Table 4.5 provides a capacity comparison of several types of body designs. C_v values shown are typical. The table is arranged generally with low capacity bodies listed at the top and high capacity bodies at the bottom of the table.

Body Materials

Valve bodies are made from most any metal that may be cast, forged or machined. The most common materials are iron, bronze, carbon steel and alloy steels. Monel, Hastelloy, aluminum and plastics are also used. Rubber and plastic liners are used on diaphragm valves, and similar materials are used for making tight closures on other body types.

Listed below are suggestions for consideration in the selection of valve body types and materials.

1. Determine the material requirements for the service in which the valve is used. Piping specifications providing this information are normally available. Valve material, including trim, must be compatible with the piping specifications for the job.
2. Corrosive or slurry services may dictate the use of a body type that exhibits poor control characteristics.
3. Economy through high capacity or simple design may outweigh a costlier design with better control capability.
4. Flashing service or other erosive service condition may indicate that the use of a valve of higher initial cost may still be an economical choice.

End Connections

Methods of connecting the valve body to adjacent piping were mentioned briefly in a previous section. Body sizes under 2 inches usually have screwed connections. Sizes 2 inches and above are normally flanged with raised-face flanges predominant. In any case, the following checks and considerations need to be made prior to selecting end connections.

1. Check the piping specifications to determine the pressure and temperature ratings for the line or service in which the valve is used. Make sure that valve connections are compatible with adjacent piping.
2. Even when a line specification calls for welded ends, the valve ends may need to be different for easier removal when maintenance is required.
3. In small fractional body sizes, tubing connections may be desirable.

Face-to-face dimensions for flanged control valves have been standardized by most valve manufacturers through cooperation with the Instrument Society of America and the Fluids Control Institute, Inc. Acceptance is not universal, however, so manufacturers' literature must be consulted for definite information. Table 4.6 lists face-to-face dimensions for valves as given in ISA Recommended Practice RP4.1. Table 4.7 is a dimension comparison table listing face-to-face dimensions of valves of several manufacturers for different pressure ratings. The ISA RP4.1 recommended dimensions are shown and manufacturer's dimensions are shown, only if they differ from those recommended.

Single Seat Versus
Double Seat Construction

The choice between single and double seated valve construction (Figure 4.28) is often a function of permissible leakage in the closed position and/or force requirements needed for valve actuation.

Leakage is defined as the quantity of fluid passing through a valve when it is in a fully closed position under stated closure forces with pressure differential and temperature as specified. It is normally expressed as a percentage of maximum capacity.

A typical manufacturer's standard for single seated valves with metal-to-metal seats might be .05% leakage or less. Claimed leakage values for double seated valves may range from 0.1 to 1%. These rates are attainable when the valves are new or are in good condition. However, the manufacturer may suggest that if leakage rates of 2% cannot be tolerated single seated valves should be used.

Table 4.6. Valve Face-to-Face Dimensions per ISA RP4.1

Pipe Size inches	125 psi Iron 150 psi Steel*	250 psi Iron 300 psi Steel†	600 psi Steel‡
½	—	7½	8
¾	—	7⅝	8⅛
1	7¼	7¾	8¼
1½	8¾	9¼	9⅞
2	10	10½	11¼
2½	10⅞	11½	12¼
3	11¾	12½	13¼
4	13⅞	14½	15½
6	17¾	18⅝	20
8	21⅜	22⅜	24

* Cast iron bodies with 125 psi ASA B16A-1939 flanges, and steel bodies with 150 psi raised face ASA-B16E-1939 flanges.
† Cast iron bodies with 250 psi ASA B16B-1944 flanges, and steel bodies with 300 psi raised face B16E-1939 flanges.
‡ Steel bodies with 600 psi raised face ASA B16E-1939 flanges.

Difficulty in getting tight closure in double seated valves results because manufacturing tolerances and alignment problems prevent two separate plugs and seats from coming together perfectly at the same time. This is particularly true when the valve is closed, and even when it is throttling near the seat, the flow tends to act as a high velocity jet which causes erosive damage to the trim and valve body. A partial solution to the problem is to size the valve so that it seldom operates near the seat.

Control valves should not be used to shut in a system. Gate or other type valves should be used for that purpose.

Single seated valves sometimes fail to provide the tight shutoff desired by the user. Tight shutoff is not an easily defined term—at least, there is no universal agreement as to what constitutes "tightness." Expressions such as "bubble tight," "dead tight" and "zero leakage" have been used. These, too, may be subject to discussion. When metal-to-metal seats do not provide an acceptable closure, special designs with resilient seats are used to give the shutoff characteristic needed.

Even though new valves possess the degree of tightness desired, service conditions may soon alter this

Table 4.7. Comparison of Manufacturer's Face-to-Face Valve Dimensions with ISA RP4.1

Face-to-Face Dimensions for Flanged Control Valves

Flange Size	125-lb. F.F. USAS Iron 150-lb. R.F. USAS Steel							250-lb. R.F. USAS Iron 300-lb. R.F. USAS Steel							600-lb. R.F. USAS						
	ISA*	Fisher	Mason-eilan	Hammel-Dahl	Honey-well	Cono-flow	Annin	ISA*	Fisher	Mason-eilan	Hammel-Dahl	Honey-well	Cono-flow	Annin	ISA*	Fisher	Mason-eilan	Hammel-Dahl	Honey-well	Cono-flow	Annin
½	NG	NA	NA	NA	NA	7¼	6¾	7½	—	NA	NA	NA	—	6¾	8	—	NA	NA	NA	—	6¾
¾	NG	7⅜	7¼	—	NA	—	6¾	7⅝	—	—	—	NA	—	6¾	8⅛	—	—	—	NA	—	6¾
1	7¼	—	—	—	—	—	8½	7¾	—	—	—	—	—	8½	8¼	—	—	—	—	—	8½
1½	8¾	—	—	—	—	—	9½	9¼	—	—	—	—	—	9½	9⅞	—	—	—	—	—	9½
2	10	—	—	—	—	—	11½	10½	—	—	—	—	—	11½	11¼	—	—	—	—	—	11½
2½	10⅞	—	—	—	—	NA	NA	11½	—	—	—	—	NA	NA	12¼	—	—	—	—	NA	NA
3	11¾	—	—	—	—	—	**14**	12½	—	—	—	—	—	14	13¼	—	—	—	—	—	14
4	13⅞	—	—	—	—	—	**17**	14½	—	—	—	—	—	17	15½	—	—	—	—	—	17
6	17¾	—	—	—	—	—	**22**	18⅝	—	—	—	—	—	22	20	—	—	—	—	NA	28
8	21⅜	—	—	—	—	NA	**32**	22⅜	—	—	—	—	NA	32	24	—	—	—	—	NA	36½
10	NG	26½	24⅝	26⅞	22	NA	**37**	NG	27⅞	26	28¼	23	NA	37	NG	29⅝	27¾	30	31	NA	NA
12	NG	NA	28¾	NA	29	NA	**56**	NG	NA	30¼	NA	30½	NA	56	NG	NA	32	NA	32¼	NA	NA

Notes: * Recommended Dimensions Per ISA RP4.1
— Means Dimension Same As ISA RP4.1
NG Dimensions Not Given In ISA RP4.1
NA Not Available From Manufacturer

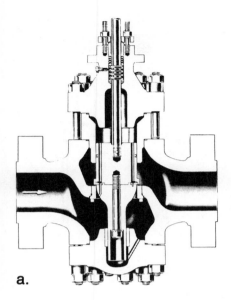
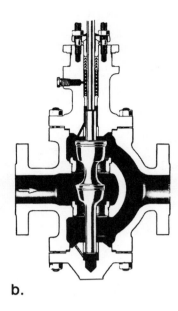

a.   b.

Figure 4.28. The choice of single port versus double port design must be decided after reviewing the desirable and undesirable characteristics of each. (Courtesy of Fisher Controls Co.)

characteristic. Thermal distortion, erosion and corrosion will cause "leak tight" valves to leak after a few months' service.

Force requirements for actuator operation of single seated valves are much greater than for double seated valves. Most of the force requirement is caused by the unbalanced force on the plug. It is the product of the pressure drop across the valve (plug seated) times the seat area, plus friction, for single port valves. For double port valves, it is the product of the pressure drop times the difference of the two port areas plus friction. Obviously, large single seated valves with high ΔPs often require prohibitively bulky and expensive actuators, especially in the spring and diaphragm type. Use of double port valves in most instances is logical and desirable.

To summarize selection criteria, single seated valves should be used where tight shutoff is required and for sizes below 2 inches. Double ported valves should be used where tight shutoff is not necessary and for sizes 2 inches and above.

Reduced Capacity Trim

Reduced trim is needed occasionally for applications where normal full flow capcity is not desirable in the valve size used. There are several valid reasons for reduced trim.

1. To obtain mechanical strength with a large body size while retaining correct trim size
2. To provide valve bodies for future requirements with plug size for current conditions
3. To provide valve bodies large enough to reduce inlet and outlet velocities, thereby reducing noise
4. To avoid use of expensive line reducers
5. To correct oversizing errors

Reduced trim is available from some manufacturers at 40% rated capacity. For example, if the full capacity C_v

rating of a 2-inch valve is 40, reduced trim C_v rating would be 40% $\times$ 40 = 16. In small valve sizes, reduced trim is available in nominal pipe sizes; for example, reduced trim of ¾-,½-, ⅜-inch would be available for 1-inch valve bodies.

Selection of Actuators

The choice of actuators for control valves depends on factors such as those listed below.

1. Pressure drop across the valve
2. Process hazards
3. Valve size
4. Response requirements
5. Valve distance from controller
6. Maintenance
7. Existing plant practice
8. Availability of operating medium

The spring and diaphragm type is the most commonly used actuator. It is simple, economical, relatively trouble-free, safe for hazardous areas and fast enough for most applications. In processes where pneumatic instruments are used primarily, air power (when required) is already distributed. When boosters or positioners are not required, no outside power is needed for valve operation—the signal from the controller provides the needed power.

The piston or cylinder type is also widely used. A built-in positioner assures a valve position exactly proportional to the instrument signal. Positioners require use of high pressure air but have the advantage of developing high stem forces needed for large valves and high pressure drop applications.

Electrohydraulic actuators are used when extremely fast actuation or maximum power is required. A typical application is the high pressure letdown valve for a high pressure polyethylene process. Electrohydraulic actuators are useful also on applications requiring fast opening valves for venting operations. Systems using external hydraulic power are available and provide full open-to-close or close-to-open operation in approximately 50 milliseconds. Such systems are expensive in comparison to other actuators. However, they are economical for some processes because of improved control characterisitcs.

Electric actuators are used in relatively few applications other than for on-off control. They are useful in remote locations where air is not available and electricity is available.

Advantages and disadvantages of the types discussed are given below.

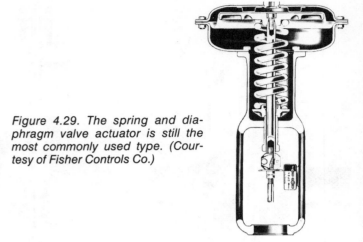

Figure 4.29. The spring and diaphragm valve actuator is still the most commonly used type. (Courtesy of Fisher Controls Co.)

Spring and Diaphragm (Figure 4.29)

Advantages

1. Simple, easy to maintain
2. Economical
3. Safe in electrically hazardous locations
4. Sufficiently fast response for a majority of applications

Disadvantages

1. Not fast enough for some applications
2. Large, cumbersome actuators required for some applications

Piston or Cylinder (Figure 4.30)

Advantages

1. Exact positioning relative to control signal
2. Relatively fast response
3. Accommodates large stem force requirements
4. Safe in electrically hazardous locations

Disadvantages

1. Requires high pressure air supply
2. More expensive than spring and diaphragm type
3. Sometimes cumbersome to achieve fail-safe condition

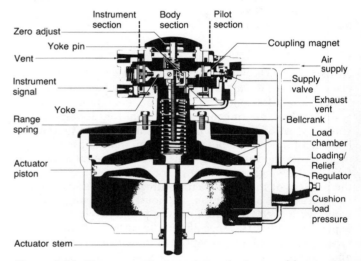

Figure 4.30. Piston or cylinder valve actuator provides greater power than the spring and diaphragm type. (Courtesy of ITT Hammel Dahl/Conoflow)

Electrohydraulic (Figure 4.31)

Advantages

1. Extremely fast response
2. Develops maximum power for actuation
3. Ensures accurate positioning of valve

Disadvantages

1. Expensive
2. Hydraulic systems require extremely clean and well-filtered piping systems
3. More difficult to maintain

Electric (Figure 4.32)

Advantages

1. Economical when air or gas is not available as a power medium

Figure 4.31. Electrohydraulic valve operators provide great power and fast response but are expensive. (Courtesy of Masoneilan Int.)

Figure 4.32. Electric valve operators are slow to respond but are economical when other power sources are not readily available. (Courtesy of Hills-McCanna)

Figure 4.33. Spring and diaphragm valve actuators sometimes need valve positioners for good control. (Courtesy of Fisher Controls Co.)

Disadvantages

1. Slow response
2. More costly compared to pneumatic actuators
3. More difficult to maintain in hazardous areas

Use of a Valve Positioner

When spring and diaphragm actuators (Figure 4.33) are used, the question often arises as to the use of valve positioners. Applications on which there is general agreement that they should be used include the following:

1. Split range operation which requires full valve stroke from some fraction of the controller signal (for example, a 3-15 psi positioner output from a 3-9 psi signal input)
2. When a maximum loading pressure greater than 20 psi is required
3. When the best possible control is required—minimum overshoot and fast recovery
4. When reversing action is necessary (this may be accomplished with a reversing relay also).

Other situations in which positioners are often used are listed below.

1. On valve sizes 6 inches and larger
2. Where long transmission lines exist
3. Where the pressure drop across the valve is high (100 psi or more)
4. On high temperature applications
5. Where excessive friction may occur at packing glands and guides
6. In sludge services which may cause sticky stems and guides

There is some disagreement to arbitary use of positioners in some of these situations, and argument is made that a control system should be analyzed to determine the need. However, a careful analysis during engineering is seldom feasible.

A positioner is used also to change the flow characteristic of a valve by contouring or characterizing the cam. This is done rather infrequently, but in special cases, it is a handy method of obtaining a desired flow characteristic.

Selection of Other Mechanical Features

Guidelines are offered below to help determine when to use hardened trim, seals and extension bonnets; when lubrication is needed; and how to select the proper plug guiding method.

Hardened Trim

Type 316 stainless steel is the most commonly used material for valve plugs, seats and stems—the moving parts of the valve that come in contact with the flowing fluid. Many applications, however, require corrosion or erosion resistant materials to meet severe operating conditions.

Hastelloy, nickel, Monel and other alloys are used when corrosion is a problem. Stellite No. 6, 440C stainless steel, 17-4PH stainless, Colmony No. 6 and carbides are used in erosive services requiring the use of hardened trim. High operating temperatures also may require its use.

Erosion occurs when high pressure drops across control valves result in turbulence and increased velocity through the valve or when particulate matter, catalyst fines, etc., pass through the valve.

Hardened seating material is recommended for the following services:

1. Steam flows with pressure drops of 150 psi or more
2. Flashing services
3. Any process above 550°F
4. Any pressure drop above 300 psi
5. Slurry flow with pressure drops of 100 psi or more

Seals

Bellows seals (Figure 4.34) are used where no leakage along the stem can be tolerated. When the flowing fluid is toxic, volatile, radioactive or highly expensive, bellows seals should be considered. Typical potential applications include:

1. HC1 and other similar vapors
2. Phosgene or other lethal gases
3. Flammable or explosive fluids

Bellows seals are available for pressures as high as 800 psig at 1,000°F. Pressure ratings generally increase as temperature decreases. Short valve strokes allow higher pressure ratings.

Extension Bonnets

Extension bonnets are used for either high or low temperature services to protect packing from extreme temperatures. Typically they are used on:

1. Cold services where temperature is below 20°F
2. Hot services where temperature exceeds 450°F (Some packing materials are suitable to considerably higher temperatures so that 450°F can be exceeded without using extension bonnets.)

Packing and Lubrication

Teflon V-ring packing is the most common type of packing material used. It has self-lubricating qualities, requiring

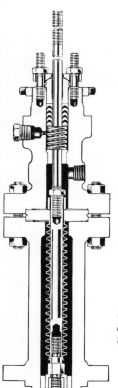

Figure 4.34. Bellows seals are used when it is essential to eliminate leakage. (Courtesy of Fisher Controls Co.)

Table 4.8. Valve Stem Packing Materials and Temperature Ranges	
Packing Type	**Allowable Temperature Range**
Solid Teflon V-rings	−20 to 465°F
Teflon-asbestos	−20 to 465°F
Graphite-asbestos	−20 to 550°F
Inconel wire reinforced graphite-asbestos	For high temperature service. When applications necessitate use of this packing, submit full details to manufacturer. In addition to selecting packing for high temperature applications, every other valve component material must be carefully considered to assure dependable service.
Courtesy of Fisher Controls Co.	

no lubrication. Other self-lubricating packing materials are also available. Packings that require lubrication include graphited asbestos, asbestos yarn impregnated with suspensoid Teflon and semimetallic packing. Table 4.8 lists commonly used packing materials and temperatures at which they may be used suitably.

Packing materials generally must meet the following physical and chemical characteristics:

1. Inertness to attack or deterioration by the flowing fluid, if it is to have normal life
2. Low friction when compressed sufficiently for a tight seal
3. Low expansion in order that clearances established for tightness at low temperatures will not result in excessive friction due to expansion when higher temperatures are encountered
4. Desired characteristics over the rated temperature ranges
5. Resiliency in order that it may conform to minute surface contours and maintain tightness
6. Resistance to cold flow or permanent deformation to prevent loss of effectiveness as a seal
7. Resistance to abrasion that leads to long packing life without excessive wear
8. Ease of replacement in the field without special skill or lengthy operations

Table 4.9 summarizes valve stem lubricants recommended by one valve manufacturer and lists temperature ranges, general and specific services for which they are suitable as well as some services for which they are unsuitable. Lubricants are not always identified by the manufacturer.

Lubricants also help to seal in the flowing fluid in some packing designs. Prefabricated packing rings (Figure 4.35) on either side of a lantern ring (a metal spacer allowing lubricant to be injected between the packings) form a seal

Table 4.9. Valve Stem Lubricants with Recommended Usages							
	No. 1	**No. 2**	**No. 3**	**No. 4**	**No. 5**	**Rockwell No. 421**	**Dow-Corning Silicone***
General Services	Water Dilute aqueous solutions, inert gases	Petroleum distillates and derivatives	Chemical solutions Foodstuffs (A white lubricant)	Coal tar solvents and similar hydro-carbons	Steam (can be used on hot oils)	Strong concentra-tions of chemicals	Steam Hot air
Packing Temperature	0° to 210°F	−20° to 150°F	0° to 150°F	−10° to 450°F	70° to 550°F	30° to 350°F	−40° to 400°F
Specific Services	Acetic acid Air Alcohol Ammonia Boric acid Na OH Creosote Glue Hydrogen HCl Lime Nitrogen Oleic acid Phenol (dil.) Soap sol. Na salts Tannic acid Zn Cl	Acetylene Nat. gas Crude oil Fuel oil Gasoline Pet. oils Waxes	Beer Chlorine Citric acid Dyes (sol.) Fatty acids Fruit juices Glycerine Helium H_2O_2 HNO_3 Lacquer solvents SO_2 SO_3 H_2SO_4 Caustic liquors	Acetone Benzol Carbon tet. Crude oil Fuel oil Gasoline Nat. gas Pet. oils Methyl Cl. Toluol Turpentine	Acetic acid Asphalt Feed water Nitrogen Steam Tar	SO_2 H_2SO_3 H_2SO_4 Chlorine Alkalies Alum Ammonia (Same as No. 1 & 3) except higher temperature	Steam Hot water Hot air Vacuum Acetic acid Alcohols Brines Chlorine Foodstuffs Animal oils Vegetable oils Digester liquors Textile liquors
Unsuitable for	Petroleum Coal tar Oxygen	Water Chemicals	Petroleum Coal tar Mineral oils Concentrated acids	Water Concentrated acids Alcohol HNO_3	Concentrated acids	Petroleum Coal tar Min. oils	Light hydrocarbons Concentrated acids Oxygen Acetone Beer Carbon tet. Ether Chloroform Benzol Toluol

Source: Masoneilan Int.

*Dow-Corning Silicone is available in 2-ounce tubes. It is not recommended, except where a wide temperature range is necessary. A lubricant ball check pressure fitting should be used in place of pressure lubricator if possible.

against the fluid. The lubricant becomes a vital part of the sealing system. Isolating valves for injecting the lubricant are used normally when lubrication is specified.

Guiding

The type of valve guiding used becomes more important as valves increase in size and as fluid velocity increases through the valve. An order of preference for guiding methods in terms of ruggedness and dependability is listed below:

1. Cage
2. Top and bottom
3. Top and port
4. Port
5. Top
6. Stem

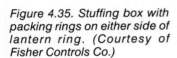

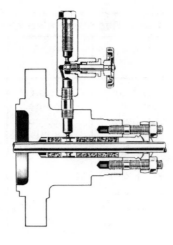

Figure 4.35. Stuffing box with packing rings on either side of lantern ring. (Courtesy of Fisher Controls Co.)

Soft Seats

Soft seat construction for tight shutoff applications is available for both single port and double port valves. Soft seating may be obtained by using the soft material as a seat ring (Figure 4.36) or by a soft insert (Figure 4.37) in a metallic seat ring. Commonly used soft materials include Teflon, Viton, Buna N, Kel-F, nylon and rubber.

Booster Relays

Booster relays (pneumatic amplifiers) are used to reduce time lags resulting from long transmission lines or where controller outputs are insufficient to meet the capacity required for high demand devices such as large diaphragm actuators. The following are typical applications for which they may be considered.

1. Pneumatic transmission lines in excess of 500 feet long
2. Large volume valve actuators
3. Fast process control systems such as (a) flow loops, (b) liquid pressure loops and (c) gas pressure loops where system capacities are relatively low

Handwheels

Handwheels are mounted on control valves for manual operation in an emergency, during startup or in the event of air failure or diaphragm rupture. They may also be used as limit stops to prevent full closure or full opening of valves.
Typical applications include:

1. Critical (but seldom used) services such as (a) reaction quenching systems and (b) emergency venting systems
2. Control loops where bypass valves are not used
3. Critical process loops where control must be maintained in emergencies such as air loss or diaphragm rupture

Control Valve Manifolds

The question of whether to use or omit bypass manifolds depends on a knowledge of plant operating philosophy and

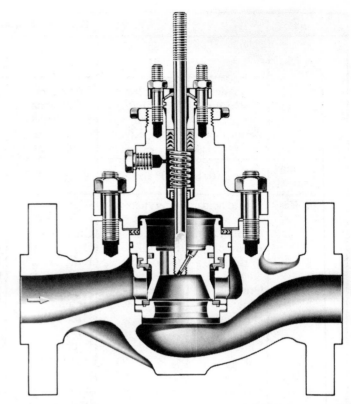

Figure 4.36. Valve plug with soft "O"-Ring seats provide tight shutoff. (Courtesy of Fisher Controls Co.)

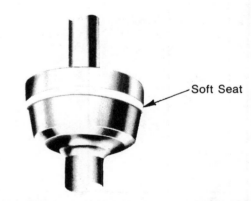

Soft Seat

Figure 4.37. Soft insert in metallic seat ring is another method of providing good seals. (Courtesy of ITT Hammel Dahl/Conoflow)

the economics involved in a particular process; it depends on the process itself. Following are some valid points for consideration about their use.

Typical applications for use of manifolds:

1. Main sources of supply—utilities or feed streams
2. To meet Boiler Code or other similar safety code requirements

3. Service where the life of the control valve is shorter than the expected period between plant turnarounds
4. Small size control valves in general process services (cost for manifold is nominal)

Typical applications where bypass manifolds may cause operating problems:

1. Solids handling streams
2. Streams whose fluids "set up" or solidify if not kept moving
3. Refrigeration systems or other services that are susceptible to ice or hydrate formation
4. Cold climates where trapped fluids present freezing problems

Omission of control valve manifolds where large or alloy valves are used offers large potential savings. It must be borne in mind, however, that loss of revenue due to downtime offsets the cost of many manifolds.

Split Ranging Control Valves

The use of two or more control valves from one controller signal is called split range operation. This operation is usually accomplished by using valve stem positioners adjusted to give full control valve stroking at a signal range less than the full range of the controller. The most commonly used split range operation consists of two valves, one operating on the 3-9 psig portion and the other on the 9-15 psig portion of a 3-15 psig control signal.
Typical applications include:

1. Services where the flow range exceeds the range of a single control valve. In such cases, the smaller valve is selected to control at the minimum flow rate and high differential pressure across the valve. The other valve is sized so that the combined rates of it and the small valve have the needed capacity at the lowest differential expected across the valves (Figure 4.38).
2. Systems in which the valves are operated by a single controller but have opposite effects on the system. This system is often used in steam power stations for "makeup" or "draw-off" of condensate in the steam cycle. It is used in pressure control of many process systems where gases or vapors must be added to or vented from the system to maintain the proper pressure. It is also used for temperature control of systems where heat must be added to or removed from a system.
3. Blending systems in which two control valves are used. With this arrangement, the total flow of the system is normally constant so that a decrease through one valve causes an increase through the other.

Valve Noise Problems

People are increasingly aware of environmental problems, and noise pollution is a problem that is getting increasing attention. Control valve noise is recognized as a contributor to the problem. Valve noise associated with fluid transmission results primarily from (a) mechanical vibration of control valve components, (b) cavitation and (c) aerodynamic noise.

Mechanical

Mechanical vibrations are caused primarily by lateral movement of the valve plug relative to guide surfaces and by valve components resonating at their natural frequencies.

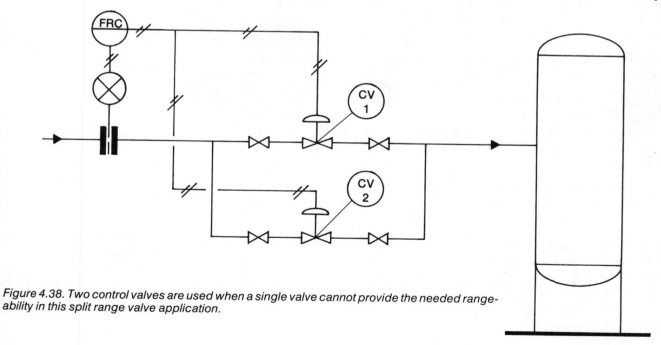

Figure 4.38. Two control valves are used when a single valve cannot provide the needed rangeability in this split range valve application.

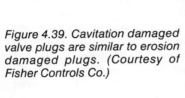
Figure 4.39. Cavitation damaged valve plugs are similar to erosion damaged plugs. (Courtesy of Fisher Controls Co.)

Noise is of secondary concern in these cases because vibrating components fatigue and fail when these conditions persist. Improved designs eliminate or reduce these problems.

Cavitation

Cavitation may be described as the formation and subsequent collapse of voids or cavities in a hydraulic system. It produces noise and results in damage to solid surfaces that confine the cavitating fluid (Figure 4.39). Cavitation is reduced through trim design by decreasing fluid velocity and turbulence. This is accomplished by causing the fluid to take relatively long paths through the valve body.

The erosive effects of cavitation can be reduced further by use of hardened trim described in a previous section of this chapter.

Aerodynamic

Aerodynamic noise is due to the Reynolds stresses or sheer forces that are inherent properties of turbulent flow. Two basic methods are used to combat the problem, (a) attenuation of the noise at the source—improved valve design—and (b) in-line muffling or silencing devices.

Typical valve designs for noise control include the Masoneilan Lo-db valve (Figure 4.40) and Fisher Controls' Whisper Trim (Figure 4.41) which can be installed in the Fisher E Series bodies.

Following are typical examples of in-line devices that are used downstream of control valve installations for noise attenuation.

1. A horizontal silencer (Figure 4.42), manufactured by Fisher Controls which uses an inlet diffuser, a silencer core assembly and acoustical insulation to absorb and reduce sound waves—all are nonmoving parts. It is available in standardized designs for 2- through 12-inch sizes at 150-, 300-, 900- and 1,500-psig ratings.
2. A method utilizing a series of multiple restriction orifices called Lo-db Expansion Plates is offered by Masoneilan (Figure 4.43). The plates are easily installed. They are de-

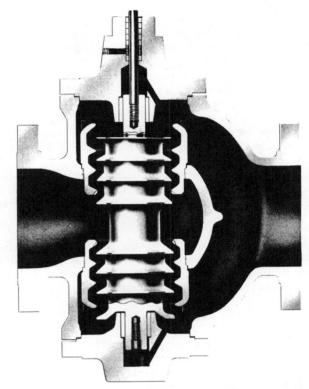
Figure 4.40. Low noise valve trim is becoming increasingly important as noise regulations increase. (Courtesy of Masoneilan Int.)

Figure 4.41. Another low noise valve trim type. (Courtesy of Fisher Controls Co.)

signed with tortuous multistep flow paths to provide maximum pressure loss with minimum fluid velocity (responsible for excessive noise). They are available in sizes 3 through 36 inches from 150 to 2,500 psig.

Other methods used to combat aerodynamic noise include (a) use of heavy wall pipe, (b) burial of downstream pipe, (c) insulation of pipe and (d) design considerations—elimination of sudden expansion and contraction of gases due to piping.

Figure 4.42. Noise is also reduced by use of packless in-line silencers. (Courtesy of Fisher Controls Co.)

Safety Considerations

In addition to the control function of control valves, they also serve as safety devices in processes. There are two basic considerations relative to the safety aspect: (a) the position of the valve when other equipment fails (or process upsets occur) and (b) the position of the valve, itself, if it fails.

Although other failures occur, the most common one in valve operation is loss of the operating medium, usually air. The choice of valve actuation, then, is made so that it fails safe. Fail-safe is defined as the position of the valve (either open or closed) following loss of its operating medium. The term fail-safe is used extensively when discussing spring-and-diaphragm actuators because the spring either closes or opens the valve on loss of air. When cylinder or piston operators are used and an air failure occurs, the valve assumes the position dictated by the forces on the valve. If flow tends to close, the valve will probably close but will remain closed only if the pressure upstream is greater than it is downstream.

Cylinder operated valves are often furnished with capacity (or volume) tanks to open or close the valve when air failure occurs. They also may be furnished with lockup valves so that they remain in the last controlled position in the event of air failure. When supply air pressure is restored, they resume normal operation.

As the term implies, fail-safe indicates the valve position that is the safest or most advantageous to the process or to plant personnel when air failure occurs. If danger is not involved, its failure position is selected to prevent wasted product, to prevent unnecessary maintenance or to protect equipment. The examples of fail-safe action listed below illustrate the principles involved in determining the desired action.

1. The bypass valve on a positive displacement pump should fail open to prevent excessive pressure buildup.
2. The steam valve to a reboiler on a distillation column should fail closed to remove the heat source to the column.
3. The reflux valve on a distillation column should fail open to allow cooling liquid to return to the column to lower the pressure.
4. The pressure control valve on an inert gas blanketing system should fail open to prevent loss of the inert gas blanket.
5. The cooling medium valve of a heat exchanger removing heat from an exothermic reactor should fail open to prevent a runaway reaction.
6. The valve controlling removal of noncondensible gas from the condensed liquid in an accumulator should fail open to avoid a pressure buildup in the system.
7. In a distillation system using hot oil as a column heat source, the temperature control valve should fail open, and the oil heater fuel valve should fail closed. Continued hot oil circulation minimizes coking of the heater tubes while the heat transferred to the tower material is removed by the overhead condenser.
8. Fuel gas valves to a furnace should fail closed to prevent gas flow when a furnace shuts down for any reason.

Special Purpose Valves

Many special purpose valves are available from various valve manufacturers. No attempt is made to provide a comprehensive survey of these. Two areas, however, that are frequently encountered in pilot plants, research laboratories and commercial process units are applications for low flow range valves and high pressure drop valves. Some of the valves available for these services are listed below.

Low Flow Applications

Precision Products & Controls, Inc. features Research control valves with C_v values ranging from 3.0 down to 0.00008. Trim sizes are designed from A ($C_v = 3.0$) to F ($C_v = 0.32$) in a ½-inch body size; and from G ($C_v = 0.2$) to P1 ($C_v = 0.002$) and P2 ($C_v = 0.0013$) to P9 ($C_v = 0.00008$)

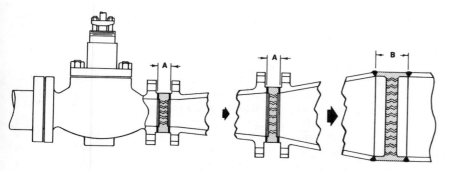

Figure 4.43. Special type multiple restriction orifices may be placed in line to reduce noise level. (Courtesy of Masoneilan, Int.)

in a ¼-inch body size (Figure 4.44). These valves are available in bronze, steel and stainless steel, from forged, cast and barstock bodies. Figure 4.45 shows a valve designed for pressures to 60,000 psig. They are available with standard flow characteristics and standard actuators.

The Annin "Wee Willie" of Masoneilan International (Figure 4.46) is a forged angle valve, made of 316 stainless steel and other forgeable alloys. It is available in ¼, ⅜ and ⁹⁄₁₆-inch sizes. It has pressure ratings up to 30,000 psig. Trim sets are designated from "AA" (C_v range from 0.000001 to 0.001) to "D" (C_v range from 0.02 to 1.0). The valve utilizes a unique linkage arrangement which provides for stroke adjustment from 0.01 to 0.15 inch. The Wee Willie uses the Annin Domotor cylinder operator.

Fisher Controls' Type 530 "Gismo" (Figure 4.47) is a low flow control valve available in ¼-, ⅜- and ½-inch sizes or ⁹⁄₁₆-inch Aminco (high pressure type) fittings. Its bar stock body is machined from carbon steel, stainless steel, Monel, nickel, Hastelloy or other alloys. The pressure rating is 10,-000 psig (15,000 psig for Type 530A). Trim sizes are available with C_v values as low as 0.006. The "Gismo" uses

Figure 4.45. Low flow, high pressure valve operates at pressures to 60,000 psig. (Courtesy of Badger Meter, Inc., Precision Products Div.)

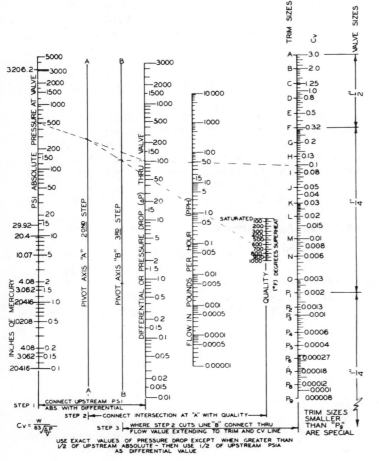

Figure 4.44. This chart of Precision Products reveals the low C_v trim available for low flow applications. (Courtesy of Badger Meter, Inc., Precision Products Div.)

Figure 4.46. Low flow, high pressure angle valve utilizes unique linkage arrangement that allows stroke adjustment from 0.01 to 0.15 inches. (Courtesy of Masoneilan Int.)

a spring and diaphragm actuator with a Fisher Series 3560 I/P positioner.

A Norriseal Uniflow bar stock valve (Figure 4.48) is made in ¼- to 1-inch sizes with reduced trim whose C_v ratings range from 0.00001 to 0.05. It is available in carbon or stainless steel, Hastelloy B or C, Monel or Duramet 20. Body materials also include brass, aluminum, penton, PVC and Teflon.

When control valves with low C_v ratings are used, filters should be placed upstream of the valves to prevent plugging of the valve orifices. Even when services are clean, line scales may cause plugging problems.

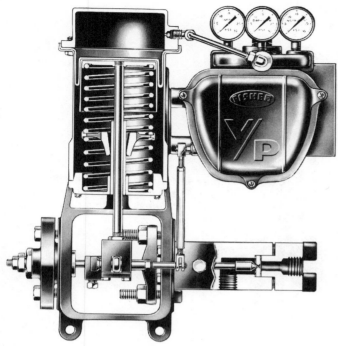

Figure 4.47. Low flow, high pressure control valve called the "Gismo." (Courtesy of Fisher Controls Co.)

Figure 4.49. High pressure drop valve with electrohydraulic operator for fast response. (Courtesy of Masoneilan Int.)

Figure 4.48. Low flow barstock needle valve. (Courtesy of Uniflow Valve, Norris Div. of Dover Corp.)

High Pressure Drop Applications

The Annin Division of Masoneilan offers a valve for high pressure drop applications (Figure 4.49). It uses an electrohydraulic actuator. The control valve system utilizes a control unit which accepts a low level DC input signal. A feedback signal is also available from the valve stem. A servoamplifier in the control unit delivers a current output to the servovalve proportional to the controller signal.

The control valve uses an auxiliary hydraulic system with pressures to 3,000 psig to provide the necessary power to overcome the stem forces involved. The valve has stroking capabilities for speeds up to 50 inches per second, providing valve operations up to 20 hz. It is available for pressures up to 60,000 psig.

A typical service for the valve is the high pressure "letdown" valve in a high pressure polyethylene unit.

Fisher Controls makes a split construction angle valve, Type 16531-CZA, for high pressure applications (Figure 4.50). Pressure rating of the valve is 50,000 psig. The body is made of ANSI Type 4340 vacuum remelted steel in 1- and 2-inch sizes.

The actuator used for the valve is a Fisher Type 16531, a unique approach using two size 30 Type 480 piston actuators on opposite ends of a steel plate framework. Each actuator strokes 4 inches independently of each other, providing 0.080-inch travel through a crank lever assembly (the total valve travel is 0.160 inch). Each piston actuator has its own Type 3570 valve positioner. The maximum thrust of the Type 16531 is 14,000 pounds.

The CZA body valve is also available with a Type 365 electrohydraulic actuator (Figure 4.51). The actuator consists of a hydraulic cylinder mounted on the yoke complete with a servovalve and a valve stem position transmitter. The actuator works in conjunction with a remote servoamplifier, Type 3200 (Figure 4.52) which controls the actuator. A separate hydraulic unit complete with pump, motor, reservoir, accumulator, filters and other accessories is necessary for power. The maximum thrust of the Type 365 actuator is 49,000 pounds.

The Type 365 unit accepts standard instrument signals of 1 to 5 ma, 4 to 20 ma, 10 to 50 ma DC or DC voltage signals.

Figure 4.51. Angle body valve with electro-hydraulic actuator for high pressure applications. (Courtesy of Fisher Controls Co.)

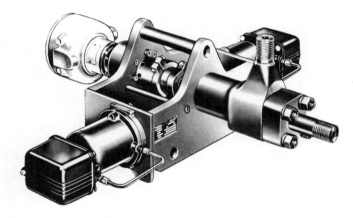

Figure 4.50. Split construction angle type valve for high pressure applications with two type 480 piston actuators—often used in high pressure polyethylene letdown service. (Courtesy of Fisher Controls Co.)

Figure 4.52. Fisher type 3200 servoamplifier for use with electro-hydraulic operator. (Courtesy of Fisher Controls Co.)

5 Control Valve Sizing

B. J. Normand

Control valve sizing requires a familiarity with sizing formulas, a knowledge of the process fluids and the service conditions for the valve. These are discussed in the following paragraphs after a brief discussion of the control valve capacity factor (C_v) by which control valves are rated.

The inherent nature of a control loop using a control valve as the final control element is such that acceptable control can be achieved in many cases, even when the wrong size valve is used. This is true because of many compensating factors:

1. High rangeabilities of control valves
2. The flexibility of controller mode adjustments
3. Stated maximum flow requirements are seldom needed
4. Available pressure drops are usually greater than the allowed design values
5. Availability of control valve bypasses

Regardless of these favorable factors, care should be taken to ensure proper sizing because these factors do not necessarily correct for sizing deficiencies. Good control can best be achieved if the control valve can handle the minimum flow and about 120% of the maximum flow required.

Definition of C_v

The concept of capacity factor (C_v) for control valves was introduced in 1944. C_v is defined as the gallons per minute of water at 60°F that a control valve will pass with 1 pound per square inch pressure drop across the valve.

For control valve sizing purposes, process fluid flow rates are converted to the equivalent flow rates of the proper reference fluids (water for liquid flows and air for vapor flows) because control valve capacities (C_v) are determined by tests using water and/or air, the reference fluids, and are charted or tabulated on this basis.

Equations

Numerous studies have been made and reported concerning control valve sizing. Because of the many approaches suggested for special cases, sizing sometimes appears to be a difficult and laborious task. Normally this is not true. Most applications are simple and straightforward. Except for cases involving flashing, cavitation, high viscosity fluids and mixed phase flow, the problems of sizing control valves are usually resolved by using simple formulas and a little logic. Equations have been developed also for the more difficult sizing problems. These are presented and discussed briefly.

From experience and a review of literature on the subject, a group of selected equations are given in the *Table of Recommended Formulas* at the end of this chapter which should cover most control valve sizing problems. The basic equations from which all the control valve sizing formulas are derived, the simplified FCI equation used on most control valve slide rules and equations for special conditions covering more exacting requirements are given below. Nomenclature for terms used in the equations is given at the end of this chapter.

Basic Equation

The basic equation used to derive control valve sizing equations for liquid, vapors and steam is the well-known fluid flow equation for flow through an orifice:

$$V = c \sqrt{2gh} \qquad (5.1)$$

In this expression, V equals velocity in feet per second, g is the gravitational constant of acceleration (32.2 ft/sec/sec), h is the head of the flowing fluid in feet and c is the orifice discharge coefficient. For valve sizing problems, the interest is in a flowing quantity and is obtained by multiplying the velocity times the area.

It should be remembered that fluid flow calculations are made from empirical solutions and that formulas and expressions seldom represent the actual conditions. For example, the theoretical C_v for a valve (or any other flow restriction) with a port or orifice of 1 square inch, a pressure drop of 1 psi and for water would be determined as follows:

velocity = $\sqrt{2gh}$
flow = $\sqrt{2gh}$ x area
C_v (for 1 square inch)
   = $[60 \sqrt{2 \times 32.2 \times 2.308} \times (12 \times 1)]/231$
$C_v = 38$

This value is theoretical and ignores such factors as jet contraction, the velocity of approach, velocity distribution effect, friction and valve geometry. The addition of the constant, c, is used in Equation 5.1 to account for these factors and is obtained by testing the valve at conditions agreed to by valve manufacturers.

ISA Equations

The control valve sizing equations found in Chapter 6 of the *ISA Handbook of Control Valves*, 2nd Ed., ISA, 1976, illustrate a highly accurate method to size control valves. Factors are included to allow for conversion to metric units, for effects of non-turbulent flow conditions, and for piping geometry differing from the standard test manifold. Limitations to the equations are discussed in detail.

The calculation of size for valves to be used in mixed phase (Liquid/Gas, or Liquid/Vapor) services is also thoroughly treated.

The ISA formulas are listed at the end of this chapter and are more precise than those given in the following paragraphs using the FCI equations, but the FCI equations are quite adequate for valve sizing.

FCI Equations

The growth of the process industries has been accompanied by a growing complexity of technological aspects relating to control valve applications. From the basic flow equation mentioned in the previous section, a diversity of formulas has been derived for obtaining valve sizes, causing confusion among manufacturers and users. The Fluid Controls Institute, Inc. (FCI) began a study in 1952 in an attempt to standardize formulas and test methods for determining valve C_v values. The Control Valve Standards Committee reached an agreement in 1961, and the published results became known as the FCI equation.

These simplified equations, given below for liquids, gas and steam, have been in use since 1961. They are the basis for most slide rules issued by control valve suppliers. They do not take into account such factors as viscosity, flashing, compressibility, high recovery characteristics, etc. The limitations of these simplified expressions are listed and discussed briefly.

Liquid Formulas (Noncompressible Fluids)

Volume basis:

$$C_v = Q \sqrt{G/\Delta P} \qquad (5.2)$$

Weight basis:

$$C_v = W/(500 \sqrt{\Delta P \times G}) \qquad (5.3)$$

Equations 5.2 and 5.3 exclude control valve calculations where flashing, high viscosity, two-phase flow or high pressure drops are involved. These are discussed in some detail in a later section of this chapter.

Gas and Vapors (Other than Steam)

Volume basis:

$$C_v = (Q/1360) \sqrt{2GT/[\Delta P(P_1 + P_2)]} \qquad (5.4)*$$

Equation 5.4 does not take into consideration gas compressibility, mixed phase flow or high pressure drop. Compressibility is considered in Equations 5.13 and 5.14. Mixed phase flow is considered later in a separate section. Usable pressure drop for valve sizing calculations for gas, vapor and steam flows is always limited to one-half of the absolute inlet pressure of the control valve. At that approximate value, critical (sonic) velocity is reached and any further reduction of the downstream pressure will not increase the velocity through a valve, thus there will be no increase in flow.

Steam Formula

$$C_v = \frac{W(1.0 + .0007Tsh)}{2.1 \sqrt{\Delta P(P_1 + P_2)}} \qquad (5.5)*$$

Special Conditions

Equations are given below to correct for the conditions mentioned previously (flashing, cavitation, high viscosity fluids and mixed phase flow) where the simplified equations may lead to inaccurate sizing. These additional expressions, together with the FCI equations, provide satisfactory solutions for most control valve sizing applications.

Liquids

Flashing Service

When flashing occurs, the simplified liquid sizing equation gives erroneous results. Flashing occurs when liquids

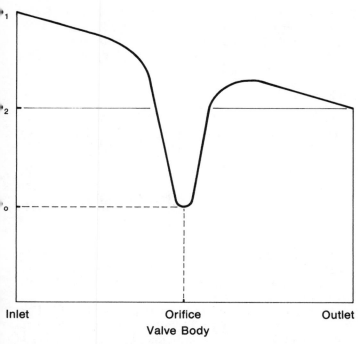

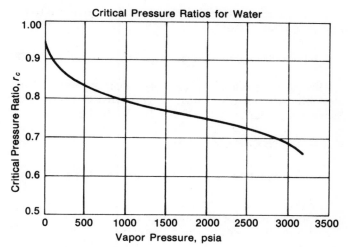

a.

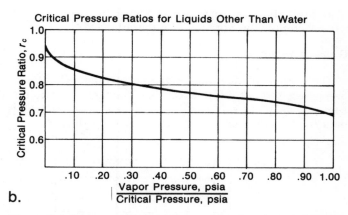

b.

Figure 5.1. Typical control valve pressure gradient. The ΔP between P₁ and P₂ is much less for high recovery valves.

Figure 5.2. These critical pressure ratios (r_c) are for use with Equation 5.7.

enter a valve at or near their boiling points, and they begin vaporizing as their pressures decrease, due to conversions of static pressure head to velocity head. With the advent of high recovery valves, this condition has been aggravated. High recovery valves take a much higher pressure drop from inlet to valve orifice than standard control valves (see Figure 5.1). When the pressure at the orifice drops to the vapor pressure of the fluid at flowing temperature, the maximum valve capacity is reached. This is true because flashing occurs across the valve orifice and a flow condition similar to the critical flow of vapors exists.

Two approaches take into account the flashing condition.

1. Hans D. Baumann, in a paper presented in *ISA Transaction,* Volume 2, Number 2 (April, 1963), introduced an equation using a C_f factor to help determine the valve capacity when flashing occurs. The equation, based also on the use of liquid vapor pressure (P_v), is

$$C_v = (Q/C_f) \sqrt{G/(P_1 - P_v)} \qquad (5.6)$$

2. C.B. Schuder, Fisher Controls, in a paper, "How to Size High Recovery Valves Correctly," proposed an equation to solve for the maximum pressure drop to be used for liquid sizing:

$$\bar{\Delta}P_m = K_m(P_1 - r_c P_v) \qquad (5.7)$$

(See Figure 5.2 for values of r_c.)

When this limiting pressure drop from Equation 5.7 is used in the liquid flow formula, Equation 5.2 gives a valve size in which flashing will not occur, unless the pressure drop increases above the calculated ΔP_m. When the ΔP exceeds ΔP_m, flashing and possible cavitation can occur. Cavitation is the effect caused by the sudden condensing of vapors when the static pressure increases to a point above the boiling point of the liquid.

When the possibility of cavitation exists, the process should be reviewed and the condition eliminated if at all possible. A high recovery valve should not be used in this application. This second approach is suggested as a good solution to the flashing problem.

Viscosity Correction

The liquid sizing formulas in general use have no factor applied to correct for viscosity. This correction is a complex function involving the Reynolds number which varies with the valve size, port area, velocity and viscosity. No correction is required for a viscosity value below 100 SSU. Above viscosity values of 100 SSU, two equations are given to obtain correction factors:

$$R = (10,000 \times q)/\sqrt{C_v} \times cs \qquad (5.8)$$

This equation is used for viscosity values between 100 and 200 SSU; the viscosity must be converted to centistokes. When viscosities exceed 200 SSU, Equation 5.9 applies:

$$R = (46,500 \times q)/\sqrt{C_v} \times SSU \qquad (5.9)$$

In either equation, the following steps should be followed:

1. Solve for C_v, assuming no viscosity effects.
2. Solve for factor R using Equation 5.8 or 5.9.
3. From the viscosity correction curve, Figure 5.3, read the correction factor at the intercept of factor R on the curve.
4. Multiply the C_v obtained in Step 1 by the correction factor obtained in Step 3 for the corrected C_v.

Gases and Vapors

The simplified equation for gases and vapors (Equation 5.4) agrees favorably with experimental data when used for conventional globe valves, but agreement with experimental data is poor when used with high recovery valves.

A universal equation proposed by C.B. Schuder, Fisher Controls, in *Instrumentation Technology,* February, 1968, reportedly showed excellent agreement with experimental results covering a wide range of body plug configurations and pressure drop ratios.

Volume basis:

$$Q = \sqrt{\frac{520}{GT}}\, C_1 C_2 C_v P_1 \, \sin \left[\frac{3417}{C_1 C_2} \sqrt{\frac{\Delta P}{P_1}} \right] \text{ Deg. } \quad (5.10)^*$$

Weight basis:

$$W = 1.06 \sqrt{dP_1 Z}\, C_1 C_2 C_v \, \sin \left[\frac{3417}{C_1 C_2} \sqrt{\frac{\Delta P}{P_1}} \right] \text{ Deg. } (5.11)^*$$

The C_1 factor in Equations 5.10 and 5.11 is currently available from Fisher Controls for their valves. Other companies presumably do not publish this factor. (The C_2 factor is shown in Figure 5.4.)

An equation for gases and vapors proposed by the Foxboro Company (Foxboro TI Bulletin 31-4K) is identical with the FCI equation, except that P_2 is used instead of $(P_1 + P_2)/2$:

$$C_v = (Q/1360) \sqrt{GT/\Delta P(P_2)} \qquad (5.12)^*$$

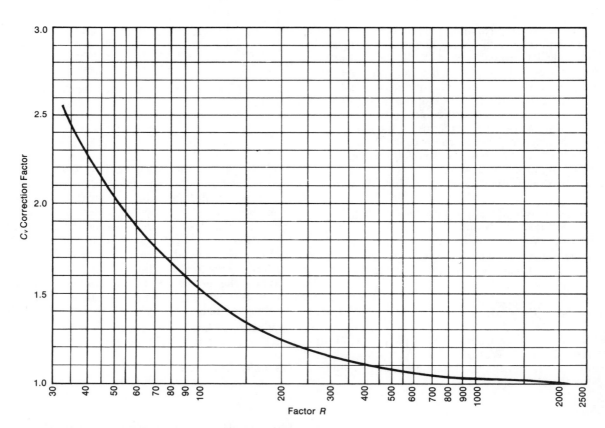

Figure 5.3. Viscosity correction curve for use with Equation 5.9.

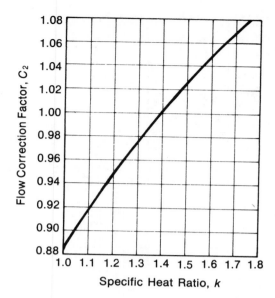

Figure 5.4. *The C₂ factor is applicable only to Equations 5.10 and 5.11, the Fisher universal equations for gas and vapor flows.*

A comparison of the Fisher Universal, FCI and Foxboro equations is made in the TI Bulletin 31-4K. A plot is given showing the results of these three equations using the same data (Figure 5.5).

The Foxboro equation gives C_v values essentially the same as the universal equation at pressure drops up to 25% of P_1 and gives 7% larger values at a pressure drop of 50% of P_1. At the higher end of $\Delta P/P_1$ ratios, a slightly larger valve should be helpful to allow for the expanding vapors. The same agreement is claimed for steam flows although no plot is given.

To correct for deviation from the universal gas law, the compressibility factor is inserted in Equation 5.12, becoming:

$$C_v = (Q/1360) \sqrt{G_b TZ/(\Delta P \times P_2)} \qquad (5.13)*$$

On a weight basis, this equation becomes:

$$C_v = (W/104) \sqrt{TZ/(\Delta P \times P_2 \times G_b)} \qquad (5.14)*$$

Mixed Phase Flow

Mixed phase flow is generally experienced when:

1. Liquid entrainment is carried by a gaseous flow.
2. Liquid at or near its boiling point vaporizes as it flows through a line and its restrictions.

In the discussion under "Flashing Service," it is explained that flashing can be reduced or eliminated by limiting the ΔP across the valve. When two-phase flow already exists or when it occurs in the control valve, two methods are given to size the valve.

1. A rule of thumb that has been used for years is to calculate the C_v based on liquid only and increase the valve body to the next size. This is satisfactory in many cases, but it obviously does not take into account varying ratios of liquid to vapor.
2. A preferred method is to determine the amount of vapor and liquid, then calculate and add the C_v of each phase to get the overall C_v requirement. The percent vaporization can be calculated by the responsible engineer if it is not furnished by the process people.

Example:

Fluid—Propane
100 gpm @ 300° psia and 130°F
s.g. @ 130°F = 0.444
s.g. @ 60°F = 0.5077
s.g. @ P_2 (200 psia) = 0.466 @ 104°F
ΔP = 100 psi
% vaporization = 15.05%
M.W. = 44
$Z = 0.8$
$G_b = 44/29 = 1.518$
Flow = 100 gpm x 0.444 = 22,200 lb/hr
Vapor = 22,200 x 0.1505 = 3,340 lb/hr
Liquid = 22,200 − 3,340 = 18,860 lb/hr

Using Equation 5.14:

$$C_v\text{(vapor)} = (W/104)\ \sqrt{(T \times Z)/(\Delta P \times G_b \times P_2)}$$
$$= (3340/104)\ \sqrt{(564)(.8)/(100(1.518)(200))}$$
$$= 3.92$$

Using Equation 5.3:

$$C_v\text{(liquid)} = (W/500)/(18{,}860/\ \sqrt{\Delta P \times G})$$
$$= 18{,}860/(500\ \sqrt{100 \times .466})$$
$$= 5.53$$
$$C_v\text{(total)} = 3.92 + 5.53$$
$$= 9.45$$

Pressure Drop

The preceding discussion has centered on formulas to be used for various applications and conditions. Little has been said about physical constants of the fluid and stream conditions. Some physical constants are well known or are easily determined. Others can be calculated or assumed without changing valve sizes appreciably.

A very important consideration affecting control valve sizing is the pressure drop available. There are two general type systems to consider when determining what it should be for good control:

1. A constant pressure drop system
2. A variable pressure drop system

In either system, the total pressure drop is absorbed by the static and dynamic losses in the system. The control valve pressure drop constitutes a part of the dynamic loss.

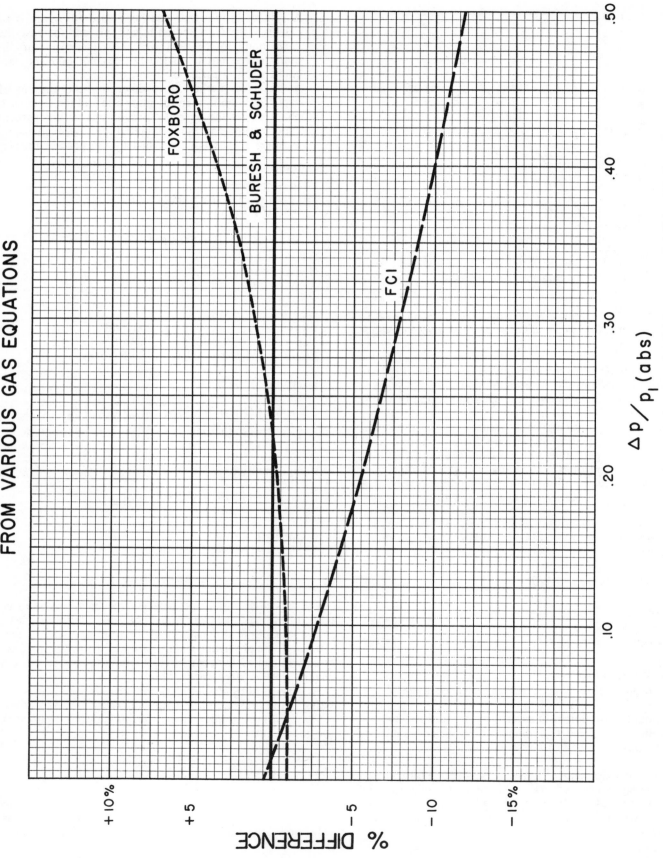

COMPARATIVE Cv VALUES
FROM VARIOUS GAS EQUATIONS

FOXBORO

BURESH & SCHUDER

FCI

% DIFFERENCE

+10%
+5
-5
-10
-15%

$\Delta p / p_1$ (abs)

.10   .20   .30   .40   .50

Figure 5.5. These curves show the close agreement (particularly for low pressure drops) of the Foxboro, FCI and Fisher universal equations for gas flows. (Courtesy of the Foxboro Co.)

Constant Pressure System

A constant pressure system (Figure 5.6) is one in which the fluid flows from one vessel to another without any means (pump, etc.) of boosting the pressure between the two. Pressures in the two vessels are fixed by the process and remain relatively constant.

The pressure drop available for a control valve in a constant pressure system is easy to determine. It is the difference between the two terminal pressures plus or minus the difference in elevation minus the dynamic (friction) losses in the line. It may be expressed as:

$$\Delta P = (P_s - P_r) \pm 0.433hG - F \qquad (5.15)$$

Variable Pressure System

A variable pressure system is similar to the constant pressure system, except that it utilizes a pump to boost the system pressure (Figure 5.7). The pressure drop available for the control valve is determined in a similar manner to the above and may be expressed as:

$$\Delta P = (P - P_r) \pm 0.433hG - F \qquad (5.16)$$

In this expression, P is the pump output pressure and replaces P_s shown in Equation 5.15. The value used for P is a function of the pump characteristic curve and is selected at the intersection of the pump curve and the maximum process flow (see Figure 5.7).

In addition to calculating C_v for the maximum flow case, checks for the C_v requirements at normal and minimum operating conditions are desirable. This will reveal how the valve should operate at these conditions. For these checks, use values for P where normal and minimum process flows intersect the pump curve. From Figure 5.7 it can be seen that the available ΔP for the valve is much greater at minimum flow than at maximum flow. This is true because line losses are lower and the pump discharge pressure is higher at minimum flow conditions.

Valve Selection Guidelines

The engineer is responsible for selecting the proper control valve size and the characteristic that best suits the requirements. Following are some guidelines for valve selection.

1. After solving the C_v required at maximum and minimum conditions, select a valve size that will handle the maximum C_v at 85 to 90% open and the minimum C_v at about 10 to 15% open.
2. Use linear trim for level control systems and on other systems when 40% or more of the system pressure drop is used by the control valve. Use equal percentage trim for other applications. For further clarification, see Chapter 4.

3. Do not use high recovery valves for flashing and cavitation services.
4. When a control valve size is the same or greater than line size, both should be reviewed since the valve size is normally less than line size.
5. Pressure drop across the valve should be 20% or greater of the total dynamic loss.

For more details on valve selection, refer to the previous chapter where guidelines relative to materials, valve rangeabilities and characteristics, body design, actuators and other mechanical features are given.

Conclusion

The formulas and procedures listed here for control valve sizing will provide an excellent tool for solving the majority of control valve problems.

While it is recognized that some of the valve sizing procedures are based on assumptions, these assumptions have been validated by much experimental data. Without the use of simplified theory and reasonable assumptions, the problem of sizing valves becomes very complicated.

Although it is based on experimental data and corrected for body losses and efficiency of the port and passages, the experimental determination of the valve C_v does not take into account (a) the variations of the characteristics of the fluid in question, (b) the changes in the state of the fluid and (c) the wide variations of pressure drop and velocity which may exist in various installations. It is in this region that the rule of thumb, based on experience and rational assumptions, is necessary. Improved techniques and more exact determinations of the behavior of fluids under various conditions may improve the theoretical accuracy, but the overall accuracy will still be limited somewhat by the difficulty of evaluating operating conditions.

Recommended Table of FCI Formulas

Liquid Sizing

Volume basis:

$$C_v = Q \sqrt{G/\Delta P} \qquad (5.2)$$

Weight basis:

$$C_v = W/500 \sqrt{\Delta P \times G)} \qquad (5.3)$$

Viscosity correction for 100 to 200 SSU:

$$R = (10,000 \times q)/(\sqrt{C_v} \times cs) \qquad (5.8)$$

Viscosity correction for 200 SSU and above:

$$R = (46,500 \times q)/(\sqrt{C_v} \times SSU) \qquad (5.9)$$

Use R value in Figure 5.3 to obtain viscosity correction factor.

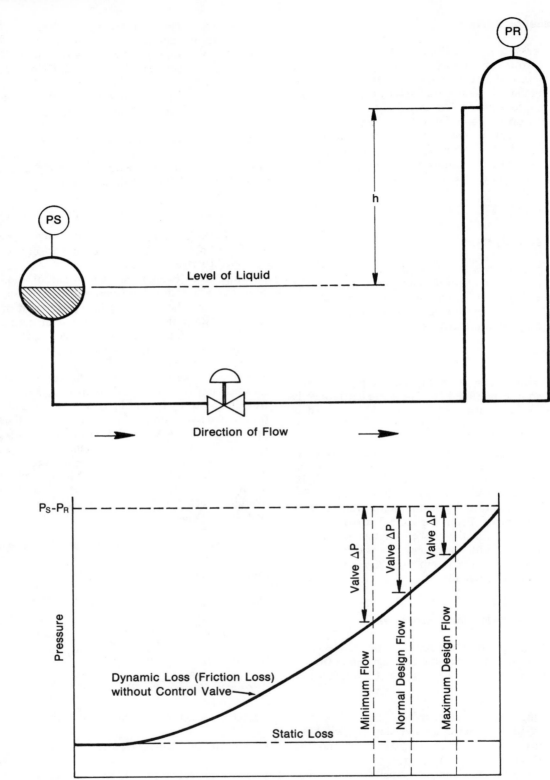

Figure 5.6. The variation of control valve pressure drop with flow in a constant pressure drop system.

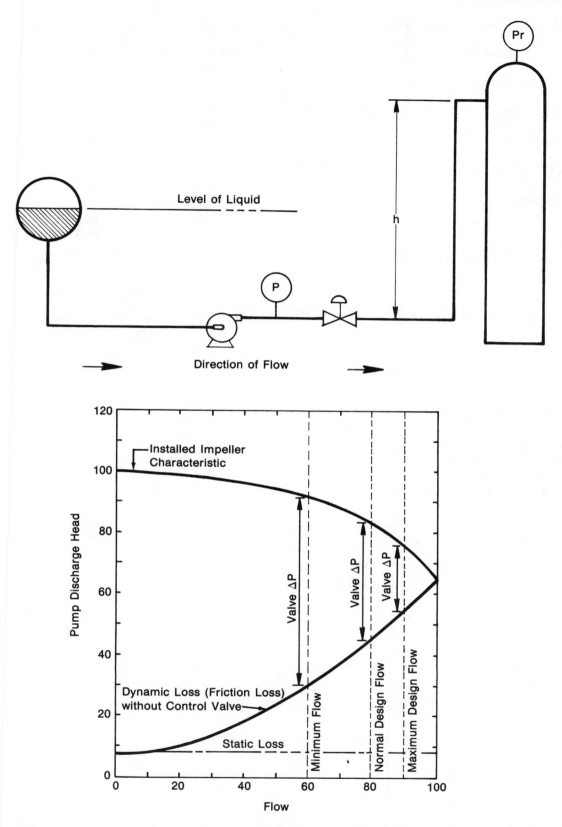

Figure 5.7. The variation of control valve pressure drop with flow in a variable pressure drop system.

Available valve ΔP for constant pressure system:

$$\Delta P = (P_s - P_r) \pm 0.433hG - F \qquad (5.15)$$

Available valve ΔP for variable pressure system:

$$\Delta P = (P - P_r) \pm 0.433hG - F \qquad (5.16)$$

Limiting pressure drop:

$$\Delta P_m = K_m (P - r_c P_v) \qquad (5.7)$$

Gases and Vapors

Volume basis:

$$C_v = (Q/1360) \ \sqrt{(T \times G_b \times Z)/(\Delta P \times P_2)} \qquad (5.12)*$$

Weight basis:

$$C_v = (W/104) \ \sqrt{(T \times Z)/(\Delta P \times G_b \times P_2)} \qquad (5.14)*$$

Steam

$$C_v = [W(1 + 0.0007Tsh)]/[2.1 \ \sqrt{\Delta P(P_1 + P_2)}] \qquad (5.5)*$$

Nomenclature for FCI Formulas

c = orifice discharge coefficient
C_1 = valve sizing factor
C_2 = specific heat correction (refer to Figure 5.4)
C_f = critical flow factor, from valve supplier

cs = viscosity at flowing temperature, centistokes
C_v = valve flow coefficient
d = density of vapors at valve inlet, lb/ft.³
d_1 = density of vapors at valve outlet, lb/ft.³
F = system dynamic loss, psig
G = specific gravity of fluid at flowing temperature, water = 1.0
G_b = specific gravity of gas or vapor at base condition
g = acceleration of gravity, 32.2 ft/sec²
h = pressure drop or change in elevation in a process system, feet of fluid
K_m = valve recovery coefficient
$M.W.$ = molecular weight of vapors
P = pump discharge
P_1 = valve upstream pressure, psia
P_2 = valve downstream pressure, psia
P_o = valve orifice pressure, psia
P_r = receiver vessel pressure, psig (refer to Figure 5.6)
P_s = source vessel pressure, psig (refer to Figure 5.6)
P_v = fluid vapor pressure at flowing temperature, psia
ΔP = pressure drop $(P_1 - P_2)$, psi
ΔP_m = maximum usable ΔP (choked flow occurs), psi
q = flow at conditions, gpm
Q = flow of vapors, SCFH
R = factor for use in Figure 5.3
r_c = theoretical critical pressure ratio at P_v given in Figure 5.2a or 5.2b
SSU = viscosity, Saybolt Seconds Universal at flowing temperature
Tsh = degrees superheat, °F
T = flowing temperature, degrees Rankine
V = velocity of fluid, ft/sec
W = flow of fluid, lb/hr
Z = compressibility factor

---

*The value of ΔP must never exceed $P_1/2$ for sizing calculation

Table of ISA Formulas

REMARKS	EQUATIONS	VALUE OF N[1] U.S.	SI
LIQUID Turbulent and Non-Cavitating	$q_f = N_1 F_p C_v \sqrt{\dfrac{\Delta p}{G_f}}$	1.00	0.0865
	$w_f = N_6 F_p C_v \sqrt{\Delta p \gamma}$	63.3	2.73
Choked	$q_f = N_1 F_{Lp} C_v \sqrt{\dfrac{p_1 - p_{vc}}{G_f}}$	1.00	0.0865
	$w_f = N_6 F_{Lp} C_v \sqrt{\dfrac{p_1 - p_{vc}}{G_f}}$	63.3	2.73
	$p_{vc} = F_F p_v$ $F_F \simeq 0.96 - 0.28 \sqrt{\dfrac{p_v}{p_c}}$		
	$F_{LP} = \left[\dfrac{1}{F_L^2} + \dfrac{K_i}{N_2}\left(C_d\right)^2 \right]^{-1/2}$	890	0.00214
	$K_i = $ (See Piping Geometry Factor)		
Laminar	$q_f = N_{10} \dfrac{\Delta p}{\mu}\left(F_s F_p C_v\right)^{3/2}$	52	173
	$F_s = \left(\dfrac{F_p F_d^2}{F_{LP}}\right)^{1/3} \left[\dfrac{(F_{LP}C_v)^2}{N_2 D^4} + 1\right]^{1/6}$	890	0.00214
Transitional	$q_f = N_1 F_R F_P C_v \sqrt{\dfrac{\Delta p}{G_f}}$	1.00	0.0865

[1] U.S. units are: pounds per hour, gallons per minute, pounds per square inch absolute, pounds per cubic foot, °R, and inches.

SI units are: kilograms per hour, cubic meters per hour, kPa, kilograms per cubic meter, °K, and millimeters.

(Reprinted with permission of Instrument Society of America; from ISA Handbook of Control Valves, 2nd Edition © Instrument Society of America, 1976.)

(table continued on next page)

Table of ISA Formulas (continued)

REMARKS	EQUATIONS	VALUE OF N U.S.	SI
Gas or Vapor - (All Equations: $x \leq F_k x_T$)			
	$w_g = N_6 F_p C_v Y \sqrt{x p_1 \gamma_1}$	63.3	2.73
	$q_g = N_7 F_p C_v p_1 Y \sqrt{\dfrac{x}{G_g T_1 Z}}$	1360	4.17
Variations for Selected units.	$w_g = N_8 F_p C_v p_1 Y \sqrt{\dfrac{xM}{T_1 Z}}$	19.3	0.948
	$q_g = N_9 F_p C_v p_1 Y \sqrt{\dfrac{x}{M T_1 Z}}$	7320	22.4
Expansion factor lower limit = 0.667	$Y = 1 - \dfrac{x}{3 F_k x_T}$		
Sp. ht. ratio factor	$F_k = k/1.40$		
Mfr's. Factors	$x_T = \dfrac{C_1^2}{1600} = 0.84 C_f^2$		
x_T with reducers	$x_{TP} = \dfrac{x_T}{F_p^2} \left[\dfrac{x_T K_i}{N_5} \left(C_d \right)^2 + 1 \right]^{-1}$ $K_i = $ (See Piping Geometry Factor)	1000	0.0024
Steam (Dry and Saturated)			
For $x < x_{Tp}$	$w = N F_p C_v p_1 \left(3 - \dfrac{x}{x_{TP}} \right) \left(\sqrt{x} \right)$	1.0	0.152
For $x \geq x_{Tp}$ (Choked Flow)	$w = N F_p C_v p_1 \sqrt{x_{TP}}$	2.0	0.304

Table of ISA Formulas (continued)

REMARKS	EQUATIONS	VALUE OF N U.S.	SI
Piping Geometry Factor For F_{LP} see "Liquid Choked Flow"	$F_p = \left[\dfrac{\sum K}{N_2} \left(C_d \right)^2 + 1 \right]^{-\frac{1}{2}}$	890	0.00214
Sum of velocity head coefficients	$\sum K = K_1 + K_2 + K_{B1} - K_{B2}$		
Bernoulli coefficient	$K_{B1} = K_{B2} = 1 - \left(\dfrac{d}{D} \right)^4$		

Resistance coefficients for abrupt transitions	$K_1 = 0.5 \left[1 - \left(\dfrac{d}{D} \right)^2 \right]^2$	
Inlet fitting co-efficient for F_{LP} and x_{TP}.	$K_2 = 1.0 \left[1 - \left(\dfrac{d}{D} \right)^2 \right]^2$ $K_i = K_1 + K_{B1}$	

Line Velocity	Feet/Second	Meters/Second	Range (Ft./Sec.)
Liquid	$U = \dfrac{q}{2.45 D^2}$	$U = 354 \dfrac{q}{D^2}$	5-10 Norm. 40-50 Max.
Gas	$U = \dfrac{qT}{695 p D^2}$	$U = 1.24 \dfrac{qT}{p D^2}$	250-400
Vapor	$U = \dfrac{w}{19.6 \gamma D^2}$	$U = 354 \dfrac{W}{\gamma D^2}$	70 Wet 300 superheated
Steam	$U = \dfrac{23w}{p D^2}$	$U = 685 \dfrac{W}{p D^2}$	

Acoustic Velocity (Mach 1.0)

Gas	$U_a = 223 \sqrt{\dfrac{kT}{M}}$	$U_a = 91 \sqrt{\dfrac{kT}{M}}$	< 0.3 Mach
Air	$U_a = 49 \sqrt{T}$	$U_a = 20 \sqrt{T}$	
Steam, Superheated	$U_a = 60 \sqrt{T}$	$U_a = 24.5 \sqrt{T}$	<0.15 Mach
Steam, Dry Saturated	$U_a = 1650$	$U_a = 500$	<0.10 Mach
Vapor	$U_a = 68.1 \sqrt{kpv}$	$U_a = 1038 \sqrt{kpv}$	

Calculated Values of F_P and x_{TP} for Valves Installed Between Short Pipe Reducers Assuming Two Reducers of the Same Size with Abrupt Change in Area

Cd	10					15						20					25					30			
xT	.40	.50	.70	.80		.40	.50	.60	.70	.80		.40	.50	.60	.70		.20	.30	.40	.50		.15	.20	.25	
d/D	x_{TP}				F_P	x_{TP}					F_P	x_{TP}				F_P	x_{TP}				F_P	x_{TP}			F_P
.80	.40	.49	.69	.78	.99	.40	.49	.58	.67	.75	.98	.39	.48	.56	.64	.96	.21	.30	.39	.47	.94	.17	.21	.26	.91
.75	.40	.50	.69	.78	.98	.40	.49	.58	.67	.75	.97	.40	.49	.57	.65	.94	.22	.31	.40	.48	.91	.18	.23	.27	.88
.67	.40	.50	.69	.78	.98	.41	.50	.59	.68	.76	.95	.42	.51	.59	.67	.91	.24	.33	.43	.51	.87	.19	.25	.30	.83
.60	.41	.51	.70	.79	.97	.42	.52	.61	.69	.78	.93	.43	.53	.61	.69	.89	.25	.36	.45	.54	.84	.21	.27	.32	.79
.50	.41	.52	.70	.80	.96	.44	.53	.63	.71	.79	.91	.46	.55	.64	.72	.85	.28	.39	.49	.58	.79	.24	.30	.36	.73
.40	.42	.52	.71	.80	.95	.44	.55	.65	.74	.82	.89	.49	.58	.67	.75	.82	.30	.42	.53	.62	.76	.26	.33	.40	.70
.33	.43	.53	.72	.81	.94	.46	.56	.66	.75	.83	.88	.50	.60	.69	.78	.81	.31	.44	.55	.64	.74	.27	.34	.40	.69
.25	.44	.53	.73	.83	.93	.48	.58	.67	.76	.85	.87	.52	.62	.71	.79	.79	.33	.46	.57	.67	.72	.27	.37	.44	.65

EXAMPLE: A 2-inch valve is rated at $Cv = 80$ and $x_T = 0.65$. Find F_P and x_{TP} if the valve is installed in a 3-inch pipe line with short reducers. $C_d = Cv/d^2$ or 20 and $d/D = 2/3$ or 0.67. Under the heading $C_d = 20$ on line $d/D = 0.67$ find $F_p = 0.91$. For x_{TP} interpolate the values given in the columns headed $x_T = 0.60$ and $x_T = 0.70$ (0.59 and 0.67 respectively). The answer is $x_{TP} = 0.63$.

(Reprinted with permission of Instrument Society of America; from ISA Handbook of Control Valves, 2nd Edition © Instrument Society of America, 1976.)

Valve Sizing Data
Representative Valve Factors

BODY & TRIM TYPE	FLOW DIRECTION	LINE SIZE BODY (D=d)						HALF SIZE (D=2d)			
		C_d	F_L	X_T	$F_d{}^1$	F_s	K_c	$\dfrac{N_3 C_v}{D^2}$	F_{LP}	x_{TP}	F_s
Single Seat Globe											
Wing Guided	Either	11	.90	.75	1.0	1.05	c	2.8	.85	.75	1.04
V-Skirt	Either	9	.90	.75	1.5	1.38	c	2.3	.86	.75	1.36
Contoured	Open	11	.90	.72	1.0	1.05	.65	2.8	.85	.73	1.04
Contoured	Close	11	.80	.55	1.0	1.09	.58	2.8	.76	.57	1.08
V-Plug	Either	9.5	.90	.75	1.0	1.05	.80	2.4	.86	.75	1.04
Cage	Open	14	.90	.75	1.0	1.06	.65	3.5	.82	.75	1.04
Cage	Close	16	.80	.70	1.0	1.11	c	4.0	.72	.71	1.08
Double Seat Globe											
Wing Guided	--	14	.90	.75	.71	0.84	c	3.5	.82	.75	0.83
V-Skirt	--	13	.90	.75	.71	0.84	c	3.3	.83	.75	0.83
Contoured	--	13	.85	.70	.71	0.85	.70	3.3	.79	.71	0.84
V-Plug	--	12.5	.90	.75	.71	0.84	.80	3.1	.83	.75	0.84
Angle											
Full Port Contour	Close	20	.80	.65	1.0	1.12	.53	5.0	.69	.68	1.08
Full Port Contour	Open	17	.90	.72	1.0	1.08	.64	4.3	.78	.73	1.04
Restricted Contour	Close	≥6	.70	.55	1.0	1.13	c	1.5	.69	.56	1.13
Restricted Contour	Open	≥5.5	.95	.80	1.0	1.02	c	1.3	.93	.80	1.02
2:1 Tapered Orif.	Close	12	.45	.15	1.0	1.31	c	3.0	.44	.17	1.31
Cage	Open	12[a]	.85	.65	1.0	1.08	c	3.0	.80	.66	1.06
Cage	Close	12[a]	.80	.60	1.0	1.10	c	3.0	.75	.62	1.08
Venturi	Close	22	.50	.20	1.0	1.29	.17	5.5	.46	.26	1.26
Ball											
Std. Bore[b]	--	30	.55	.15	1.0	1.28	.25	7.5	.47	.24	1.22
Characterized	--	25	.57	.25	1.0	1.25	.22	6.3	.50	.33	1.21
Butterfly											
60-Deg. Open	--	17	.68	.38	.71	0.92	.3	4.3	.63	.43	0.91
90-Deg. Open	--	>30	.55	.20	.71	1.01	c	>7.5	.45	.33	0.97

[1]The values of F_d are based on limited test data which have not been corroborated by independent laboratories. F_s is computed from F_d. Key: a = Variable, b = Orif. ≈ 0.8d, c = Unavailable.

(Reprinted with permission of Instrument Society of America; from ISA Handbook of Control Valves, 2nd Edition © Instrument Society of America, 1976.)

Reference Data for Steam and Gases

	SP. GRAVITY G	SP. HEATS RATIO k	FACTOR F_k
Acetylene	0.897	1.28	0.914
Air	1.000	1.40	1.00
Ammonia	0.587	1.29	0.921
Argon	1.377	1.67	1.19
Carbon Dioxide	1.516	1.28	0.914
Carbon Monoxide	0.965	1.41	1.01
Ethylene	0.967	1.22	0.871
Helium	0.138	1.66	1.19
Hydrogen Chloride	1.256	1.40	1.00
Hydrogen	0.0695	1.40	1.00
Methane	0.553	1.26	0.900
Methyl Chloride	1.738	1.20	0.857
Nitrogen	0.966	1.40	1.00
Nitric Oxide	1.034	1.40	1.00
Nitrous Oxide	1.518	1.26	0.900
Oxygen	1.103	1.40	1.00
Sulphur Dioxide	2.208	1.25	0.893
Steam (dry saturated) P_1			
0-80		1.32	0.94
80-245		1.30	0.93
245-475		1.29	0.92
475-800		1.27	0.91
800-1050		1.26	0.90
1050-1250		1.25	0.89
1250-1400		1.23	0.88

(Reprinted with permission of Instrument Society of America; from ISA Handbook of Control Valves, 2nd Edition © Instrument Society of America, 1976.)

Nomenclature of ISA Formulas

a	Area of orifice or valve opening, in.^{2}
C	Coefficient of discharge, dimensionless. Includes effect of jet contraction and Reynolds number, mach number (gas at high velocities), turbulence.
C_d	Relative capacity (at rated C_v) $C_d = N_3 C_v/d^2$.
c_p	Specific heat at constant pressure.
c_v	Specific heat at constant volume.
C_v	Valve coefficient, $38\,a\bar{K}/F_L$.
d	Valve inlet diameter, inches or mm.
D	Pipe diameter, inches or mm.
F	Velocity of approach factor $= \dfrac{1}{\sqrt{1-m^2}}$.
F_d	Experimentally determined factor relating valve C_v to an equivalent diameter for Reynolds number. (See Table IV)
f	Weight fraction
F_F	Liquid critical pressure ratio factor, $F_F = p_{vc}/p_v$
F_k	Ratio of specific heats factor.
F_L	Pressure recovery factor. When the valve is not choked: $$F_L = \sqrt{(p_1 - p_2)/(p_1 - p_{vc})}$$
F_{LP}	Combined pressure loss and piping geometry factors for valve/fitting assembly.
F_P	Correction factor for piping around valve (e.g. reducers) $F_P C_v =$ effective C_v for valve/fitting assembly.
F_R	Correction factor for Reynolds number, where $F_R C_v =$ effective C_v.
F_s	Laminar, or streamline, flow factor.
g	Acceleration due to gravity.
G_f	Specific gravity of liquids at flowing temperature relative to water at 60°F or 15°C.
G_g	Specific gravity of gas relative to air with both at standard temperature and pressure.
h	Effective differential head, height of fluid.
$\bar{K}$	Flow coefficient $= CF$, dimensionless.
ΣK	Sum total of effective velocity head coefficients where $K(U^2/2g) = h$.
K_B	Bernoulli coefficient $= 1 - (d/D)^4$.

(Reprinted with permission of Instrument Society of America; from ISA Handbook of Control Valves, 2nd Edition © Instrument Society of America, 1976.)

(continued on next page)

Nomenclature of ISA Formulas (continued)

K_c Cavitation index. Actually the ratio $\Delta p/(p_1 - p_v)$ at which cavitation measurably affects the value of C_v.

K_i Inlet velocity head coefficient, $K_1 + K_{B1}$

K_1 Resistance coefficient for inlet fitting.

K_2 Resistance coefficient for outlet fitting.

k Ratio of specific heats of a gas $= c_p/c_v$, dimensionless.

M Molecular weight.

m Ratio of areas.

N Numerical constant (See Table I)

p Absolute static pressure.

p_c Thermodynamic critical pressure

p_r Reduced pressure, p/p_c

P_v Vapor pressure of liquid at inlet.

q Volume rate of flow.

Re_v Reynolds number for a valve.

T Absolute temperature.

T_c Thermodynamic critical temperature

T_r Reduced temperature, T/T_c

U Average velocity.

v Specific volume ($1/\gamma$).

w Weight rate of flow.

x Ratio of differential pressure to absolute inlet static pressure, $x = (p_1 - p_2)/p_1$

x_T Terminal or ultimate value of x, used to establish expansion factor, Y.

x_{TP} Value of x_T for valve/fitting assembly.

Y Expansion factor. Ratio of flow coefficient for a gas to that for a liquid at the same Reynolds number (includes radial as well as longitudinal expansion effects).

Nomenclature of ISA Fomulas (continued)

Z   Compressibility factor. (See Nelson-Obert Compressibility Charts in Appendix.)

γ (gamma)   Specific weight ($1/v$).

Δ (delta)   Difference, (eg., $\Delta p = p_1 - p_2$).

μ (mu)   Viscosity, centipoise.

v (nu)   Kinematic viscosity, centistokes

$\sum$ (sigma)   Summation

SUBSCRIPTS

1	Upstream
2	Downstream
e	Effective value
f	Liquid
g	Gas
t	Theoretical
T	Terminal or ultimate value
vc	Vena contracta (point of minimum jet stream area)

6 Pressure Relief Systems

Andrew Jackson Stockton, William G. Andrew

The purpose of pressure-relieving devices is to protect personnel and property. This has been discussed in some detail in Chapter 12, Volume 1, where the types and functions of relieving devices are also described, and design features, code requirements and construction materials are discussed.

Relieving devices are usually classified as relief valves, safety valves, safety-relief valves, pilot-operated relief valves and rupture discs. *Pressure-relief valve* is the generic term that encompasses all of these devices except the rupture disc. The distinction between relief and safety valves is that relief valves open in proportion to the increase in pressure over the set or opening pressure, whereas safety valves exhibit a rapid full opening or pop action after their set pressure is exceeded. Relief valves are used primarily for liquid service; safety valves, for gas or vapor service.

Safety-relief valves are suitable for either relief or safety valve applications, depending on the setting of the blowdown ring. When the blowdown ring is backed off so that the huddling chamber effect (which produces the pop action of safety valves) is absent, the valve functions as a relief valve. When the blowdown ring is set to produce pop action, the valve functions as a safety valve.

Pilot-operated relief valves may function as relief, safety or safety-relief valves, but each operates differently. Whereas the latter three are spring operated, the pilot-operated valve uses a floating piston as a main valve with process pressure on both sides of the piston and utilizes a pilot valve to initiate opening or closing action. This feature allows quick and positive opening and closing of the valve; it also allows process operation much closer to the set pressure.

Rupture discs are used for applications where pressure-relief valve operation is questionable (because of plugging, freezing and similar problems). They are used as secondary relief devices for explosion protection, in series with relief valves to prevent plugging or freezing or because they are much more economical.

Figure 6.1 shows a conventional safety-relief valve, a balanced valve, a pilot-operated valve and a rupture disc. The purpose of the balanced valve is to eliminate the effect of varying back pressure on the set pressure of the valve. This is accomplished by using a bellows so proportioned that the area open to the atmosphere is exactly equal to the area of the disc exposed to the process pressure. Theoretically this eliminates the effect of back pressure on the set pressure. Back pressures are often present in discharge systems common to several valves or where valves discharge to lower pressure systems.

Orifice sizes for pressure-relief valves have been standardized for full-nozzle safety-relief valves as listed in Table 6.1. Orifice diameters vary from 0.110 inches (Size D) to 26.0 inches (Size T) as standard but are available in smaller and larger sizes by some manufacturers.

Another type relieving device whose use is confined essentially to near-atmospheric conditions is the pressure-vacuum relief valve. Used normally on atmospheric storage vessels rather than on process equipment, it is designed to relieve both pressure and vacuum. Its operating mechanism usually consists of weight-loaded pallets or springs.

To ensure a proper understanding of the various pressure-relieving devices and their associated terms, the following definitions are given. The quoted definitions are taken directly from API RP 520, Part 1, Paragraph 3.1.

Relief valve: "A relief valve is an automatic pressure relieving device actuated by the static pressure up-stream of the valve, and which opens in proportion to the increase in

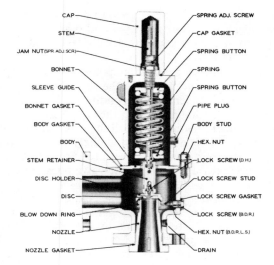

CAP
STEM
JAM NUT (SPR ADJ SCR)
BONNET
SLEEVE GUIDE
BONNET GASKET
BODY GASKET
BODY
STEM RETAINER
DISC HOLDER
DISC
BLOW DOWN RING
NOZZLE
NOZZLE GASKET

SPRING ADJ. SCREW
CAP GASKET
SPRING BUTTON
SPRING
SPRING BUTTON
PIPE PLUG
BODY STUD
HEX. NUT
LOCK SCREW (D.H.)
LOCK SCREW STUD
LOCK SCREW GASKET
LOCK SCREW (B.D.R.)
HEX. NUT (B.D.R.L.S)
DRAIN

a.

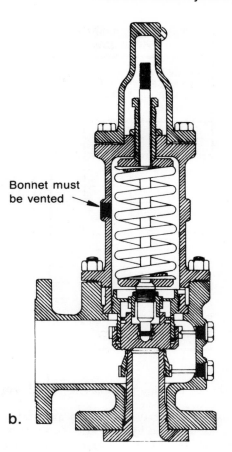

Bonnet must
be vented

b.

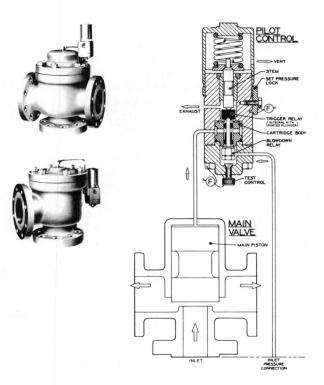

PILOT
CONTROL
VENT
STEM
SET PRESSURE
LOCK
EXHAUST
TRIGGER RELAY
(INTEGRAL WITH
PORTED PLUNGER)
CARTRIDGE BODY
BLOWDOWN
RELAY
TEST
CONTROL

MAIN
VALVE
MAIN PISTON

INLET

INLET
PRESSURE
CONNECTION

c.

d.

Figure 6.1. Pressure relieving devices may be classified as (a) conventional safety relief valves, (b) balanced safety relief valves, (c) pilot-operated relief valves and (d) rupture discs. The first two devices may operate as relief or safety valves, depending on the blowdown setting. (Courtesy of Teledyne Farris Engineering and BS&B Safety Systems)

Table 6.1. Standard Orifice Sizes	
EFFECTIVE ORIFICE AREA SQ. IN.	ORIFICE DESIGNATION
0.110	D
0.196	E
0.307	F
0.503	G
0.785	H
1.287	J
1.838	K
2.853	L
3.60	M
4.340	N
6.380	P
11.050	Q
16.0	R
26.0	T

pressure over the opening pressure. It is used primarily for liquid service."

Safety valve: "A safety valve is an automatic pressure-relieving device actuated by the static pressure up-stream of the valve and characterized by rapid full opening or pop action. It is used for gas or vapor service. (In the petroleum industry, it is used normally for steam or air.)"

Safety-relief valve: "A safety relief valve is an automatic pressure-relieving device suitable for use as either a safety or relief valve, depending on application. (In the petroleum industry, it is normally used in gas and vapor service or for liquid.)"

Pressure-relief valve: "A generic term applying to relief valves, safety valves, or safety relief valves."

Rupture disc: "A rupture disc consists of a thin diaphragm held between flanges. Its purpose, of course, is to fail at a predetermined pressure, serving essentially the same purpose as a pressure-relief valve."

Maximum allowable working pressure: "As defined in the construction codes for unfired pressure vessels, the maximum allowable working pressure depends on the type of material, its thickness and the service conditions set as the basis for design. The vessel may not be operated above this pressure or its equivalent at any metal temperature. It is the highest pressure at which the primary pressure-relief valve is set to open."

Operating pressure: "The operating pressure of a vessel is the pressure, in pounds per square inch gage, to which the vessel is usually subjected in service. A processing vessel is usually designed for a maximum allowable working pressure, in pounds per square inch gage, which will provide a suitable margin above the operating pressure in order to prevent an undesirable operation of the relief device. (It is

suggested that this margin be as great as possible consistent with economical design of the vessel and other equipment, system operation and the performance characteristics of the pressure-relieving device.)"

Set pressure: "The set pressure, in pounds per square inch gage, is the inlet pressure at which the pressure-relief valve is adjusted to open under service conditions. In a relief or safety-relief valve on liquid service, the set pressure is to be considered the inlet pressure at which the valve starts to discharge under service conditions. In a safety or safety-relief valve on gas or vapor service, the set pressure is to be considered the inlet pressure at which the valve pops under service conditions."

Cold differential test pressure: "The cold differential test pressure, in pounds per square inch gage, is the pressure at which the valve is adjusted to open on the test stand. The cold differential test pressure includes the corrections for service conditions of back pressure and/or temperature."

Accumulation: "Pressure increase over the maximum allowable working pressure of the vessel during discharge through the relief valve, expressed as a percent of that pressure, or in pounds per square inch, is called accumulation."

Overpressure: "Pressure increase over the set pressure of the primary relieving device is overpressure. It is the same as accumulation when the relieving device is set at the maximum allowable working pressure of the vessel. NOTE: From this definition, it will be observed that when the set pressure of the first (primary) safety or relief valve to open is less than the maximum allowable working pressure of the vessel, the overpressure may be greater than 10 percent of the set pressure of the safety valve."

Blowdown: "Blowdown is the difference between the set pressure and the reseating pressure of a relief valve, expressed as percent of the set pressure, or in pounds per square inch."

Lift: "The rise of the disc in a pressure-relief valve is called lift."

Superimposed back pressure: "Superimposed back pressure is the pressure in the discharge header before the safety-relief valve opens."

Built-up back pressure: "Built-up back pressure is the pressure in the discharge header which develops as a result of flow after the safety-relief valve opens."

Figure 6.2 shows the relationship between several terms defined above and helps to clarify their meanings.

Codes and Recommended Requirements for Protective Devices

State and local regulatory bodies usually require compliance with some or most of the following codes and recommended practices. The designer of pressure-relief systems should be familiar with their contents.

1. *ASME Boiler and Pressure Vessel Codes,* Section I, Power Boilers

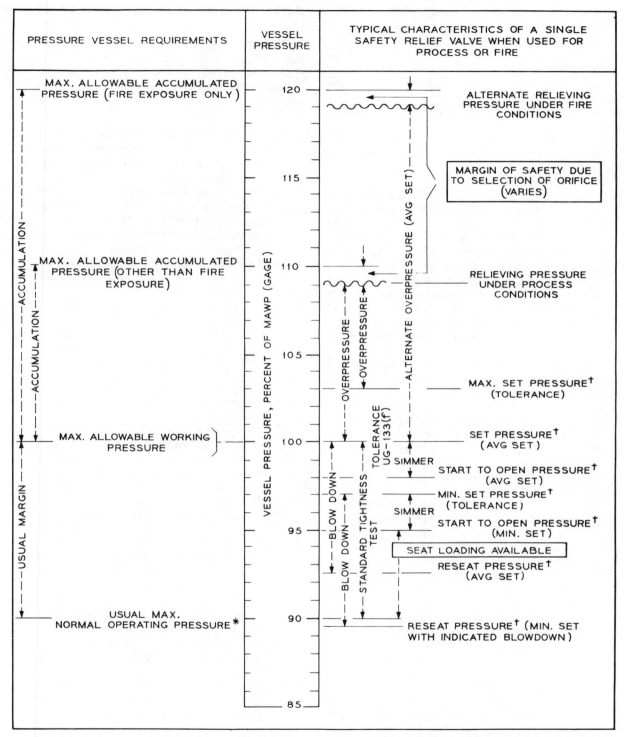

*Operating pressure may be any lower pressure required.
†Set pressure, and all other values related to it (noted with †), may be moved downward if operating pressure permits.

Note: This valve meets the requirements of *ASME Boiler and Pressure Vessel Code,* Sect. VIII, "Unfired Pressure Vessels," Par. UG-125(c and d) and Par. UG-133(a).

Figure 6.2. This chart shows typical pressure conditions that may exist for a single relief valve installed on a pressure vessel. It illustrates pictorially the relationship that exists among various relief terms. (Courtesy of the American Petroleum Institute)

2. *ASME Boiler and Pressure Vessel Codes,* Section VIII, Pressure Vessels
3. API RP 520, *The Design and Installation of Pressure-Relieving Systems in Refineries*—Part I: Design; Part II: Installation
4. API RP 521, *Guide for Pressure Relief and Depressuring Systems*
5. API Standard 2000, *Venting Atmospheric and Low-Pressure Storage Tanks*
6. NFPA 30, *Flammable and Combustible Liquids Code*
7. NFPA 58, *Liquefied Petroleum Gases, Storage and Handling*

The paragraphs of the ASME Codes, Section I, Power Boilers, which are directly pertinent to safety valves are: (a) PG-67, Boiler Safety Valve Requirements; (b) PG-68 Superheater Safety Valve Requirements; (c) PG-69, Testing; (d) PG-70, Capacity; (e) PG-71, Mounting; (f) PG-72, Operation; and (g) PG-105, Code Stamp Symbols.

That portion of the ASME Code, Section VIII, Pressure Vessels, which relates to relief devices includes UG-125 through UG-134. They cover requirements for safety and relief valves and rupture discs, the determination of capacity, their certification and stamping, pressure setting and installation.

Excerpts of the above codes of Section I and Section VIII are given at the end of the chapter as reference sources in designing and installing relief systems.

API RP 520, Parts 1 (Design) and 2 (Installation), provides more complete information on relieving devices, including the determination of relief requirements, disposal systems, piping and installation practices, than was previously available from published sources. It supplements the information in Sections I and VIII of the ASME Boiler and Pressure Vessel Codes.

API RP 521 provides an additional source of knowledge and experience gained by the petroleum processing and related industries. It recommends economically sound and safe practices for pressure relief. This source considers the complex design and operation of modern processing plants and the vast amount of energy stored. It suggests analysis guides for determining capacity requirements and design concepts for discharge and disposal systems.

API Standard 2000, *Venting Atmospheric and Low-Pressure Storage Tanks,* covers relief requirements for vessels and tanks 15 psig and under. Negative as well as positive pressures are considered. The determination of capacity requirements, the methods of venting and the testing of relief devices are included.

The National Fire Protection Association Code NFPA 30, *Flammable and Combustible Liquids Code,* paragraph 2150 (and Appendix A), closely parallels the information in API Standard 2000.

NFPA 58, *Liquefied Petroleum Gases, Storage and Handling,* also has sections pertinent to safety devices. Paragraph B.10, Basic Rules, contains information on use, sizing, location and installation. There are other references to safety devices under various container systems such as cylinder systems, cargo tanks, propelled vehicles, etc.

Pertinent Facts Concerning Codes and Guides

The following information taken from code and guide requirements and from experience in designing relief systems is offered for ease in the sizing and selection of individual relieving devices and their installation in relieving systems.

1. The ASME Codes, Sections I and VIII, are considered the authority for minimum safety and performance by virtue of their broad use as reference or base codes.
2. The ASME Boiler and Pressure Vessel Code, Sections I and VIII, and API RP 520 and 521 apply to relieving devices and their discharge systems on vessels and equipment designed for maximum allowable working pressure of more than 15 psig.
3. API Standard 2000 applies to tanks and vessels designed for operation from ½ ounce per square inch vacuum through 15 psig.
4. The maximum accumulated pressure allowed for a relief device other than for fire exposure is 110% of the maximum allowable working pressure.
5. Twenty percent accumulation is allowed relief devices used for fire exposure service.
6. Pressure-relief valves are not required on every vessel—only on each vessel or group of vessels (system) that may be isolated by control valves or other valves.
7. The spring in a safety valve in service for pressures up to 250 psig should not be reset for any pressure 10% above or 10% below that for which the valve is marked. At set pressures higher than 250 psig, the spring should not be reset 5% above or below that for which the safety valve is marked.
8. Vessel connections for rupture discs should have cross-sectional areas equal to or greater than the relief area of the rupture disc.
9. All pipe and fittings between a vessel and its pressure-relieving device should have openings with at least the area of the relieving device inlet.
10. Fail-safe protective devices and automatic startup equipment should not replace relief valves as protection for individual process equipment.
11. On equipment such as pumps and piping, it is customary to base pressure-relief sizing on 25% accumulation for liquids.
12. Oversized valves have a tendency to chatter, thereby presenting operational and maintenance problems.
13. When multiple valves are required to obtain the required relief capacity, staggered pressure settings may be used to reduce chatter and leakage problems and minimize reaction forces to relief valve openings. The smallest valve should be used for the lowest set value.
14. Balanced relief valves are suggested as a means to minimize the effect of flashed vapor on valve capacity.

15. Where balanced safety valves are used, a back pressure equal to 30 to 50% of set pressure may be used before their capacity is affected. (This value varies with pressure settings. Manufacturers' curves should be checked for more exact figures).

16. In calculating fire relief requirements, wetted surfaces higher than 25 feet above grade (or other levels where fire could be maintained) need not be considered. In the case of spheres or spheroids, the elevation should be at least to that where the maximum horizontal diameter occurs.

17. Though water applied to vessels exposed to fire is effective and highly recommended, no credit is recommended in reducing heat input because of the uncertainty of such systems—freezing weather, high winds, clogged systems, undependable water supply, etc.

18. No reduction in relief capacity should be allowed for control devices that are designed to remain stationary in the last "controlled" position when their power source fails. They should be considered fully open or closed, whichever is most conservative.

19. When limit stops are installed on valves, no credit for the stops should be taken in calculating relief requirements.

20. Relief protection for tube rupture in a heat exchanger or a reboiler is not necessary unless the design of the low pressure side is less than two-thirds of the operating pressure of the high side. This is true primarily because the low pressure side has been hydrostatically tested to one and one-half times its operating pressure.

21. Flashing conditions of relieving fluids should be considered and capacities of relief devices reduced accordingly. The effect of solids formation should be evaluated; one approach used is to increase the relief device by one size.

22. The determination of a relief requirement need not be based on the simultaneous occurrence of two or more conditions that could result in overpressure if the causes are unrelated.

23. Relief valve manifolds should be designed without pockets to avoid trapping liquids.

24. The length of inlet pipe between a relief valve and the equipment it protects should be as short as possible. In no case should the pressure drop due to piping friction and entrance losses be greater than 3% of the set pressure.

25. Exercise caution to prevent the mixing of chemicals which may react in flare headers or other disposal systems.

26. Vapor depressuring streams may use the same header system used for pressure-relief valves.

27. Where conventional relief valves are used, the relief manifolds should be sized to limit back pressure to 10% of the set pressure.

28. Where several valves discharge into the same manifold, make sure that the pressure in the manifold will be satisfactory for all relief devices discharging

into it. In other words, be sure that the imposed or built-up back pressure does not limit the required discharge capacity of the valve.

29. When redundant valves are installed and used so that one valve can be removed for maintenance while the other remains in service, block valves underneath the relief valves must be mechanically linked so that they both cannot be closed at the same time.

Determining Relieving Requirements for Various Overpressure Causes

Overpressure in process systems results when normal flows of material and energy become unbalanced, producing a pressure buildup in some part of the system. The possible causes of overpressure must be analyzed and their magnitudes determined to ensure that the process system and all of its components are not subjected to pressures exceeding the maximum allowable accumulated pressure.

It would be difficult to list all the possible causes of overpressure. The more common ones include external fires, closed outlets on vessels, utility failures, electrical or mechanical failures, reflux failures, abnormal heat input, liquid expansion, accumulation of noncondensible gases, entrance of volatile materials and overfilling. To determine the relieving capacity needed for a vessel or system, all these individual causes must be examined and calculations made. The cause that produces the greatest requirement prevails in selecting a valve size.

The simultaneous occurrence of two or more conditions that could result in overpressure is not considered if the causes are unrelated. If mechanical or electrical linkages do exist in possible causes, however, they should be considered as occurring at the same time and sized accordingly.

Another consideration in determining relief requirements is the time required for the heat input to raise the fluid in the system to relieving pressure. If the time is sufficiently long, operator action should be able to counteract the cause and avoid overpressuring the system. Ten to 30 minutes is a commonly accepted time interval for operator response, depending on the complexity of the process.

External Fires

Fire is one of the most prevalent and devastating hazards in the processing industry. Fires are always possible around equipment and systems that contain flammable liquids or gases. Leakage, operational mishaps, equipment failures and transfer operations may release flammable mixtures that are easily ignited. When liquid-containing equipment is exposed to fire, pressures beyond the rupture point are reached because heat absorbed by the vessel causes the liquid to boil and the gas to expand. Rupture is likely to occur if adequate relief is not provided.

The relief requirement for vessels exposed to fire is determined by calculating the amount of heat absorbed by the vessel. Tests have confirmed that heat absorption is proportional to the wetted vessel surface area exposed to the fire. For example, if only 50% of a vessel surface area is wetted, only half of its surface area is used in computing heat input. Appendix A of API RP 520, Part 1, contains information on tests that confirm the heat calculation method commonly used. A discussion of test conditions is also included.

The formula arrived at and used for the calculation is

$$Q = 21,000 \, FA^{0.82} \qquad (6.1)$$

where Q = total heat input to the wetted surface in British thermal units (btu) per hour
  A = total wetted surface in square feet
  F = environment factor (values shown in Table 6.2)

Since vessel liquid levels vary under operating conditions, it is necessary to assume a level for determining A, the wetted surface area. For example, surge drums normally operate about half full so the wetted surface area used would be 50%; knockout pots usually operate at a small fraction of their possible level, hence a proportionate low percent of their total surface area is used; other vessels operate essentially full so 100% surface area may be used.

Consideration is also given to the effective height a source of fire reaches. It is recommended that only the wetted surface within 25 feet of grade (or other level at which a fire could be sustained) be used in computing area A.

As shown in Table 6.2, the environment factor, F, is 1.0 for a bare vessel, 0.3 for an insulated vessel whose insulation

Table 6.2. Environment Factors Used in Calculating Tank Venting Requirements for Fire Exposure

Type of Installation	Factor F*
1. Bare vessel	1.0
2. Insulated vessels† (these arbitrary insulation conductance values are shown as examples and are in British thermal units per hour per square foot per degree fahrenheit):	
a. 4.0	0.3
b. 2.0	0.15
c. 1.0	0.075
3. Water-application facilities on bare vessel ‡ . .	1.0
4. Depressurizing and emptying facilities §	1.00
5. Underground storage	0.0
6. Earth-covered storage above grade	0.03

*These are suggested values for the conditions assumed in Par. 6.2. When these conditions do not exist, engineering judgement should be exercised either in selecting a higher factor or in providing means of protecting vessels from fire exposure as suggested in Par. 7.1.

† Insulation shall resist dislodgement by fire hose streams. For the examples a temperature difference of 1,600°F was used. In practice it is recommended that insulation be selected to provide a temperature difference of at least 1,000°F and that the thermal conductivity be based on a temperature that is at least the mean temperature.

‡ See Par. 7.3(c) for recommendations regarding water application.

§ Depressurizing will provide a lower factor if done promptly, but no credit is to be taken when safety valves are being sized for fire exposure.
Note: Paragraphs referred to are from API RP 520.

(Courtesy of the American Petroleum Institute)

conductance value is 4.0 and 0.075 for an insulation conductance value of 1.0. When credit is taken for insulation, it should be a fire-resistant type. Some engineers prefer not to take credit for insulation since there is a possibility it might be dislodged when a fire hose is turned on the vessel during an emergency.

Conductance values for various thicknesses of insulation are available from manufacturers of insulating materials.

It should be noted that the F factor is pertinent to steel vessel surfaces. If vessels are made of plastic, glass or any material other than steel, the conductance value changes, altering factor F.

Figure 6.3 shows a chart used for heat absorption determination for various surface areas and F values. It provides a quick and easy solution to Equation 6.1.

No credit is given (in computing heat input) for water-application facilities (for they might fail), and no credit is given for depressuring and emptying facilities. It is also evident from the table that underground storage requires *no* fire relief capacity for the F factor is zero.

After the heat absorption rate is determined by the chart or the formula, capacity is determined (if pressure and temperature conditions are below the critical point) by dividing the total rate of heat absorption by the latent heat of vaporization,

$$W = Q/\lambda \qquad (6.2)$$

where W = relieving rate in pounds per hour
  λ = latent heat of vaporization in btu per pound

The determination is easy when the latent heat of vaporization is known. When pressure-relief conditions are above the critical point, the relief rate depends on the rate at which the fluid expands as a result of the heat absorption rate. The valve size can be calculated from Equation 6.3.

Figure 6.4 is a chart giving the latent heat of vaporization for pure single-component paraffin hydrocarbon liquids. It is directly applicable only to those liquids. It applies approximately to paraffin hydrocarbon mixtures whose molecular weights are approximately the same. Other sources of latent heat data should be used if Figure 6.4 does not apply. (Volume 3 contains some sample calculations for determining latent heat values for multicomponent mixtures.)

For vessels containing only gas or when pressure-relief conditions are above the critical point, relief areas must be calculated on the basis of gas expansion due to the external fire. The following equation applies:

$$A = F'A_s/\sqrt{P_1} \qquad (6.3)$$

where A = effective discharge area of the valve, in square inches
  F' = an operating factor determined from Figure 6.5
  A_s = exposed surface area of the vessel, in square feet
  P_1 = upstream pressure, in psia. This is the set pressure multiplied by 1.10 or 1.20 (depending on the accumulation permissible) plus the atmospheric pressure, in psia.

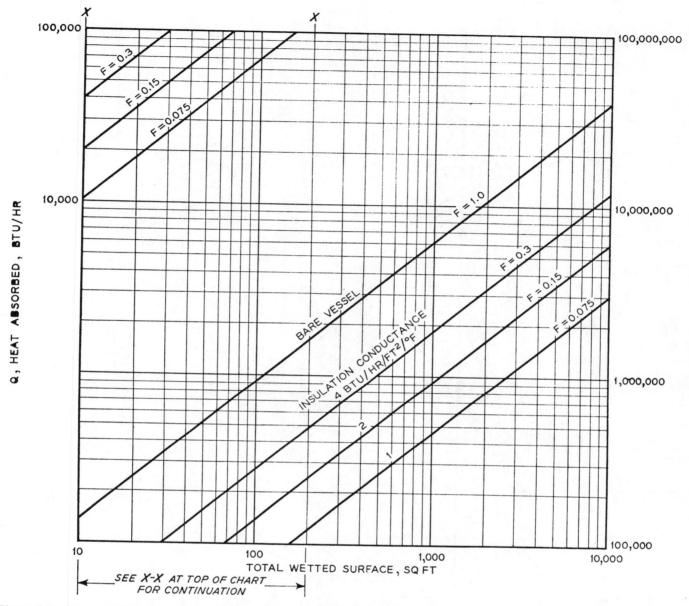

Figure 6.3. API formula for heat absorbed from fire on wetted surface of pressure vessel, $Q = 21,000\ FA^{0.82}$. (Courtesy of the American Petroleum Institute)

Closed Outlets

When all outlets on a vessel or system are blocked, the pressure-relieving capacity must be as great or greater than can be fed from a higher pressure source (liquid or gas pump-in rate, steam, etc.) plus that generated under normal operation. It is permissible to take into account the effect of the increase in pressure, when relieving, in suppressing the vapor generation.

If all outlets are not blocked, credit is given for the capacity of the unblocked outlets.

Cooling Failure in Condenser

The loss of cooling water supply to a condenser results in a capacity requirement equal to the total entering vapor. In a fractionating tower the required capacity would be equal to the total overhead vapor rate from the tower. In any case, the requirement for a system would be recalculated at a temperature corresponding to the vapor condition at set pressure plus accumulation.

If partial condensing occurs, credit is taken for that which occurs, and the relieving requirement is based on the

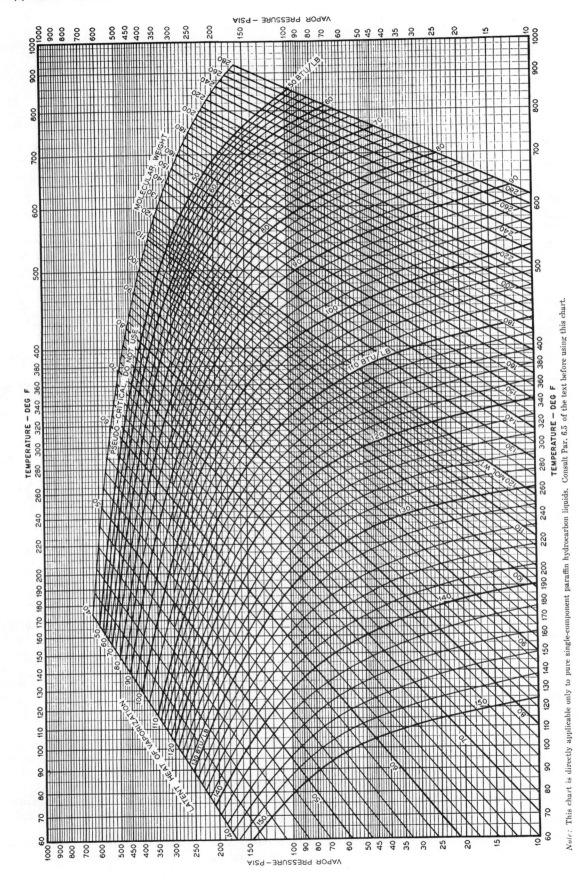

Figure 6.4. Vapor pressure and heat of vaporization of pure single-component paraffin hydrocarbon liquids. (Courtesy of the American Petroleum Institute)

Note: This chart is directly applicable only to pure single-component paraffin hydrocarbon liquids. Consult Par. 6.5 of the text before using this chart.

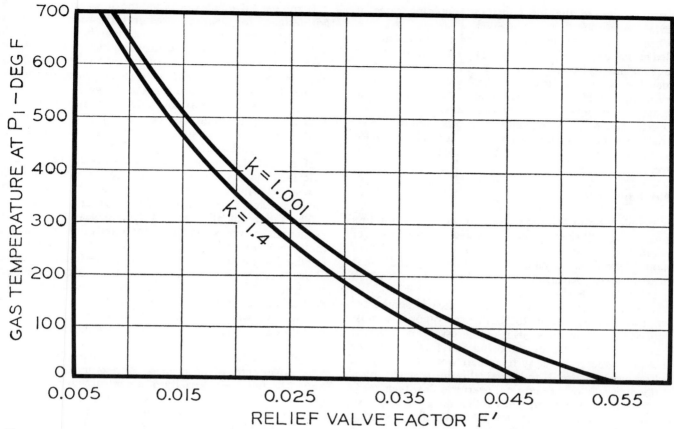

Notes:

1. Table 1 gives k values for some gases; for others, the values can be determined from the properties of gases as presented in any acceptable reference work.

2. These curves are for vessels of carbon steel.

3. These curves conform to the relationship $F' = \left(\dfrac{0.1406}{CK}\right)\left(\dfrac{\Delta T^{1.25}}{T_n^{0.6506}}\right)$

Where:

C = coefficient which is determined by the ratio of the specific heats of the gas at standard conditions. This can be obtained from Fig. 2, Fig. C-2, or Table D-1.

K = coefficient of discharge, which value is obtainable from the valve manufacturer. The K for a number of nozzle-type valves is 0.975.

T_1 = gas temperature, absolute, in degrees fahrenheit + 460, at the upstream pressure, and is determined from the relationship:

$$T_1 = \left(\frac{P_1}{P_n}\right)T_n$$

T_n = normal operating gas temperature, in degrees fahrenheit + 460.

P_n = normal operating gas pressure, in pounds per square inch absolute.

P_1 = upstream pressure, in pounds per square inch absolute. This is the set pressure multiplied by 1.20 (or less, depending on the overpressure permissible) plus the atmospheric pressure, in pounds per square inch absolute.

$\Delta T = T_w - T_1$. Difference between wall temperature and the temperature of the gas at P_1.

T_w = vessel wall temperature, in degrees fahrenheit + 460.

The curves are drawn using 1,100 F as the vessel wall temperature. This value is a recommended maximum temperature for the usual carbon steel plate materials whose physical properties at temperatures in excess of 1,100 F show signs of undesirable tendencies. Where vessels are fabricated from alloy materials, the value for T_w should be changed to a more proper recommended maximum.

It is recommended that the minimum value of $F' = 0.01$.

Figure 6.5. Relief valve factors for noninsulated vessels in gas service exposed to open fires. (Courtesy of the American Petroleum Institute)

difference between the incoming and outgoing rates at relieving conditions.

When electrical power failures cause fan losses at air-cooled exchangers, credit may still be taken for heat transfer because of natural convection; 20 to 30% of normal duty condensing capacity is generally acceptable. Steam driven spare pumps should be considered only if they idle on line and start up automatically on power failure.

Louver closure on air-cooled condensers should be considered as total coolant failure unless credit can be taken for partial condensing as indicated by the second paragraph above.

Reflux Failure

A complete reflux failure is likely to occur on pump loss due to power failure or on control valve closure, either of which may cause flooding of the condenser and total loss of coolant. The relieving capacity required for this condition is difficult to assess because there are so many factors which may affect the relieving rate.

A conservative approach, and one which is sometimes taken, is to assume that the total overhead vapor flow must be relieved. In many cases this is not necessary, however. If the system is capable of removing the total condenser overhead product from the system, there would be negligible effect on overpressure. If full removal is not possible, eventual flooding may occur and total overhead flow then needs to be considered. However, when reflux fails, the composition of the overhead stream changes, normally producing different vapor properties which affect capacity. Each case must be examined individually for the particular components and their effect.

The relieving rate for sidestream reflux failure or failure of the pump-around circuit should be equal to the difference in the amount of vapor entering and leaving the section in question.

Failure of Automatic Controls

Automatic control failures may stem from the loss of a detecting element, a transmission signal or the operating medium of the final control element such as a valve operator. The most significant effect is likely to come from the loss of the operating medium—instrument air in most cases.

The valve or other control element should be considered to be fully open or fully closed, depending on its action, unless there is some specific reason to prevent those actions. Some control devices are designed to remain in the last control position on air failure. However, since that position cannot be determined, the valve is commonly assumed to be fully open or closed, a conservative approach to that particular valve and/or occurrence.

On air failure two conditions need to be examined: (a) the failure of a single control valve and (b) the failure of all controls or valves of the system in question.

When considering a single valve failure, an open inlet and a closed outlet need to be considered.

If an inlet valve fails open (and the downstream system is not designed for upstream pressure), the relieving requirement should be equal to the difference in the capacity of the wide open inlet valve less the normal outlet flow at relieving conditions. The upstream pressure of the inlet valve should be the normal pressure. The downstream pressure should be the relief valve set pressure plus the allowable overpressure. If a downstream valve is controlled by the pressure of the system itself, it would soon be fully open. If it were controlled by another system pressure or another process variable, its position is more questionable.

If an outlet control device fails closed, the pressure-relieving capacity should be equal to the maximum inlet flow at relieving conditions less any flow that might occur through other outlets at relieving conditions.

The failure of all control valves in a system must be evaluated as well as independent failures. There may be multiple valves on the inlet, outlet or both. Analyzing relief requirements is similar to the single valve failure discussed above. Failure positions of all valves are ascertained, and the difference in inlet and outlet flow at relieving conditions determine the needed relieving capacity.

The following are some additional considerations sometimes evaluated on control valve failures:

1. Manual bypass valves could be partially open, inadvertently or on purpose, during startup.
2. The full open capacity of a control valve need not be used if the primary source of power (compressor, for example) is unable to deliver that quantity.
3. The valve capacity must be corrected to relieving conditions rather than design conditions.
4. Controller response times such as proportional band, reset and rate affect valve responses for abnormal operating conditions, but no credit should be taken other than for normal operation.
5. Wide open inlet valves may produce sufficient flow to reduce the normal inlet pressure. The capacity of the inlet source should be checked.
6. Fluid may vaporize as it passes through a valve—flashing should be considered.

Accumulation of Noncondensibles

Under normal circumstances noncondensible vapors are released with process streams and do not accumulate in the system. It is possible, however, with certain piping configurations for noncondensibles to accumulate and, in effect, "block" a condenser. The effect is the same as a total loss of coolant, although the rate of buildup is usually very slow for most systems.

Heat Exchanger Tube Failure

Thermal shock, vibration, corrosion and other causes may produce tube failures in shell and tube heat exchangers. When either the shell or tube side operates at a lower pressure than the normal operating pressure of the other side, it must be protected from overpressure as a result of tube failure if that possibility exists. The possibility of tube rupture is considered to be sufficiently remote so that protection of the low side is required only if the hydrostatic test pressure would be exceeded.

Several formulas have been proposed for calculation of relieving capacity for a split tube. One is given below:

$$W = 2{,}404 \, KA \sqrt{P_2 (P_1 - P_2)} \tag{6.4}$$

where W = flow in lb/hr

K = coefficient of discharge whose value is usually considered between 0.60 and 0.65 (use of the higher value is suggested)

A = split-tube flow area $= 2[(\pi/4)(\text{Tube } ID)^2]$, sq in.

P_1 = high-pressure-side pressure, psia.

P_2 = low-pressure-side pressure, psia.

For vapors, P_2 depends on critical flow pressure, P_{CF}.

1. If P_2 is less than or equal to P_{CF}, use P_{CF} as P_2.
2. If P_2 is greater than P_{CF}, use P_2 as P_2.

3. $P_{CF} = P_1 \left[\dfrac{2}{K+1} \right]^{k/k-1}$

ρ_2 = fluid density at the vena contracta in lb/cu ft. If the fluid is a liquid at the high pressure, use the liquid density. If the fluid is a vapor at the high pressure, the density, ρ_2, is determined as follows:

$$\rho_2 = \rho_1 / (P_1/P_2)^{1/k} \tag{6.5}$$

where k = specific heat ratio, C_p/C_v

$\rho_1 = P_1 M / [10.73 \, ZT]$ = density high-pressure side, lb/cu ft.

where M = molecular weight

Z = compressibility factor (at P_1 and T_1)

T = temperature at high-pressure side, in °R

Equation 6.4 is a simplified form of Equation 2 of API RP 520, Part I, Section 3.3. It is useful for liquid and vapor calculations.

Hydraulic Expansion

Vessels, long pipe lines and other liquid filled systems should be protected from thermal expansion if the vessels or systems can be closed off. Systems particularly subject to thermal expansion include blocked-in vessels and piping which have been filled when cold but are subsequently heated, the blocked-in cold side of an exchanger with flow on the hot side, and closed systems that are filled when cold, then later are subjected to solar radiation.

Since these applications relieve liquid, the required capacity is usually small. A ¾- by 1-inch relief valve is commonly used. On large diameter pipe lines where there is reason to believe that ¾- by 1-inch valves are inadequate, the relieving capacity can be calculated if the contained

volume and the thermal expansion coefficient of the fluid are known. The following formula may be used:

$$Q = \beta H / 500 \, Sc \tag{6.6}$$

where Q = required capacity, gpm

H = heat input, btu/hr

β = coefficient of volumetric expansion per °F

= 0.0010 for light hydrocarbons

= 0.0008 for gasoline

= 0.0006 for distillates

= 0.0004 for residual fuel oil

= 0.001 for water

S = specific gravity (water = 1.0 at 60°F)

c = specific heat, btu/lbm/°F

Chemical Reaction

The determination of relieving rates for chemical reactions is often difficult if it can be done at all. It requires a knowledge of reaction kinetics that frequently is not available. Exothermic reactions in particular are unpredictable because of their accelerating reaction rates with increasing temperature. In many cases, large volumes of noncondensibles may be produced at high temperature levels.

It is generally agreed at this time that normal relief measures are often inadequate to prevent overpressures in runaway reactions. The more obvious methods of preventing them include rapid depressuring of the system and the addition of volatile fluids to counteract or stop the reaction. To be effective, either of these measures is dependent on an early detection technique that provides sufficient time to react to the problem.

Abnormal Heat Input

To be consistent with the general philosophy of analyzing all the possible causes of overpressure, abnormal situations must also be investigated. There are many indeterminate aspects of process design that result in overcapacity systems because high safety factors are used. In other instances overcapacity is intentionally built in as an economical factor related to future expansion, changes in feed composition or other factors.

Typical examples include overcapacity of a heater burner, limit stops on valves and use of "fouled" rather than "clean" conditions in calculating heat transfer rate on shell and tube heat exchangers.

Overfilling Storage or Surge Vessels

This condition is obvious, simply requiring that relieving capacity be equal to pump-in or delivery rate at relieving conditions.

Opening Manual Valves

Pressure buildup may occur by the inadvertent opening of manually operated valves separating systems of different pressures. The relief requirement must be equal to the flow through the open valve at relieving conditions. Credit may be taken for flows into alternate systems if they can reasonably be expected to be in operation. Only one inadvertently opened valve needs to be considered at a time.

Entrance of Highly Volatile Materials

The entrance of highly volatile materials, such as water or a light hydrocarbon, into hot oil can produce such an overpressure that normal relieving methods are inadequate. The situation is similar to that of chemical reactions discussed previously. Even if the relief quantity could be calculated, it is questionable whether the relief devices would operate fast enough for the almost instantaneous relief requirement. For example, the water vapor-to-liquid volume ratio is approximately 1,400 to 1 at atmospheric pressure.

Normally no pressure-relieving device is provided for this contingency. Instead, other precautions in process design and operation should be observed to minimize such occurrences. In piping arrangements, pockets where water might collect should be avoided, proper steam condensate traps should be used, and double block and bleed configurations should be provided where water or other volatile materials connect to hot process lines.

Internal Explosions

There is no widely accepted method for calculating the relief requirements for systems where internal explosions are likely to occur. The inability to determine gaseous mixtures and the temperature and pressure rise rates leave so much uncertainty concerning the thermodynamics involved that pressure relief is questionable.

It is generally agreed that where protection is required, rupture discs should be used instead of relief valves because relief valves react too slowly to protect the system from the pressure buildup resulting from an explosion.

When rupture discs are used, a suggested method for determining the relief area is to use a ratio from 1 to 2 square feet for each 100 cubic feet of vessel volume. (API RP 521 suggests that 2 square feet per 100 cubic feet provide adequate protection for normal refinery applications involving air-hydrocarbon mixtures.)

Power Failure

Partial power failures have been considered in some of the previously mentioned causes of overpressure—for example, the loss of pumps in producing cooling water and reflux failures and the loss of instrument air in producing automatic control failures. In addition to considering the power loss as it affects a single system, a careful analysis needs to be made to evaluate how it affects the entire plant, both for partial and complete failures.

Special consideration must be given to the effect of simultaneous operation of several relief valves, particularly if they discharge into a closed header system. This is discussed in greater detail in a later section on disposal systems.

Instrument air failure has been discussed previously as it relates to control valves and transmitter and controllers in the process. Like electrical power failures, total instrument air loss affects operation of several relief systems simultaneously.

The loss of steam power used on turbine drives for pumps, compressors and blowers, or in any other equipment service where its loss affects the process, must be evaluated in the same manner as other power sources.

Following is a convenient check list of equipment that may be affected by the different power sources—electrical, instrument air and steam. Fuel gas might also be used occasionally as a convenient power source for some applications.

Electric

1. Cooling water pumps
2. Boiler feed pumps
3. Reflux pumps
4. Air-cooled heat exchanger fans
5. Cooling tower fans
6. Process vapor compressors
7. Instrument air compressors
8. Refrigeration compressors
9. Instrumentation
10. Motor-operated valves

Instrument Air

1. Control valves
2. Transmitters and controllers
3. Alarm and shutdown systems

Steam

1. Turbine drivers for pumps, compressors or blowers
2. Direct use of steam for the process
3. Boiler fans
4. Cooling water pumps
5. Reciprocating pumps
6. Steam heating where its loss allows freezing that affects the process or control devices

Fuel Gas (or Oil)

1. Boilers for process on utility steam
2. Engine drivers for pumps or electric generators

Evaluation of Relief Device Types and Features

The designer of pressure-relief systems is confronted not only with determining relieving quantities for the various systems of a process but also with selection of the proper device for the system. Factors which affect the choice include the nature of the fluid, the corrosive effect of the fluid on construction materials, the set pressure, the proximity of the operating pressure to the set pressure, back pressure in the disposal system and the ultimate economy of the device. These factors are discussed below, and advantages and disadvantages of the various relief devices are listed and evaluated.

Conventional relief and safety valves are more widely used than any other pressure-relieving device but possess some inherent disadvantages in operating features and characteristics.

Conventional Valves

Advantages

1. Simple
2. Wide use in general services
3. Relatively low cost for standard materials, especially in small sizes

Disadvantages

1. Limited pressure
2. Tendency to leak at high pressures and when operating pressure approaches set pressure
3. High cost for exotic materials, especially in large sizes
4. Back pressure variations may present problems
5. Can be costly to maintain

Spring-operated relief and safety valves are characterized by simmering action near the set point. Simmering occurs as the set pressure is reached and fluid flows through the valve. As pressure is relieved, the valve tries to reseat but may seat improperly, allowing leakage of additional fluid through the valve. The action occurs because the forces (process pressure versus spring action) are so nearly equal that the valve trim vacillates between the open and closed positions.

Variable back pressure also presents problems for conventional relief valves. These valves operate when the differential pressure for which they are set is exceeded. If the valves exhaust to atmosphere or to closed systems whose back pressures are constant, relief settings present no problems. However, when back pressures are not constant, the opening pressure varies in proportion to the back pressure variation. In such instances, balanced valves should be used. Compensation for back pressures within 10% of the set pressure can be accommodated.

Relief valves are available for pressure settings to 20,000 psig. However, such high settings are unusual and may present operating and maintenance problems. When valves at these extended pressures fail to seat properly, dangerous conditions exist, expecially if the service is flammable or toxic. Rupture discs are normally preferred for these services.

As relief valve sizes increase, particularly if standard steel material is not used, valve costs become excessive. Rupture discs may need to be considered because of their initial low cost.

Balanced Valves

Advantages

Balanced valves overcome back pressure problems.

Disadvantages

1. All the disadvantages of conventional relief valves, except the back pressure problem
2. Higher cost
3. More limited on pressure ceiling

Balanced relief valves have all the inherent disadvantages of conventional valves except their near-immunity to back pressure variations—within limits. They may be used with back pressures to 50% of set pressure without significant effect on pressure settings.

Balanced valves are more limited on pressure ceilings, cost more than conventional valves and have the same leakage and maintenance difficulties. Leakages result from improper reseating (simmering), scratching of metal-to-metal seats, corrosion, piping strains and vibration. Having set pressures too near the operating pressure can also cause leakage.

Pilot-Operated Valves

Advantages

1. High operating pressure in relation to valve set pressure
2. Ease of setting pressure
3. Short blowdown
4. Higher capacities for the same size body
5. Reduction of product losses with improved leakage characteristics
6. Less atmospheric contamination

Disadvantages

1. Not as widely accepted
2. More parts that can give trouble
3. Pilot device not always acceptable

The pilot-operated relief valve is particularly useful for operation close to the set pressure. A differential piston in the main valve is loaded through a pilot operator by process pressure. Closure is maintained by a larger piston area on top of the piston, holding the valve closed. Tight seating is maintained even when process pressure approaches the set point. The pilot can be set to close at values just below the set pressure. Positive opening and closing actions avoid the simmering and chattering characteristics of conventional and balanced valves. Product losses are reduced, and maintenance costs to repair damaged seats are considerably lower.

Other significant factors include the reduction of atmospheric contaminants (increasingly important) and minimal repair cost (compared to chattering valves).

Because there are more parts to a pilot-operated valve that are subject to failure and because pilot lines might plug and become inoperative, acceptance of these valves has been slow. Back pressures may also present problems. It acts against the overhang of the main valve piston, causing the main valve to open. This can be avoided by a piping arrangement that directs this pressure to the top of the main valve piston to hold it closed.

Rupture Discs

Advantages

1. Good for slurries and viscous services where relief valves are impractical
2. Can be used as explosion protectors
3. Very economical, especially where high cost alloys must be used
4. Available in any size and in a wide variety of materials
5. No leakage until failure occurs
6. Can be used on high pressure applications beyond the range of relief valves
7. Good as a secondary relieving device such as a reaction runaway
8. May be used upstream of relief valves to protect against corrosion, freezing, etc., allowing the use of less expensive relief valves
9. May be used downstream of relief valves to protect from corrosive materials in discharge system

Disadvantages

1. Greater loss of material when failure occurs
2. Process downtime required to replace ruptured discs
3. Cannot be set close to operating pressure, especially where pressures are cyclic
4. Settings are affected by temperature and corrosive characteristics
5. Life expectancy short for many applications
6. Accuracy poor compared to relief valve settings

The regulating codes refer to rupture discs primarily in terms of secondary relief devices and, by implication, discourage their use except for that purpose or in conjunction with spring-operated relief valves. The complexities of modern processes, however, with the wide assortment of viscous, corrosive and slurry-like materials, have resulted in an increasing use of rupture discs as primary relief devices. Some high pressure processes require relief beyond the practical range of spring-actuated devices at their present state of technology. Rupture discs are the logical solution to these applications. They are economical; they can be made in any practical size; there is a wide variety of materials available; and they are not subject to the leakage problems that are characteristic of spring-actuated relief valves.

On disc failure, however, rupture discs do present problems. Greater product loss is incurred unless other means are used to shut down process flows automatically. Process downtime is required to install new rupture discs.

Figure 6.6. The reverse buckling rupture disc design allows burst pressures to be set close to operating pressures and still retain relatively long life expectancy. (Courtesy of BS&B Safety Systems)

These two features are very objectionable for most processes, and rupture discs are used only if other relief devices are impractical.

Until recent years, rupture discs could not be set close to the operating pressure, especially if process pressures were cyclic or pulsating in nature. Recently, however, rupture disc fabrication and installation techniques (such as the reverse buckling disc shown in Figure 6.6) have allowed operating pressure much closer to the set pressure and increased the life expectancy of the disc significantly. The accuracy of disc settings has also increased. An accuracy of ±2% of set pressure is not an uncommon requirement, and this degree of accuracy is achieved.

Sizing Formulas

Formulas are given below for sizing relief valves for gas, steam and liquid services. Calculation examples for these are given in Section 8, Volume 3.

Nonviscous Liquids

The formula for sizing valves for liquid service is taken from Appendix D of API RP 520.

$$\text{gpm} = 27.2\, A K_p K_w K_v \sqrt{(p-p_b)/G}$$
or
$$A = \text{gpm}/27.2\, K_p K_w K_v \text{x} \sqrt{[G/(p-p_b)]} \qquad (6.7)$$

where gpm = flow rate at the selected percentage overpressure, in U.S. gallons per minute.
A = effective discharge area in square inches.

K_p = capacity correction factor due to overpressure. Many, if not most, relief valves in liquid service are sized on the basis of 25% overpressure, in which case $K_p = 1.00$. The factor for other percentages of overpressure can be obtained from Figure 6.7.

K_w = capacity correction factor due to back pressure. If the back pressure is atmospheric, the factor can be disregarded, or $K_w = 1.00$. Conventional valves in back pressure service require no special correction: $K_w = 1.00$. Balanced bellows valves in back pressure service will require the correction factor as determined from Figure 6.8.

K_v = capacity correction factor due to viscosity. For most applications, viscosity may not be significant, in which case $K_v = 1.00$.

p = set pressure at which relief valve is to begin opening, in psig.

p_b = back pressure, in psig.

G = specific gravity of the liquid at the flowing temperature referred to water = 1.00 at 70°F.

Viscous Liquids

When sizing a relief area for a viscous liquid service, it is necessary to determine the Reynolds number, R, so that the correction factor, K_v, can be determined. R is calculated from Equation 6.8 using a preliminary area calculated from the nonviscous formula. The orifice size is selected from the first size greater than the calculated area. When R is found and K_v is obtained from Figure 6.9, the area is then calculated on the basis of the viscous fluid. If the calculated area is larger than the chosen area used to determine R, the next larger standard orifice size is used to recalculate the area until the correct Reynolds number is used for the orifice size eventually used.

The Reynolds number, R, is determined from either of the following relationships:

$$R = \text{gpm} \, (2{,}800 \, G)/\mu \sqrt{A} \qquad (6.8)$$

$$R = 12{,}700 \, \text{gpm}/U\sqrt{A} \qquad (6.9)$$

where gpm = flow rate at the flowing temperature, in U.S. gallons per minutes

G = specific gravity of the liquid at the flowing temperature referred to water = 1.00 at 70°F

μ = absolute viscosity at the flowing temperature, in centipoises

U = viscosity at the flowing temperature, in Saybolt Universal Seconds

A = effective discharge area, in square inches (from manufacturers' standard orifice areas)

Gases or Vapors

For gas or vapor service, any one of the following formulas may be used. Equation 6.10, based on a weight rather than a volume flow, is most commonly used.

$$W = (CKAP_1K_b \sqrt{M})/TZ$$

$$or \quad A = (W \sqrt{TZ})/(CKP_1K_b \sqrt{M}) \qquad (6.10)$$

$$V = (6.32 \, CKAP_1K_b)/\sqrt{TZM}$$

$$or \quad A = V\sqrt{TZM}/ (6.32 \, CKP_1K_b) \qquad (6.11)$$

$$V = (1.175 \, CKAP_1K_b)/\sqrt{TZG}$$

$$or \quad A = V\sqrt{TZG}/(1.175 \, CKP_1K_b) \qquad (6.12)$$

where W = flow through valve, in pounds per hour.

V = flow through valve, in standard cubic feet per minute at 14.7 psia and 60°F.

C = coefficient determined by the ratio of the specific heats of the gas or vapor at standard conditions. This can be obtained from Figures 6.10 and 6.11 or Table 6.3.

K = coefficient of discharge, which value is obtainable from the valve manufacturer. The K for a number of nozzle-type valves is 0.975. For exact coefficient values, the authoritative source is *Safety Valve and Safety Relief Valve Relieving Capacities,* National Board of Boiler and Pressure Vessel Inspectors, 1055 Crupper Ave., Columbus, OH 43229.

A = effective discharge area of the valve, in square inches.

P_1 = upstream pressure, in psia. This is the set pressure multiplied by 1.10 or 1.20 (depending on the amount of accumulation permissible) plus the atmospheric pressure, in psia.

K_b = capacity correction factor due to back pressure. This can be obtained from Figure 6.12, which applies to conventional safety relief valves, or from Figure 6.13, which applies to balanced bellows valves. The correction factor value should be read from the curve specifically applying to the type of valve under consideration. In reference to Figure 6.13 for set pressures lower than 50 psig, the valve manufacturer should be consulted for the proper value of correction factor K_b.

M = molecular weight of the gas or vapor. Various handbooks carry tables of molecular weights of materials, but the composition of the flowing gas or vapor is seldom the same as that listed in the tables. This value should be obtained from the process data. An example of how to calculate the average molecular weight of a mixture of gases is given in *Applied Instrumentation in the Process Industries,* Volume 3.

T = absolute temperature of the inlet vapor, in degrees Fahrenheit + 460.

Z = compressibility factor for the deviation of the actual gas from a perfect gas, a ratio evaluated at inlet conditions. This can be obtained from Figure 6.14 or Figure 6.15.

(text continued on page 148)

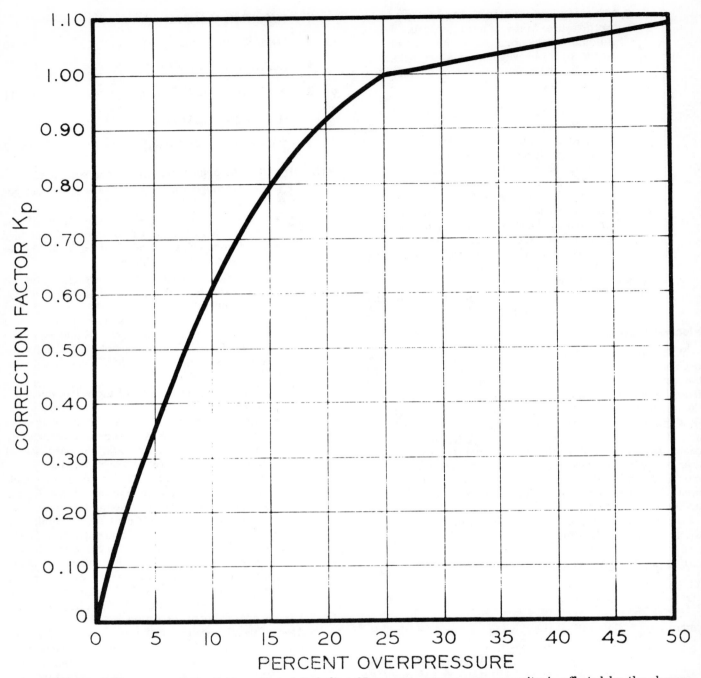

Note: The above curve shows that up to and including 25 percent overpressure, capacity is affected by the change in lift, the change in orifice discharge coefficient, and the change in overpressure. Above 25 percent, capacity is affected only by the change in overpressure.

Valves operating at low overpressures tend to "chatter"; therefore, overpressures of less than 10 percent should be avoided.

Figure 6.7. Capacity correction factors due to overpressure for relief and safety relief valves in liquid service. (Courtesy of the American Petroleum Institute)

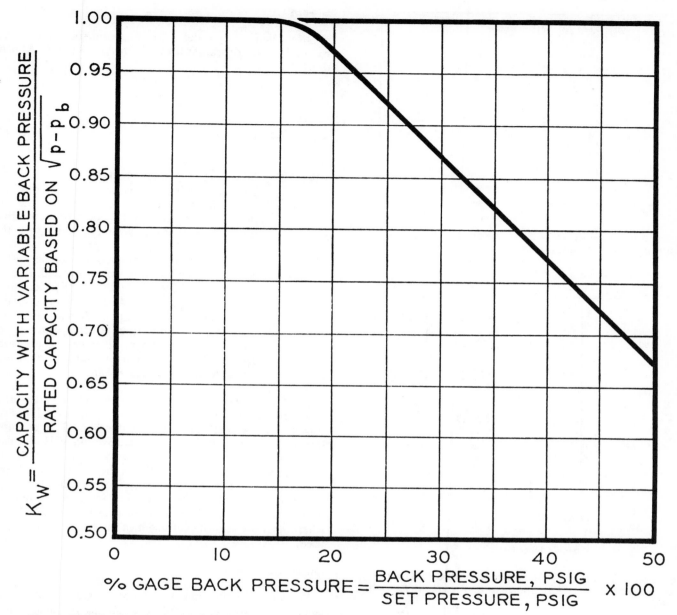

Note: The above curve represents a compromise of the values recommended by a number of relief valve manufacturers. This curve may be used when the make of the valve is not known. When the make is known, the manufacturer should be consulted for the correction factor.

Figure 6.8. Variable or constant back pressure sizing factor, K , for 25% overpressure on balanced bellows safety relief valves (liquid only). (Courtesy of the American Petroleum Institute)

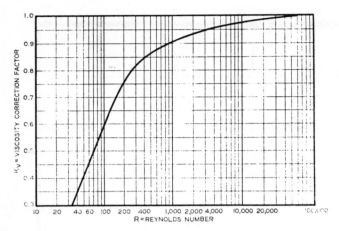

Figure 6.9. Capacity correction factor due to viscosity K_v. (Courtesy of the American Petroleum Institute)

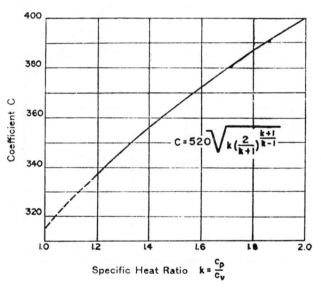

$$C = 520 \sqrt{k \left(\frac{2}{k+1}\right)^{\frac{k+1}{k-1}}}$$

Figure 6.10. Curve for evaluating coefficient C in flow formula from specific heat ratio. (Courtesy of American Petroleum Institute)

G = specific gravity of gas referred to air = 1.00 at 60°F and 14.7 psia.
  = molecular weight of the gas/28.93.

Table 6.3 complements Figure 6.10 where

$k = C_p/C_v$

the ratio of specific heats of an ideal gas, and Figure 6.16 when n = isentropic expansion coefficient of an actual gas (such as a paraffin hydrocarbon).

When k or n cannot be determined, it is suggested that c = 315.

Steam

Safety and safety relief valves in steam service are sized by the following formula taken from API RP 520, Appendix D:

$$W = 50 \, A P_1 K_{sh}$$

$$or \quad A = W / (50 \, P_1 K_{sh}) \tag{6.13}$$

where W = flow rate, in pounds per hour.
  A = effective discharge area, in square inches.
  P_1 = upstream pressure, in psia. This is the set pressure multiplied by 1.03 or 1.10 (depending on the permissible accumulation) plus the atmospheric pressure, in psia. The ASME power boiler code applications are permitted only 3% accumulation and are rated at only 90% of actual capacity. Other applications may need to conform to the ASME unfired pressure vessel code, which permits 10% accumulation.
  K_{sh} = correction factor due to the amount of superheat in the steam. This can be obtained from Table 6.4. For saturated steam at any pressure, the factor $K_{sh} = 1.0$.

The factor 50 is a constant used in the API equation that is slightly different from the one given in Section VIII of the Power Boiler Code, which uses a value of 51.5. In conjunction with the value of 51.5, however, Section VIII also uses a coefficient of discharge, K, whose value is variable, depending on the manufacturer. When the two factors of Section VIII are combined, their value is sometimes lower and sometimes higher than the value of 50 in the API equation. The factor of 50 may be replaced with the term "51.5K" where K is the known coefficient of discharge for the valve used.

Other Calculations

Section 8 of Volume 3 contains several sample relief calculation problems including examples for determining the average molecular weight of gas mixtures, the calculation of fire loads for vessels exposed to fire and the determination of various other factors used in calculating relief requirements. Calculations for thermal relief and for rupture disc sizing are also included in Section 8, Volume 3.

Depressuring Systems

Relief devices are used to protect vessels and equipment from rupture. Protection is based on their maximum allowable working pressure (MAWP) of the vessel. Under fire conditions, unwetted portions of vessels and equipment can locally overheat so that they fail at pressures below their relief settings. In such cases the equipment is no longer

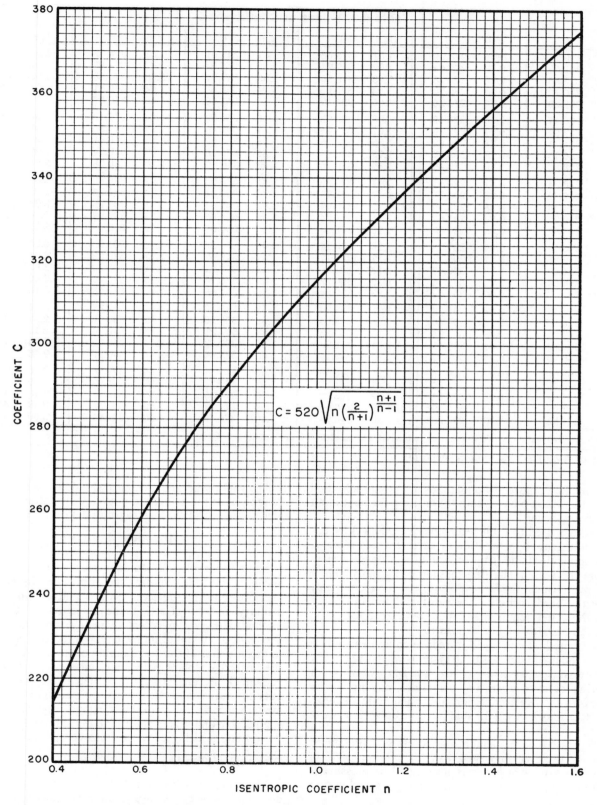

Figure 6.11. Curve for evaluating coefficient C in safety valve flow formula from isentropic coefficient, n. (Courtesy of American Petroleum Institute)

Table 6.3. Values of Coefficient C											
k, n	C	k, n	C	k, n	C	k, n	C	k, n	C	k, n	C
0.41	219.28	0.71	276.09	1.01	316.56*	1.31	347.91	1.61	373.32	1.91	394.56
0.42	221.59	0.72	277.64	1.02	317.74	1.32	348.84	1.62	374.09	1.92	395.21
0.43	223.86	0.73	279.18	1.03	318.90	1.33	349.77	1.63	374.85	1.93	395.86
0.44	226.10	0.74	280.70	1.04	320.05	1.34	350.68	1.64	375.61	1.94	396.50
0.45	228.30	0.75	282.20	1.05	321.19	1.35	351.60	1.65	376.37	1.95	397.14
0.46	230.47	0.76	283.69	1.06	322.32	1.36	352.50	1.66	377.12	1.96	397.78
0.47	232.61	0.77	285.16	1.07	323.44	1.37	353.40	1.67	377.86	1.97	398.41
0.48	234.71	0.78	286.62	1.08	324.55	1.38	354.29	1.68	378.61	1.98	399.05
0.49	236.78	0.79	288.07	1.09	325.65	1.39	355.18	1.69	379.34	1.99	399.67
0.50	238.83	0.80	289.49	1.10	326.75	1.40	356.06	1.70	380.08	2.00	400.30
0.51	240.84	0.81	290.91	1.11	327.83	1.41	356.94	1.71	380.80	2.01	400.92
0.52	242.82	0.82	292.31	1.12	328.91	1.42	357.81	1.72	381.53	2.02	401.53
0.53	244.78	0.83	293.70	1.13	329.98	1.43	358.67	1.73	382.25	2.03	402.15
0.54	246.72	0.84	295.07	1.14	331.04	1.44	359.53	1.74	382.97	2.04	402.76
0.55	248.62	0.85	296.43	1.15	332.09	1.45	360.38	1.75	383.68	2.05	403.37
0.56	250.50	0.86	297.78	1.16	333.14	1.46	361.23	1.76	384.39	2.06	403.97
0.57	252.36	0.87	299.11	1.17	334.17	1.47	362.07	1.77	385.09	2.07	404.58
0.58	254.19	0.88	300.43	1.18	335.20	1.48	362.91	1.78	385.79	2.08	405.18
0.59	256.00	0.89	301.74	1.19	336.22	1.49	363.74	1.79	386.49	2.09	405.77
0.60	257.79	0.90	303.04	1.20	337.24	1.50	364.56	1.80	387.18	2.10	406.37
0.61	259.55	0.91	304.33	1.21	338.24	1.51	365.39	1.81	387.87	2.11	406.96
0.62	261.29	0.92	305.60	1.22	339.24	1.52	366.20	1.82	388.56	2.12	407.55
0.63	263.01	0.93	306.86	1.23	340.23	1.53	367.01	1.83	389.24	2.13	408.13
0.64	264.72	0.94	308.11	1.24	341.22	1.54	367.82	1.84	389.92	2.14	408.71
0.65	266.40	0.95	309.35	1.25	342.19	1.55	368.62	1.85	390.59	2.15	409.29
0.66	268.06	0.96	310.58	1.26	343.16	1.56	369.41	1.86	391.26	2.16	409.87
0.67	269.70	0.97	311.80	1.27	344.13	1.57	370.21	1.87	391.93	2.17	410.44
0.68	271.33	0.98	313.01	1.28	345.08	1.58	370.99	1.88	392.59	2.18	411.01
0.69	272.93	0.99	314.19*	1.29	346.03	1.59	371.77	1.89	393.25	2.19	411.58
0.70	274.52	1.00	315.38*	1.30	346.98	1.60	372.55	1.90	393.91	2.20	412.15

* Interpolated values, since C becomes indeterminate as either n or k approaches 1.00.

(Courtesy of American Petroleum Institute)

protected against failure. Two methods are commonly used to reduce the failure possibilities: (a) limiting the heat input and (b) depressuring.

The methods used to limit heat input include: (a) insulation, (b) underground installation or earth-covered storage, (c) water application and (d) provision of a drainage system away from the vessel. The first method can greatly reduce the relief requirement for fire exposure, and the second may eliminate the need entirely. The third and fourth do not decrease the relief requirement, but they are effective in reducing the hazards of fire exposure.

The other method employed to reduce equipment failure potential is the use of depressuring systems when fire occurs. Most depressuring systems utilize remote controls for depressuring because operating personnel may no longer have access to the fire area. Hydraulic, pneumatic and electric operators may be used for that purpose on various valve types.

The depressed level sought when the depressuring technique is used is suggested as 100 psig or 50% of operating pressure, whichever is lower. A low pressure is used to limit the potential energy stored in the vessel in case of failure.

Additional guides on the protection of vessels from fire exposure are given in API RP 520, Part 1, paragraph 7.

The vapor load to be removed should be calculated for the conditions at the depressured level. Calculation methods are given in API RP 521, paragraph 3.17.

Disposal Systems

Some basic rules and guidelines are given in API RP 520 and API RP 521 governing the disposal of discharges from pressure-relieving devices. Safety and economy are the primary factors affecting the selection of the disposal method. However, in recent years governmental regulations pertinent to air pollution have reoriented concepts on safety to include that of the surrounding communities rather than the immediate plant personnel and adjacent property and personnel. Because of more stringent requirements from outside sources and because of the increased complexity of modern processes, disposal systems require a thorough engineering analysis to determine the maximum loading and the optimum sizing of the system components.

Disposal Methods

There are three basic methods used for relieved material disposal: (a) direct release to the atmosphere, (b) discharge to closed systems for flaring and (c) discharge to systems of lower pressure.

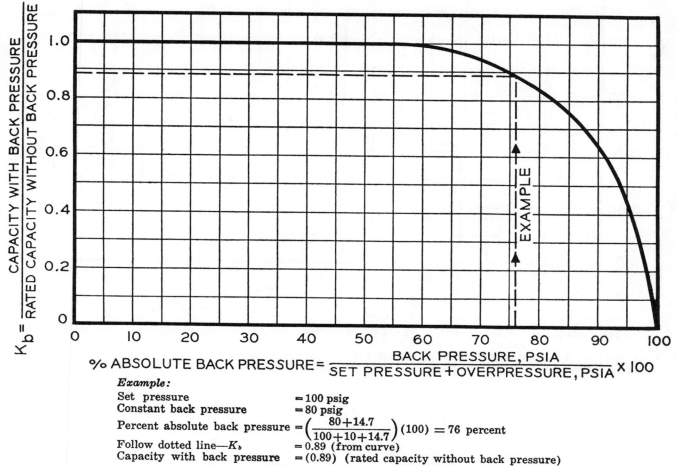

$$\% \text{ ABSOLUTE BACK PRESSURE} = \frac{\text{BACK PRESSURE, PSIA}}{\text{SET PRESSURE} + \text{OVERPRESSURE, PSIA}} \times 100$$

Example:

Set pressure $= 100$ psig
Constant back pressure $= 80$ psig
Percent absolute back pressure $= \left(\dfrac{80+14.7}{100+10+14.7}\right)(100) = 76$ percent
Follow dotted line—K_b $= 0.89$ (from curve)
Capacity with back pressure $= (0.89)$ (rated capacity without back pressure)

Note: This chart is typical and suitable for use only when the make of valve or the actual critical flow pressure point for the vapor or gas is unknown; otherwise, the valve manufacturer should be consulted for specific data.

Figure 6.12. Constant back pressure sizing factor, K , for conventional safety relief valves (vapors and gases). (Courtesy of American Petroleum Institute)

Atmospheric Discharge

Direct discharge of relieved material to the atmosphere is usually the most economical discharge method because of its simplicity and dependability. The potential hazards which may be created must be evaluated, however, to determine that compliance with safety and other requirements is assured.

Potential hazards include (a) the ignition of the relieved material at the point of emisson, (b) the formation of flammable mixtures after emission, (c) exposure of personnel to toxic vapors, (d) corrosive effect of material on equipment and piping, (e) excessive noise and (f) pollution effects. The potential for ignition of flammable materials discharged directly into the atmosphere is great. However, the turbulent action of the normal jet stream resulting from a relieved vapor causes rapid dilution of the material with the surrounding air. Tests have indicated that at discharge velocities of 500 feet per second, a typical hydrocarbon stream is diluted to its lower flammable limit at a distance of approximately 120 diameters from the end of the discharge pipe. The "typical" material referred to is a light hydrocarbon in the methane/hexane range, and flammability properties differing from that would require further analyzing. Consideration must be given to the likelihood that the 500-foot per second velocity will not always apply. Calculation of the distance at which dilution below the lower explosive limit occurs is made by the formula:

$$W/W_o = 0.264(X/D) \tag{6.14}$$

where W = vapor-air mixture, in pounds per hour, at distance X from the end of the tailpipe
W_o = relief device discharge, in pounds per hour
X = distance along tailpipe axis, at which W is to be calculated
D = tailpipe diameter, in the same units as X

(text continued on page 154)

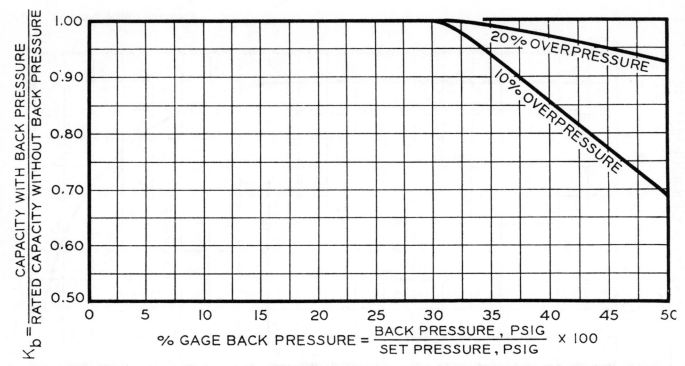

Note: The above curves represent a compromise of the values recommended by a number of relief valve manufacturers and may be used when the make of valve or the actual critical flow pressure point for the vapor or gas is unknown. When the make is known, the manufacturer should be consulted for the correction factor.

These curves are for set pressures of 50 psig and above; for set pressures lower than 50 psig, the manufacturer should be consulted for the values of K_b.

Figure 6.13. Variable or constant back pressure sizing factor, K , for balanced bellows safety relief valves (vapors and gases). (Courtesy of the American Petroleum Institute)

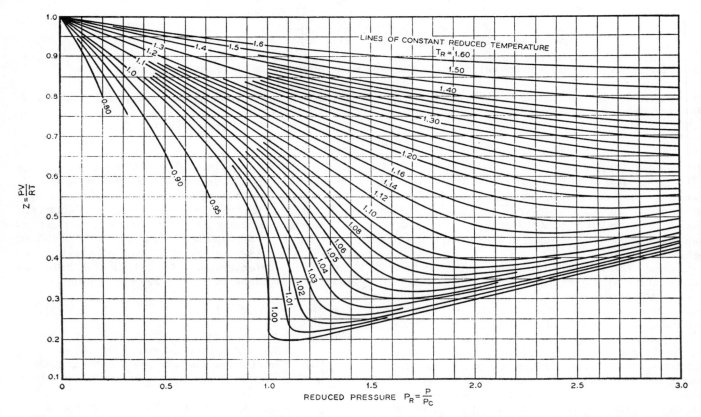

Figure 6.14. P-V-T Data correlation for hydrocarbons: PV/RT versus Pr and Tr. (Courtesy of American Petroleum Institute)

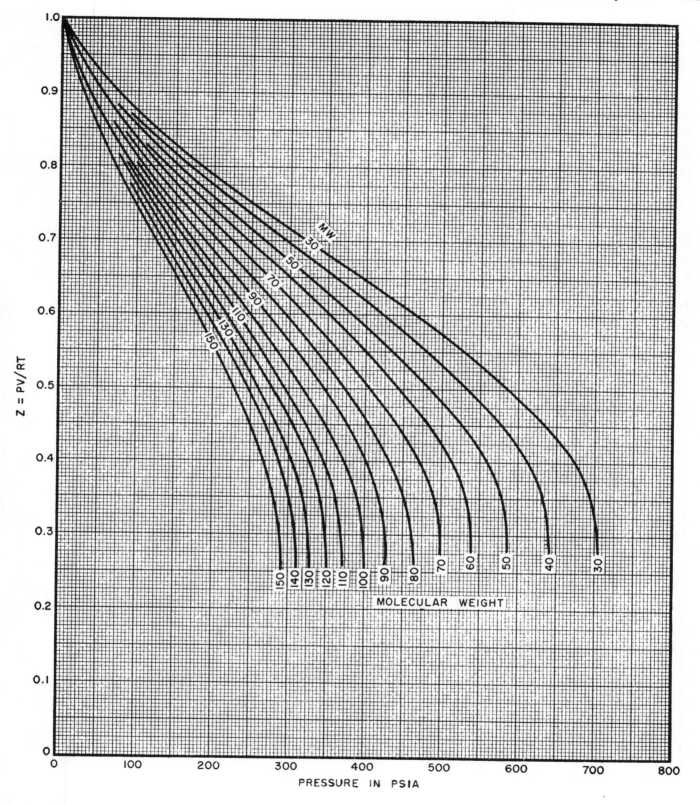

Figure 6.15. *Compressibility factor for paraffin hydrocarbons at saturated vapor conditions. (Courtesy of American Petroleum Institute)*

The release of liquids or vapors that may produce flammable mixtures after release should be carefully analyzed. Release points should be located to avoid open flames and hot surfaces where ignition may easily occur.

Consideration must be given to toxic vapors or liquids that may form toxic vapors, corrosive chemicals that present hazards to people and equipment and air pollutants that obviously must be kept at low levels of concentration. Noise emission levels are now more closely regulated, requiring closer scrutiny of potential offending sources. All these factors point to the need of careful analysis even for the simple atmospheric relief applications.

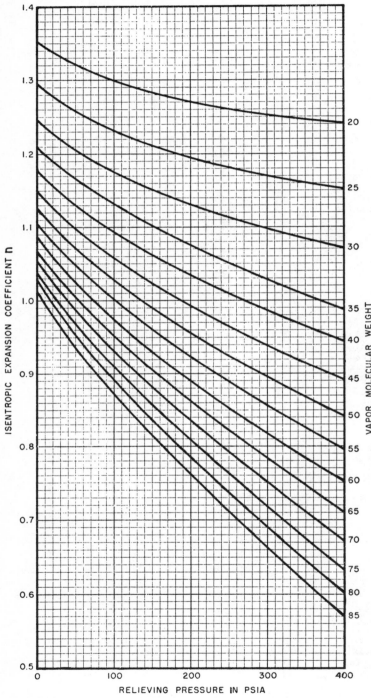

Figure 6.16. Curves for paraffin hydrocarbons expanding from relieving pressures at or near saturation to critical flow pressures. (Courtesy of the American Petroleum Institute)

Flares

The primary purpose of flares is to convert flammable, toxic or corrosive vapors to less objectionable compounds by combustion. Ground or elevated flares may be used, depending on available space, flare gas characteristics (composition, quantity and pressure level), economics and public relations.

Flare systems should be designed to prevent flashback, the entrance of oxygen (air) back into the system. Flame arrestors are not very practical because of maintenance problems, so the most obvious solution is to maintain a purge of fuel or other gas into the system. The cost of maintaining a purge flow is appreciable and may not be considered necessary if the normal venting velocity can maintain a sufficient flame. Steam is also used for purging, but it presents some additional problems if flow is cut off or reduced too much.

The flare should also be designed to operate smokelessly, at least throughout most of the range of rates it is expected to cover. This may require a considerable amount of steam which eliminates or reduces the generation of smoke. Other requirements for flares include positive pilot ignition, flame stability and acceptable noise and luminosity levels.

Elevated flares consist of a guyed stack with a supporting structure, or self-supported stacks with burner tip, pilot burners and associated fuel systems. Compared to ground flares, elevated flares conserve space, shorten flare lines and give greater flexibility in location. They have the disadvantages of higher initial cost and difficulty of maintenance. Ground flares are similar in design to elevated flares, requiring essentially the same auxiliaries. The major advantages include the absence of supporting structures and the ease of maintenance. Disadvantages include the need for isolation from other facilities, long flare lines and greater space requirements. Depending on the heat release, the suggested minimum unobstructed distance surrounding a flare may vary from 250 to 500 feet.

Discharge to Lower Pressure Systems

API RP 520 states simply that relieved material may be disposed to another system at lower pressure provided the receiving system can tolerate the additional load.

Past practice has included the use of sewer systems for nonvolatile liquids where capacity was adequate. Recent pollution control regulations make this practice questionable unless the sewers discharge to ponds for special treatment.

In some cases lower pressure process systems may be capable of handling relieving quantities, particularly in the case of liquids.

Design

The design of disposal systems requires a careful analysis of all conditions under which fluids will enter the system. One of the first steps is to use a layout (a scaled plot plan) for location of all pressure-relieving devices including those from depressuring systems, if used. Pertinent physical data such as molecular weight, discharge pressure and temperature, flow rates and load-causing information should be available, preferably on the plot plan.

Single-valve systems discharging to the atmosphere are so simple that little attention is given to them. Excessive back pressure is avoided by using the shortest practical downstream piping. They are often used at elevated locations or for piping ground level device discharges to high emission points.

Single-valve closed systems are more complex, usually discharging to a flare, a stack or to a lower pressure system. The problem is not greatly different, however, except for the back pressure of the discharge system.

Multiple valve systems are more complex. The maximum load on a multiple-valve system (or systems if more than one header is used) must be determined by establishing the abnormal condition(s) that produces the greatest pressure drop in the system. It may be a power failure, cooling water failure, steam failure or a combination of these circumstances which produces the greatest load. (The greatest pressure loss does not necessarily result from the highest pounds per hour flow rate; it is a function of the density also. For example, a flow of 100,000 pounds per hour of a 19 molecular weight vapor at 300°F develops more head loss than 150,000 pound per hour rate of a 44 molecular weight vapor at 100°F.)

If vapor depressuring systems dump into pressure-relief systems, their effect may add to the total system requirement. The fact that depressuring systems usually are voluntarily controlled whereas relief valves function automatically should be borne in mind.

For economy and safety it is often desirable to isolate certain relief streams or groups of streams. The most common causes of separation are (a) significant differences in pressure levels, (b) corrosive effects and (c) abnormally high or low temperatures.

Segregation by pressure levels is a generally accepted rule. It is usually applicable because back pressures from high pressure systems often become excessive for low-pressure relief valves, restricting their capacity and requiring increased header size or balanced valves or both. It often becomes economical to use two systems of moderate size rather than the one large system. This separation is not essential, however, and mixed levels should be used if location suggests an economy and capacities are not adversely affected.

Corrosive fluids are normally collected in separate headers or subheaders to minimize the need for large header sizes requiring special alloy material. They may be run entirely separate from the noncorrosive materials or routed through neutralizing systems prior to entry into a main or common header.

Occasionally separate systems are warranted because unusually high or low temperatures require alloy material of significantly higher cost. Small separate headers may be much more economical than using the required alloy in a larger header.

These problems and possible solutions are not intended to limit the design approaches to be used, but rather to illustrate the principle of separation of discharges. The final layout and arrangement should reflect a thorough analysis that considers all the factors affecting overall economy and safety.

Pressure drop calculations must be made for the entire disposal system. The loads and conditions for each segment of the system should be tabulated for easy reference. Equivalent pipe lengths are determined for each segment; maximum allowable pressures are listed at each node of the system starting at the discharge of the valves. (This will assure that valve capacities are not reduced by excessive back pressure.) The final calculation for pressure drop should start at the discharge end and work back toward the valves. The flow at the discharge end is the total relieving load at the maximum condition for the group of valves assigned to the system. (If water sprays or other quenching methods are used to condense part of the load, the condensed load is subtracted. Similarly, purge gas or steam injection add to the load.)

When pressure drops are determined for each segment, the final layout is checked to determine that back pressures do not exceed 10% of the set pressure for conventional valves or 50% (many companies allow only 40% or less) for balanced valves. A check should also be made to determine that individual relieving rates are adequate and that the MAWP of the individual vessel or system it protects is not exceeded.

Some notable points should be kept in mind in designing disposal systems.

1. Piping systems are complex because of the wide variety of fluids handled and the wide range of temperatures that may be encountered—thermal stresses may be severe.
2. Shock and vibration from the reactive forces of relief discharges may be severe, creating high mechanical stresses. Liquid slugs also may produce shock loading.
3. Alignment of relieving devices and discharge piping is important to prevent unnecessary mechanical stress.
4. Knockout drums, quench drums or seals may be needed.
5. Winterizing to prevent freezing of condensed liquids may be needed.
6. In checking back pressure requirements, be sure to check the valve with the lowest set pressure in the system.
7. No more than 10% variable back pressure is recommended for conventional relief valves.
8. No more than 50% variable back pressure is recommended for balanced valves.

Table 6.4. Superheat Correction Factors for Safety Valves in Steam Service

| Set Pressure (Pounds per Square Inch Gage) | Saturation Temperature (Degrees Fahrenheit) | Correction Factor K_{SH} | | | | | | | | | | | |
| | | 0.99 | 0.98 | 0.97 | 0.96 | 0.95 | 0.94 | 0.93 | 0.92 | 0.91 | 0.90 | 0.89 | 0.88 |
		Total Temperature (Degrees Fahrenheit)											
10	240	269	305	335	368	400	428	460	492	520	545	570	595
20	259	286	315	343	375	405	433	463	492	518	542	565	590
40	287	310	335	357	382	410	440	467	493	515	540	561	585
60	308	330	350	370	390	422	450	472	495	515	537	560	580
80	324	345	365	385	405	432	460	478	497	515	535	556	580
100	338	360	375	395	415	440	466	485	500	515	535	555	580
120	350	370	388	405	425	450	475	490	505	520	537	557	581
140	361	...	398	415	435	455	480	497	510	525	540	560	585
160	370	...	405	425	443	463	487	502	516	530	545	565	586
180	379	...	415	432	450	470	492	508	523	535	550	570	590
200	388	...	420	440	456	475	497	513	527	540	555	575	592
220	396	...	430	445	463	480	502	517	532	546	560	577	596
240	403	...	435	452	470	485	507	522	537	550	565	583	600
260	409	...	440	460	475	490	512	526	541	555	569	586	603
280	416	...	447	465	480	495	516	531	545	558	573	590	606
300	422	...	452	470	485	500	520	535	550	562	577	593	610
350	433	...	465	480	496	512	530	545	558	572	586	602	618
400	448	...	475	492	508	523	540	553	566	580	595	610	626
500	470	...	495	513	526	543	557	568	582	597	610	625	646
600	489	...	512	530	543	556	570	585	596	610	625	638	655
800	520	...	545	558	570	585	597	610	625	635	650	665	680
1,000	546	...	567	582	595	608	620	633	645	660	675	688	705
1,250	574	...	593	605	620	630	640	655	668	681	696	710	725
1,500	597	...	...	630	642	653	664	676	688	702	715	728	744
1,750	619	...	...	647	660	670	680	692	704	717	730	743	759
2,000	637	...	...	665	675	685	696	708	719	732	745	757	773
2,500	670	...	...	690	702	712	723	733	742	755	766	780	795
3,000	697	...	...	713	723	733	742	751	762	773	785	795	812

(Courtesy of the American Petroleum Institute)

9. Separate discharge headers may be more economical when special alloys become necessary in any system.
10. Relief valve discharge piping should not be smaller than the relief valve outlet to which it connects.
11. Special consideration must be given to liquids or vapors that could form solids or heavily viscous material such as gums, polymers, coke or ice that could prevent safe operation of the disposal system. Steam tracing may be needed to prevent such formation.
12. The basic criterion for sizing discharge piping is to prevent developed back pressure from reducing the relieving capacity of any of the pressure-relief devices below the amount required to protect the vessel or system from overpressure.
13. Give special attention to discharge of low-boiling liquids where reduced pressures cause autorefrigeration.

14. Discharge systems should be self-draining toward the discharge end; pockets should be avoided. If pockets are unavoidable, drain pots or drip legs may be necessary at low points in lines that are not sloped continuously. A suggested slope is ¼ inch per 10 feet.
15. Laterals from individual relief devices should enter a header from above.
16. Avoid the location of safety valves below a header elevation, if possible.
17. The use of angle (30° or 45°) entry of laterals into disposal headers reduces head losses and reaction forces in the system.

Flares

The sizing of flares requires the determination of stack diameter and height. Flare stack sizing is beyond the scope of this chapter, but the basis for sizing and general con-

Table 6.4 continued

Set Pressure (Pounds per Square Inch Gage)	Saturation Temperature (Degrees Fahrenheit)	Correction Factor K_{SH}											
		0.87	0.86	0.85	0.84	0.83	0.82	0.81	0.80	0.79	0.78	0.77	0.76
		Total Temperature (Degrees Fahrenheit)											
10	240	618	645	670	695	725	755	783	817	850	885	920	955
20	259	613	640	665	690	720	748	780	813	847	885	918	954
40	287	610	635	660	685	715	742	775	810	845	880	916	952
60	308	607	630	655	683	710	740	770	807	840	880	915	951
80	324	605	630	653	680	708	736	770	805	840	878	915	950
100	338	605	628	652	680	706	735	768	805	840	877	914	950
120	350	605	630	652	680	705	733	767	804	838	876	913	949
140	361	607	630	654	678	705	732	765	804	838	876	912	949
160	370	610	632	655	678	703	730	765	803	837	875	912	949
180	379	612	635	656	680	702	730	764	802	837	875	911	948
200	388	615	636	658	680	703	729	764	801	835	875	911	948
220	396	617	640	660	682	705	730	763	800	835	875	911	947
240	403	620	641	664	685	706	730	763	799	835	875	910	947
260	409	623	645	666	686	710	731	763	800	835	875	910	946
280	416	626	647	668	690	712	734	764	800	835	874	910	946
300	422	630	650	670	692	715	735	765	800	835	873	910	946
350	433	637	657	678	700	722	741	770	804	835	874	915	948
400	448	645	665	685	707	730	750	775	808	840	876	918	950
500	470	660	680	700	722	743	763	788	820	849	885	925	957
600	489	675	693	715	735	756	776	799	830	858	893	935	965
800	520	700	718	738	758	780	800	820	850	877	910	950	981
1,000	546	723	740	760	778	800	820	840	867	893	925	965	994
1,250	574	738	762	780	800	820	840	860	885	910	940	975	...
1,500	597	762	780	798	817	838	857	878	902	927	955	985	...
1,750	619	777	795	812	832	852	871	892	915	938	965	993	...
2,000	637	790	810	825	845	865	885	905	928	950	975	1,000	...
2,500	670	815	830	848	866	887	906	927	948	968	992	...	...
3,000	697	832	850	865	885	905	925	945	965	983	1,005	...	...

Note: Correction factors for pressure and temperature conditions not tabulated may be determined by interpolation if desired. However, it is practical to select the correction factor according to the next higher temperature tabulated at the closest pressure level listed.

siderations involved in the calculation of these requirements are given in API RP 521, Sections 4 and 5. A nomograph for sizing is also given in the section on Nomographs in Volume 3.

The flare stack diameter is determined primarily by the velocity requirement which ranges normally between 0.2 mach and 0.5 mach. (Mach is the ratio of vapor velocity to sonic velocity for that vapor.)

Stack height is generally based on the radiant heat intensity at grade produced by the flare flame. A formula for determining radiant heat is given in the referenced guide, API RP 521.

Smokeless burning at flares is achieved by steam injection, by operating the flares as premix burners or by using a multijet burner design. Steam injection is quite common, but operating costs are relatively high. Multijet designs have a high initial cost, but operating cost is low; they also provide a relatively quiet, nonluminous flare.

When steam is used, manual or automatic control may be used to adjust the steam load as flaring rate changes. The choice depends on load changes expected and the expected consequence of short-term smoking when upsets occur, and adjustment of steam rate is necessary.

Knockout Drums and Seals

When there is a probability of liquid discharges into a flare system or if discharge vapors are likely to condense, knockout drums should be used to remove the liquid. The capacity of the drums should provide a holdup time of 10 to 30 minutes. Drums may be vertical or horizontal. Criteria for sizing is given in API RP 521, Section 5.3.

The automatic transfer of accumulated material from the knockout drums is highly recommended because of the uncertainty of material releases and the leakage possibilities from unmonitored relief valves. In some cases, cooling of transferred material may be necessary.

Knockout drums are usually located near the area they serve. An additional seal is needed at the base of the flare stack to trap any liquid formed downstream of knockout drums.

Quench drums are used to cool liquid streams that may be released at high temperatures. A typical application is that of quenching the blowdown of a hydrocarbon furnace to reduce temperature sufficiently to allow the use of less expensive materials in the downstream piping. Drum sizing depends on the design of the drum internals as well as the amount of heat that must be removed. A common criterion for sizing is to reduce the exit stream temperature (liquid and vapor) to a range of 150° to 200°F, assuming that about 40 to 50% of the inlet stream will be vaporized.

Pipe seals are used to prevent flashback into flare headers. Flashback is the combustion of material in the header itself rather than at the flare exit. It occurs when oxygen (air) enters the header, mixes with combustible material and is ignited at the flare, propagating backward into the system.

Pipe seals consist of a trap built at the flare stack base and filled with water to provide a seal between the header and the outside atmosphere. Figure 6.17 shows a simplified seal arrangement used when condensation of header materials is not likely to present problems. Figure 6.18 shows a more complex seal system that allows the removal of light material condensates by a skimming removal method.

Pipe seals have the disadvantages of causing pulsating flows in flare headers at low flow conditions and of the possibility of blowing the liquid out of the seals when flare velocities are high.

In reference to Figures 6.17 and 6.18, the inlet flare line is sloped to provide a volume of water (below the normal

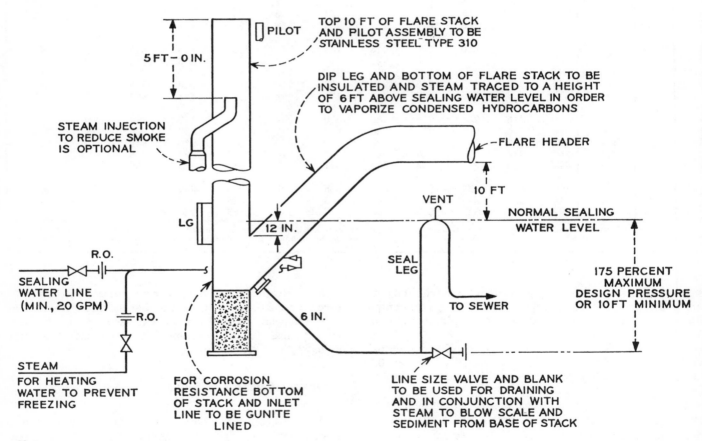

Notes:
1. Elevation of flare header should be 10 ft above water seal.
2. Slope flare header toward blowdown drum.
3. Insulate steam line to top of flare.
4. Provide handhole at bottom of stack for cleanout.

Figure 6.17. A simplified seal leg arrangement of a pipe seal placed at a flare stack to prevent flashback into flare headers. The trap is filled with water to provide the seal between the header and outside atmosphere. (Courtesy of the American Petroleum Institute)

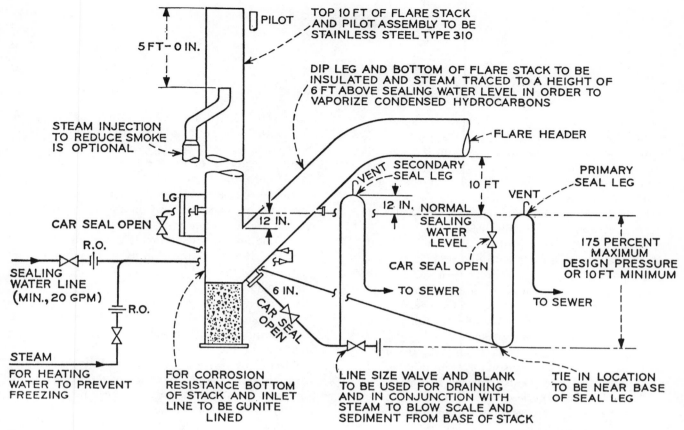

5 FT-0 IN.

PILOT

TOP 10 FT OF FLARE STACK
AND PILOT ASSEMBLY TO BE
STAINLESS STEEL TYPE 310

DIP LEG AND BOTTOM OF FLARE STACK TO BE
INSULATED AND STEAM TRACED TO A HEIGHT OF
6 FT ABOVE SEALING WATER LEVEL IN ORDER TO
VAPORIZE CONDENSED HYDROCARBONS

STEAM INJECTION
TO REDUCE SMOKE
IS OPTIONAL

FLARE HEADER

LG

CAR SEAL OPEN

R.O.

SEALING
WATER LINE
(MIN., 20 GPM)

R.O.

STEAM
FOR HEATING
WATER TO PREVENT
FREEZING

FOR CORROSION
RESISTANCE BOTTOM
OF STACK AND INLET
LINE TO BE GUNITE
LINED

CAR SEAL
OPEN

12 IN.

6 IN.

VENT SECONDARY
SEAL LEG

12 IN. NORMAL
SEALING
WATER
LEVEL

CAR SEAL OPEN

TO SEWER

LINE SIZE VALVE AND BLANK
TO BE USED FOR DRAINING
AND IN CONJUNCTION WITH
STEAM TO BLOW SCALE AND
SEDIMENT FROM BASE OF STACK

10 FT

VENT PRIMARY
SEAL LEG

175 PERCENT
MAXIMUM
DESIGN PRESSURE
OR 10 FT MINIMUM

TO SEWER

TIE IN LOCATION
TO BE NEAR BASE
OF SEAL LEG

Notes:
1. Elevation of flare header should be 10 ft above water seal.
2. Slope flare header toward blowdown drum.
3. Insulate steam line to top of flare.
4. Provide handhole at bottom of stack for cleanout.

Figure 6.18. A skimming seal leg arrangement of a pipe seal, more complex than the one shown in Figure 6.17, allows the removal of light condensates from the seal water. (Courtesy of the American Petroleum Institute)

water level) equivalent to a 10-foot volume of the line. The depth of the water seal should not exceed 12 inches since greater depths may cause gas pulsations. Seal water level is maintained by a continuous water flow of about 20 gpm. Overflow from the nonskimming seal is taken off the base of the seal through a seal leg whose height is equivalent to about 175% of the pressure at the base of the stack during maximum blowing conditions. In the skimming seal, the seal leg is taken from the top of both sides of the seal. Overflow from the stack side of the seal is tied into the base of the seal leg, though, to prevent gas bypassing through the skimming overflow connections.

Appendix*

The American Society of Mechanical Engineers set up a committee in 1911 for the purpose of formulating standard rules for the construction of steam boilers and other pressure vessels. This committee is now called the Boiler and Pressure Vessel Committee.

The Committee's function is to establish rules of safety governing the design, the fabrication, and the inspection during construction of boilers and unfired pressure vessels, and to interpret these rules when questions arise regarding their intent. In formulating the rules, the Committee considers the needs of users, manufacturers, and inspectors of pressure vessels. The objective of the rules is to afford reasonably certain protection of life and property- and to provide a margin for deterioration in service so as to give a reasonably long safe period of usefulness. Advancements in design and material, and the evidence of experience, have been recognized.

The Boiler and Pressure Vessel Committee deals with the care and inspection of boilers and pressure vessels in service only to the extent of providing suggested rules of good practice as an aid to owners and their inspectors.

The rules established by the Committee are not to be interpreted as approving, recommending, or endorsing any proprietary or specific design or as limiting in any way the manufacturer's freedom to choose any method of design or any form of construction that conforms to the Code rules.

The Boiler and Pressure Vessel Committee meets regularly to consider requests for interpretations and revisions of the rules. Inquiries must be in writing and must give full particulars in order to receive consideration. Requests for interpretations which are of a routine nature may be executed by the Secretary of the Boiler and Pressure Vessel Committee without reference to a subcommittee. All other requests are first referred to the proper subcommittee for consideration and for recommendation of action by the Main Committee. The action of the Main Committee becomes effective only after confirmation by letter ballot of the Committee and approval by the Council of the Society.

Interpretations of general interest are published in *Mechanical Engineering* as Code Cases, and inquiries are advised of the action taken. Code revisions approved by the Committee are published in *Mechanical Engineering* as proposed addenda to the Code to invite comments from all interested persons. After final approval by the Committee and adoption by the ASME Council, they are printed in the addenda supplements to the Code.

Code Cases (interpretations) may be used in the construction of vessels to be stamped with the ASME Code symbol beginning with the date of their approval by the ASME Council.

After Code revisions are approved by council they may be used beginning with the date of issuance shown on the addenda. Revisions become mandatory as minimum requirements six months after such date of issuance, except for boilers or pressure vessels contracted for prior to the end of the six-month period.

Manufacturers and users of pressure vessels are cautioned against making use of revisions and Cases that are less restrictive than former requirements without having assurance that they have been accepted by the proper authorities in the jurisdiction where the vessel is to be installed.

Each state and municipality in the United States and each province in the Dominion of Canada that adopts or accepts one or more Sections of the Boiler and Pressure Vessel Code is invited to appoint a representative to act on the Conference Committee to the Boiler and Pressure Vessel Committee. Since the members of the Conference Committee are in active contact with the administration and enforcement of the rules, the requirements for inspection in this Code correspond with those in effect in their respective jurisdictions. The required qualifications for an authorized inspector under these rules may be obtained from the administrative authority of any state, municipality, or province which has adopted these rules.

The Boiler and Pressure Vessel Committee in the formulation of its rules and in the establishment of maximum design and operating pressures considers

---

*The Appendix consists of excerpts from the ASME Boiler and Pressure Code Sections I and VIII by permission of the ASME.

materials, construction, method of fabrication, inspection, and safety devices. Permission may be granted to regulatory bodies and organizations publishing safety standards to use a complete Section of the Code by reference. If usage of a Section, such as Section I, involves exceptions, omissions, or changes in provisions, the intent of the Code might not be attained.

Where a state or other regulatory body, in the printing of any Section of the ASME Boiler and Pressure Vessel Code, makes additions or omissions, it is recommended that such changes be clearly indicated.

The National Board of Boiler and Pressure Vessel Inspectors is composed of chief inspectors of states and municipalities in the United States and of provinces in the Dominion of Canada that have adopted the Boiler and Pressure Vessel Code. This Board, since its organization in 1919, has functioned to uniformly administer and enforce the rules of the Boiler and Pressure Vessel Code. The cooperation of that organization with the Boiler and Pressure Vessel Committee has been extremely helpful. Its function is clearly recognized and, as a result, inquiries received which bear on the administration or application of the rules are referred directly to the National Board. Such handling of this type of inquiry not only simplifies the work of the Boiler and Pressure Vessel Committee, but action on the problem for the inquirer is thereby expedited. Where an inquiry is not clearly an interpretation of the rules, nor a problem of application or administration, it may be considered both by the Boiler and Pressure Vessel Committee and the National Board.

It should be pointed out that the state or municipality where the Boiler and Pressure Vessel Code has been made effective has definite jurisdiction over any particular installation. Inquiries dealing with problems of local character should be directed to the proper authority of such state or municipality. Such authority may, if there is any question or doubt as to the proper interpretation, refer the question to the Boiler and Pressure Vessel Committee.

The specifications for materials given in Section II of the Code are identical with or similar to those of the American Society for Testing and Materials as indicated, except in those cases where that organization has no corresponding specification.

SAFETY VALVES AND SAFETY RELIEF VALVES
PG-67[1] BOILER SAFETY VALVE REQUIREMENTS

67.1 Each boiler shall have a least one safety valve or safety relief valve and if it has more than 500 sq ft of water-heating surface or if an electric boiler has a power input more than 500 kw it shall have two or more safety valves or safety relief valves. The method of computing the steam-generating capacity of the boiler shall be as given in A-12.

67.2 The safety valve or safety relief valve capacity for each boiler (except as noted in PG-67.4) shall be such that the safety valve, or valves will discharge all the steam that can be generated by the boiler without allowing the pressure to rise more than 6 percent above the highest pressure at which any valve is set and in no case to more than 6 percent above the maximum allowable working pressure. The safety valve or safety relief valve capacity shall be in compliance with PG-70 but shall not be less than the maximum designed steaming capacity as determined by the manufacturer. The required steam relieving capacity, in pounds per hour, of the safety relief valves on a high-temperature water boiler shall be determined by dividing the maximum output in Btu at the boiler nozzle obtained by the firing of any fuel for which the unit is designed by 1000.

Any economizer which may be shut off from the boiler, thereby permitting the economizer to become a fired pressure vessel, shall have one or more safety relief valves with a total discharge capacity, calculated from the maximum expected heat absorption in Btu/hr, as determined by the manufacturer, divided by 1000. This absorption shall be stated in the stamping (PG-106.4).

The required relieving capacity in lb/hr of the safety or safety relief valves on a waste heat boiler shall be determined by the manufacturer. When auxiliary firing is to be used in combination with waste heat recovery, the maximum output shall include the effect of such firing in the total required capacity. When auxiliary firing is to be used in place of waste heat recovery, the required relieving capacity shall be based on auxiliary firing or waste heat recovery, whichever is the higher.

67.3 One or more safety valves on the boiler proper shall be set at or below the maximum allowable working pressure (*except as noted in PG-*

---

[1]Safety Valve: An automatic pressure relieving device actuated by the static pressure upstream of the valve and characterized by full-opening pop action. It is used for gas or vapor service.

Relief Valve: An automatic pressure relieving device actuated by the static pressure upstream of the valve which opens further with the increase in pressure over the opening pressure. It is used primarily for liquid service.

Safety Relief Valve: An automatic pressure-actuated relieving device suitable for use either as a safety valve or relief valve, depending on application.

67.4). If additional valves are used the highest pressure setting shall not exceed the maximum allowable working pressure by more than 3 percent. The complete range of pressure settings of all the saturated steam safety valves on a boiler shall not exceed 10 percent of the highest pressure to which any valve is set. Pressure setting of safety relief valves on high-temperature water boilers[1] may exceed this 10 percent range.

67.4 For a forced-flow steam generator with no fixed steam and water line, equipped with automatic controls and protective interlocks responsive to steam pressure, safety valves may be provided in accordance with the above paragraphs or the following protection against overpressure shall be provided:

67.4.1 One or more power-actuated pressure-relieving valves[2] shall be provided in direct communication with the boiler when the boiler is under pressure and shall receive a control impulse to open when the maximum allowable working pressure at the superheater outlet, as shown in the master stamping (PG-106.3), is exceeded. The total combined relieving capacity of the power-actuated relieving valves shall be not less than 10 percent of the maximum design steaming capacity of the boiler under any operating condition as determined by the manufacturer. The valve or valves shall be located in the pressure part system where they will relieve the overpressure.

An isolating stop valve of the outside-screw-and-yoke type may be installed between the power-actuated pressure relieving valve and the boiler to permit repairs provided an alternate power-actuated pressure relieving valve of the same capacity is so installed as to be in direct communication with the boiler in accordance with the requirements of this paragraph.

Power-actuated pressure relieving valves discharging to intermediate pressure and incorporated into bypass and/or startup circuits by the boiler manufacturer need not be capacity certified. Instead, they shall be marked by the valve manufacturer with a capacity rating at a set of specified inlet pressure and temperature conditions. Power-actuated pressure relieving valves discharging directly to atmosphere shall be capacity certified. This capacity certification shall be conducted in accordance with the provisions of PG-69.1.2.

67.4.2 Spring loaded safety valves shall be provided, having a total combined relieving capacity, including that of the power-actuated pressure relieving capacity installed under PG-67.4.1, of not less than 100 percent of the maximum designed steaming capacity of the boiler, as determined by the manufacturer, except when the alternate provisions of PG-67.4.3 are satisfied. In this total no credit in excess of 30 percent of the total required relieving capacity shall be allowed for the power-actuated pressure relieving valves actually installed. Any or all of the spring loaded safety valves may be set above the maximum allowable working pressure of the parts to which they are connected, but the set pressures shall be such that when all of these valves (together with the power-actuated pressure relieving valves) are in operation the pressure will not rise more than 20 percent above the maximum allowable working pressure of any part of the boiler, except for the steam piping between the boiler and the prime mover.

67.4.3 The total installed capacity of spring loaded safety valves may be less than the requirements of PG-67.4.2 provided all of the following conditions are met:

67.4.3.1 The boiler shall be of no less steaming capacity than 1,000,000 pounds per hour and installed in a unit system for power generation (i.e., a single boiler supplying a single turbine-generator unit).

67.4.3.2 The boiler shall be provided with automatic devices, responsive to variations in steam pressure, which include no less than all the following:

67.4.3.2.1 A control capable of maintaining steam pressure at the desired operating level and of modulating firing rates and feedwater flow in proportion to a variable steam output; and

67.4.3.2.2 A control which overrides PG-67.4.3.2.1 by reducing the fuel rate and feedwater flow when the steam pressure exceeds the maximum allowable working pressure as shown in the master stamping PG-106.3 by 10 percent; and

67.4.3.2.3 A direct-acting overpressure-trip-actuating mechanism, using an independent pressure sensing device, that will stop the flow of fuel and feedwater to the boiler, at a pressure higher than

---

[1]Safety relief valves in hot water service are more susceptible to damage and subsequent leakage, than safety valves relieving steam. It is recommended that the maximum allowable working pressure of the boiler and the safety relief valve setting for high-temperature water boilers be selected substantially higher than the desired operating pressure so as to minimize the times the safety relief valve must lift.

[2]The power-actuated pressure relieving valve is one whose movements to open or close are fully controlled by a source of power (electricity, air, steam or hydraulic). The valve may discharge to atmosphere or to a container at lower pressure. The discharge capacity may be affected by the downstream conditions, and such effects shall be taken into account. If the power-actuated pressure relieving valves are also positioned in response to other control signals, the control impulse to prevent overpressure shall be responsive only to pressure and shall override any other control function.

the set pressure of PG-67.4.3.2.2, but less than 20 percent above the maximum allowable working pressure as shown in the master stamping PG-106.3.

67.4.3.3 There shall be no less than two (2) spring-loaded safety valves and the total rated relieving capacity of the spring loaded safety valves shall be no less than 10 percent of the maximum designed steaming capacity of the boiler as determined by the manufacturer. These spring loaded safety valves may be set above the maximum allowable working pressure of the parts to which they are connected but shall be set such that the valves will lift at a pressure no higher than 20 percent above the maximum allowable working pressure as shown in the master stamping PG-106.3.

67.4.3.4 At least two (2) of these spring loaded safety valves shall be equipped with a device that directly transmits the valve stem lift action to controls that will stop the flow of fuel and feedwater to the boiler. The control circuitry to accomplish this shall be arranged in a "fail-safe" manner (see Note).

67.4.3.5 The power supply for all controls and devices required by PG-67.4.3 shall include at least one source contained within the same plant as the boiler and which is arranged to actuate the controls and devices continuously in the event of failure or interruption of any other power sources.

67.4.4 When stop valves are installed in the water-steam flow path between any two sections of a forced-flow steam generator with no fixed steam and water line:

67.4.4.1 The power-actuated pressure relieving valve(s) required by PG-67.4.1 shall also receive a control impulse to open when the maximum allowable working pressure of the component, having the lowest pressure level upstream to the stop valve, is exceeded; and

67.4.4.2 The spring loaded safety valves shall be located to provide the pressure protection requirements in PG-67.4.2 or 67.4.3.

Note: "Fail-safe" shall mean a circuitry arranged as either of the following:

1. *Energize to trip*: There shall be at least two separate and independent trip circuits served by two power sources, to initiate and perform the trip action. One power source shall be a continuously charged DC battery. The second source shall be an AC-to-DC converter connected to the DC system to charge the battery and capable of performing the trip action. The trip circuits shall be continuously monitored for availability.

It is not mandatory to duplicate the mechanism that actually stops the flow of fuel and feedwater.

2. *De-energize to trip*: If the circuits are arranged in such a way that a continuous supply of power is required to keep the circuits closed and operating and such that any interruption of power supply will actuate the trip mechanism, then a single trip circuit and single power supply will be enough to meet the requirements of this subparagraph.

67.4.5 A reliable pressure-recording device shall always be in service and records kept to provide evidence of conformity to the above requirements.

67.5 All safety valves or safety relief valves shall be so constructed that the failure of any part cannot obstruct the free and full discharge of steam and water from the valve. Safety valves shall be of the direct spring loaded pop type, with seat inclined at any angle between 45 and 90 deg, inclusive, to the centerline of the spindle. The coefficient of discharge of safety valves shall be determined by actual steam flow measurements at a pressure not more than 3 percent above the pressure at which the valve is set to blow and when adjusted for blowdown in accordance with PG-72. The valves shall be credited with capacities as determined by the provisions of PG-69.2.

Safety valves or safety relief valves may be used which give any opening up to the full discharge capacity of the area of the opening of the inlet of the valve (see PG-69.5), provided the movement of the steam safety valve is such as not to induce lifting of water in the boiler.

Deadweight or weighted lever safety valves or safety relief valves shall not be used.

For high-temperature water boilers safety relief valves shall be used. Such valves shall have a closed bonnet. For purposes of selection the capacity rating of such safety relief valves shall be expressed in terms of actual steam flow determined on the same basis as for safety valves. In addition the safety relief valves shall be capable of satisfactory operation when relieving water at the saturation temperature corresponding to the pressure at which the valve is set to blow.

67.6 A safety valve or safety relief valve over 3 in. in size, used for pressures greater than 15 psi gage, shall have a flanged inlet connection or a weld-end inlet connection. The dimensions of flanges subjected to boiler pressure shall conform to the applicable American National Standards as given in PG-42. The facing shall be similar to those illustrated in the Standard.

67.7 Safety valves or safety relief valves may have bronze parts complying with either Specifications SB-61 or SB-62, provided the maximum allowable working stresses and temperatures do not exceed the values given in Table PG-23.2 and shall be marked to indicate the class of material used. Such valves shall not be used on superheaters delivering steam at a temperature over 450 F and 406 F respectively, and shall not be used for high-temperature water boilers.

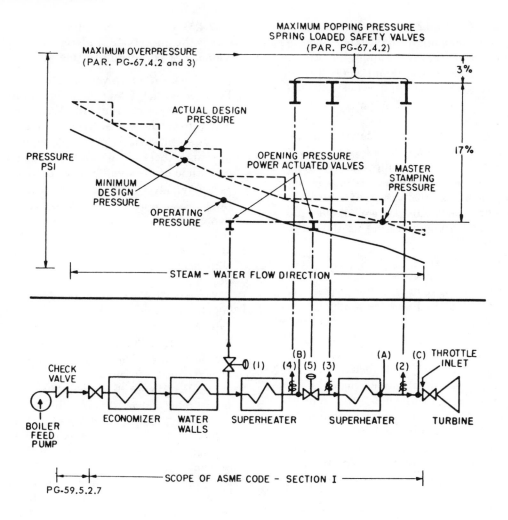

PG-59.5.2.7

FIG. PG-67.4 REQUIREMENTS FOR PRESSURE RELIEF FORCED-FLOW STEAM GENERATOR

PRESSURE
(A) = Master Stamping — (PG-106.3)
(B) = Component design at inlet to stop valve (5) — (PG-67.4.4.1)
(C) = Turbine throttle inlet — (PG-58.1.2.4)

PRESSURE RELIEF VALVES
(1) = Power actuated — (PG-67.4.1)
(2), (3) and (4) = Spring loaded safety — (PG-67.4.2)
(5) = Superheater stop — (PG-67.4.4)

RELIEF VALVE FLOW CAPACITY — (minimum, based on rated capacity of boiler)
(1) = 10 − 30% (PG-67.4.1)
(2) = minimum of one valve — (PG-68.1)
(2) + (3) when downstream to stop valve (5) = 6#/sq ft superheater surface — (PG-68.3)
(2) + (3) + (4) = 100% minus (1) — (PG-67.4.2)

RELIEF VALVE OPENING PRESSURE — (maximum)
(1) = (A), and (B) when there is stop valve (5) — (PG-67.4.1)
(2), (3) and (4) = (A) + 17% — (PG-67.4.2)
(5) = (A) — (PG-67.4.1)

"Alternate" Requirements for Safety Valves

RELIEF VALVE FLOW CAPACITY — (minimum, based on rated capacity of boiler)
(1) = 10 − 30% — (PG-67.4.1)
(2) = one valve minimum — (PG-68.1)
(2) + (3) when downstream to stop valve (5) = 6#/sq ft superheater surface — (PG-68.3)
(4) = 10% total with minimum of 2 valves when there is a stop valve (5) — (PG-67.5.3.3)
(2) + (4) = 10% total with minimum of 2 valves when there is no stop valve (5) — (PG-67.4.3.3)

RELIEF VALVE OPENING PRESSURE — (maximum)
(1) = (A), and (B) when there is stop valve (5) — (PG-67.4.1)
(2), (3) and (4) = (A) + 20% — (PG-67.4.3.3)
(5) = (A) — (PG-67.4.1)

AUTOMATIC PRESSURE CONTROLS — (PG-67.4.3)
1) At (C) for normal operation under load — (PG-67.4.3.2.1)
2) At (A) + 10% to override control (I) — (PG-67.4.3.2.2)
3) At (A) + 20% to shut off flow of fuel and feedwater — (PG-67.4.3.2.3)
4) Safety valves at (4) to shut off flow of fuel and feedwater by "fail-safe" power circuit — (PG-67.4.3.4)

PG-68   SUPERHEATER   SAFETY   VALVE REQUIREMENTS

68.1 Every attached superheater shall have one or more safety valves near the outlet. If the superheater outlet header has a full, free, steam passage from end to end and is so constructed that steam is supplied to it at practically equal intervals throughout its length so that there is a uniform flow of steam through the superheater tubes and the header, the safety valve, or valves, may be located anywhere in the length of the header.

68.2 The discharge capacity of the safety valve, or valves, on an attached superheater may be included in determining the number and size of the safety valves for the boiler, provided there are no intervening valves between the superheater safety valve and the boiler, and provided the discharge capacity of the safety valve, or valves, on the boiler, as distinct from the superheater is at least 75 percent of the aggregate valve capacity required.

68.3 Every independently fired superheater which may be shut off from the boiler and permit the superheater to become a fired pressure vessel shall have one or more safety valves having a discharge capacity equal to 6 lb of steam per square foot of superheater surface measured on the side exposed to the hot gases. The number of safety valves installed shall be such that the total capacity is at least equal to that required.

68.4 Every reheater shall have one or more safety valves, such that the total relieving capacity is at least equal to the maximum steam flow for which the reheater is designed. At least one valve shall be located on the reheater outlet. The relieving capacity of the valve on the reheater outlet shall be not less than 15 percent of the required total. The capacity of reheater safety valves shall not be included in the required relieving capacity for the boiler and superheater.

68.5 A soot blower connection may be attached to the same outlet from the superheater or reheater that is used for the safety valve connection.

68.6 Every safety valve used on a superheater or reheater discharging superheated steam at a temperature over 450 F shall have a casing, including the base, body, and bonnet and spindle, of steel, steel alloy, or equivalent heat resisting material.

The valve shall have a flanged inlet connection, or a weld-end inlet connection. It shall have the seat and disk of suitable heat erosive and corrosive-resisting material, and the spring fully exposed outside of the valve casing so that it shall be protected from contact with the escaping steam.

PG-69   TESTING

69.1.1 The manufacturer of safety valves that are to be stamped with the Code symbol shall first submit at least three valves of each of three representative sizes of each design and for three different pressures for testing, together with drawings, at a place where adequate equipment and personnel are available to conduct pressure and relieving capacity tests which shall be made in the presence of an authorized observer. The place, personnel, and the authorized observer shall be approved by the Boiler and Pressure Vessel Committee. Laboratory approval is subject to review within each five year period.

69.1.2 Tests shall be made to determine the lift, popping, and blowdown pressures and capacities at three different pressures for each of three representative sizes of each design. A coefficient shall be established for each test as follows:

$$K_D = \frac{\text{Actual steam flow}}{\text{Theoretical steam flow}} = \text{coefficient of discharge}$$

The average coefficient, K, of the nine tests required shall be taken as the coefficient of the design and shall be used for determining the relieving capacity of all sizes and pressures of the design in the following formula:

For 45 deg seat

$$W = (51.45 \times \pi DLP \times K)\, 0.90$$

For flat seat

$$W = (51.45 \times \pi DLP \times K)\, 0.90$$

For nozzle

$$W = (51.45 \times AP \times K)\, 0.90$$

where

W = weight of steam per hour, pounds
D = seat diameter, inches
L = lift, inches at 103 percent of set pressure
$P = (1.03 \times \text{set pressure}) + 14.7 =$ absolute pressure, pounds per square inch
K = average coefficient of discharge
A = nozzle throat area, square inches

The individual coefficient K_D as determined by nine tests shall fall within the range of plus or minus 5 percent of the average coefficient K.

69.2 The tests shall be made with steam and in a manner closely approximating actual operating conditions of steam boilers. The relieving capacity shall be measured by condensing the steam or with a calibrated steam flowmeter.

69.3 If a manufacturer wishes to apply the Code symbol to a power-actuated pressure relieving valve as applied under PG-67.4.1, four valves of each combination of inlet pipe size and maximum orifice size to be used with that inlet pipe size shall be tested. These four valves shall be tested at pressures approximately covering the range of the certified test facility on which the tests are conducted. The capacities, as determined by these four tests, shall be plotted against the absolute flow test pressure and a line drawn through these four test points. All points must lie within plus or minus five percent in capacity value of the plotted line and must pass through 0-0. From the plotted line, the slope of the line, $\dfrac{dw}{dP}$ shall be determined and a factor of $\dfrac{.90}{51.45} \times \dfrac{dw}{dP}$ shall be applied to capacity computations in the supercritical region at elevated pressures by means of the isentropic flow equation.

$W=$ capacity, pounds of steam per hour
$P=$ absolute inlet pressure, psia
$v=$ inlet specific volume, cubic feet per pound
$\dfrac{dw}{dP}=$ rate of change of measured capacity with respect to absolute pressure

$$W = 1135.8 \frac{.90}{51.45} \times \frac{dw}{dP} \sqrt{\frac{P}{v}}$$

Note: The constant 1135.8 is based on a γ factor of 1.30 which is accurate for superheated steam at temperature above approximately 800 F. In the interest of accuracy, other methods of capacity computations must be used at temperatures below 800 F at supercritical pressures.

Capacity of power-actuated pressure relieving valves determined as specified in this paragraph shall be computed at a pressure 3 percent above the set pressure of the valves.

69.4 The relieving capacity that may be stamped on the safety valves or the safety relief valves shall not exceed 90 percent of the value determined by the witnessed tests.

69.5 When the valve casing is marked as required by this paragraph, it shall be the guarantee by the manufacturer that the valve also conforms to the details of construction herein specified.

PG-70 CAPACITY

70.1 The minimum safety valve or safety relief valve relieving capacity for other than electric boilers, waste heat boilers and forced-flow steam generators with no fixed steam and water line when provided in accordance with PG-67.4.3, shall be determined on the basis of the pounds of steam generated per hour per square foot of boiler heating surface and

waterwall heating surface, as given in the following table:

TABLE PG-70
MINIMUM POUNDS OF STEAM PER HR PER SQ FT OF SURFACE

	Fire-tube Boilers	Water-tube Boilers
Boiler heating surface		
Hand fired	5	6
Stoker fired	7	8
Oil, gas, or pulverized fuel fired	8	10
Waterwall heating surface		
Hand fired	8	8
Stoker fired	10	12
Oil, gas, or pulverized fuel fired	14	16

Note: When a boiler is fired only by a gas having a heat value not in excess of 200 Btu per cu ft, the minimum safety-valve or safety relief valve relieving capacity may be based on the values given for hand-fired boilers above.

The minimum safety valve or safety relief valve relieving capacity for electric boilers shall be $3\frac{1}{2}$ lb per hr per kw input.

In many cases a greater relieving capacity of safety valves or safety relief valves will have to be provided than the minimum specified by this rule, and in every case the requirements of PG-67.2 shall be met.

70.2 The heating surface shall be computed as follows:

70.2.1 Heating surface, as part of a circulating system in contact on one side with water or wet steam being heated and on the other side with gas or refractory being cooled, shall be measured on the side receiving heat.

70.2.2 Boiler heating surface and other equivalent surface outside the furnace shall be measured circumferentially plus any extended surface.

70.2.3 Waterwall heating surface and other equivalent surface within the furnace shall be measured as the projected tube area (diameter × length) plus any extended surface on the furnace side. In computing the heating surface for this purpose, only the tubes, fireboxes, shells, tube sheets, and the projected area of headers need be considered, except that for vertical firetube steam boilers, only that portion of the tube surface up to the middle gage cock is to be computed. The minimum number and size of safety valves or safety relief valves required shall be determined on the basis of the aggregate relieving capacity and the relieving capacity marked on the valves by the manufacturer. Where the operating conditions are changed, or additional heating surface such as water screens or waterwalls is connected to the boiler circulation, the safety valve or safety relief valve capacity shall be increased, if necessary, to meet the new conditions and be in accordance with PG-67.2. The additional valves required on account of changed conditions may be installed on the steam or water line between the boiler and the main stop valve except when the boiler is equipped with a superheater or other piece of apparatus, in which case they may be installed on the steam pipes between the boiler drum and the inlet to the superheater or other apparatus, provided that the steam main between the boiler and points where a safety valve or valves may be attached has a cross-sectional area at least 3 times the combined areas of the inlet connections to the safety valves applied to it.

70.3 If the safety valve or safety relief valve capacity cannot be computed or if it is desirable to prove the computations, it may be checked in any one of the three following ways, and if found insufficient, additional capacity shall be provided.

70.3.1 By making an accumulation test, that is, by shutting off all other steam discharge outlets from the boiler and forcing the fires to the maximum. The safety valve equipment shall be sufficient to prevent an excess pressure beyond that specified in PG-67.2. This method should not be used on a boiler with a superheater or reheater or on a high temperature water boiler.

70.3.2 By measuring the maximum amount of fuel that can be burned and computing the corresponding evaporative capacity upon the basis of the heating value of the fuel (see A-12 through A-17).

70.3.3 By determining the maximum evaporative capacity by measuring the feedwater. The sum of the safety valve capacities marked on the valves shall be equal to or greater than the maximum evaporative capacity of the boiler. This method shall not be used on high-temperature water boilers.

PG-71   MOUNTING

71.1 When two or more safety valves are used on a boiler, they may be mounted either separately or as twin valves made by placing individual valves on Y-bases, or duplex valves having two valves in the same body casing. Twin valves made by placing individual valves on Y-bases, or duplex valves having two valves in the same body, shall be of approximately equal capacity.

When not more than two valves of different sizes are mounted singly the relieving capacity of the smaller valve shall be not less than 50 percent of that of the larger valve.

71.2 The safety valve or safety relief valve or valves shall be connected to the boiler independent of any other connection, and attached as close as possible to the boiler, without any unnecessary intervening pipe or fitting. Such intervening pipe or fitting shall be not longer than the face-to-face dimension of the corresponding tee fitting of the same diameter and pressure under the applicable American National Standard listed in PG-42 and shall also comply with PG-8 and PG-39. Every safety valve or safety relief valve shall be connected so as to stand in an upright position, with spindle vertical. On high-temperature water boilers of the watertube forced-circulation type, the valve shall be located at the boiler outlet.

71.3 The opening or connection between the boiler and the safety valve or safety relief valve shall have at least the area of the valve inlet. No valve of any description shall be placed between the required safety valve or safety relief valve or valves and the boiler, nor on the discharge pipe between the safety valve or safety relief valve and the atmosphere. When a discharge pipe is used, the cross-sectional area shall be not less than the full area of the valve outlet or of the total of the areas of the valve outlets, discharging thereinto. It shall be as short and straight as possible and so arranged as to avoid undue stresses on the valve or valves.

All safety valve or safety relief valve discharges shall be so located or piped as to be carried clear from running boards or platforms. Ample provision for gravity drain shall be made in the discharge pipe at or near each safety valve or safety relief valve, and where water of condensation may collect. Each valve shall have an open gravity drain through the casing below the level of the valve seat. For iron-and-steel-bodied valves exceeding 2-in. size, the drain hole shall be tapped not less than $^3/_8$-in. pipe size.

Discharge piping from safety relief valves on high-temperature water boilers shall be provided with adequate provisions for water drainage as well as the steam venting.

The installation of cast iron bodied safety relief valves for high-temperature water boilers is prohibited.

71.4 If a muffler is used on a safety valve or safety relief valve, it shall have sufficient outlet area to prevent back pressure from interfering with the proper operation and discharge capacity of the valve. The muffler plates or other devices shall be so constructed as to avoid a possibility of restriction of the steam passages due to deposit. Mufflers shall not be used on high-temperature water boiler safety relief valves.

When a safety valve or safety relief valve is exposed to outdoor elements which may affect operation of the valve, it is permissible to shield the valve with a satisfactory cover. The shield or cover shall be properly vented and arranged to permit servicing and normal operation of the valve.

71.5 When a boiler is fitted with two or more safety valves or safety relief valves on one connection, this connection to the boiler shall have a cross-sectional area not less than the combined areas of inlet connections of all the safety valves or safety relief valves with which it connects and shall also meet the requirements of PG-71.3.

71.6 Safety valves may be attached to drums or headers by welding provided the welding is done in accordance with Code requirements.

71.7 Every boiler shall have proper outlet connections for the required safety valve, or safety relief valve, or valves, independent of any other outside steam connection, the area of opening to be at least equal to the aggregate areas of inlet connections of all of the safety valves or safety relief valves to be attached thereto. An internal collecting pipe, splash plate, or pan may be used, provided the total area for inlet of steam thereto is not less than twice the aggregate areas of the inlet connections of the attached safety valves. The holes in such collecting pipes shall be at least $1/4$ in. in diameter and the least dimension in any other form of opening for inlet of steam shall be $1/4$ in.

Such dimensional limitations to operation for steam need not apply to steam scrubbers or driers provided the net free steam inlet area of the scrubber or drier is at least 10 times the total area of the boiler outlets for the safety valves.

71.8 If safety valves are attached to a separate steam drum or dome, the opening between the boiler proper and the steam drum or dome shall be not less than required by PG-71.7.

PG-72 OPERATION

72.1 Safety valves shall be designed and constructed to operate without chattering and to attain full lift at a pressure no greater than 3 percent above their set pressure. After blowing down, all valves shall close at a pressure not lower than 96

percent of their set pressure, except that all drum valves installed on a single boiler may be set to reseat at a pressure not lower than 96 percent of the set pressure of the lowest set drum valve. The minimum blowdown in any case shall be 2 psi. For spring loaded pop safety valves for pressures between 100 and 300 psi, both inclusive, the blowdown shall be not less than 1 percent of the set pressure. To insure the guaranteed capacity and satisfactory operation, the blowdown as marked upon the valve (PG-69.5) shall not be reduced.

Safety valves used on forced-flow steam generators with no fixed steam and water line, and safety relief valves used on high-temperature water boilers may be set and adjusted to close after blowing down not more than 10 percent of the set pressure. The valves for these special uses must be so adjusted and marked by the manufacturer.

72.2 The blowdown adjustment shall be made and sealed by the manufacturer.

72.3 The popping point tolerance plus or minus shall not exceed the following: 2 psi for pressures up to and including 70 psi, 3 percent for pressures from 71 to 300 psi, 10 psi for pressures over 301 psi to 1000 psi, and 1 percent for pressures over 1000 psi.

72.4 To insure the valve being free, each safety valve or safety relief valve shall have a substantial lifting device by which the valve disk may be positively lifted from its seat when there is at least 75 percent of full working pressure on the boiler. The lifting device shall be such that it cannot lock or hold the valve disk in listed position when the exterior lifting force is released.

72.4.1 Safety relief valve disks used on high-temperature water boilers shall not be lifted while the temperature of the water exceeds 200 F. If it is desired to lift the valve disk to assure that it is free, this shall be done when there is at least 75 percent of full working pressure on the boiler. For high-temperature water boilers, the lifting mechanism shall be sealed against leakage.

72.5 The seats and disks of safety valves or safety relief valves shall be of suitable material to resist corrosion. The seat of a safety valve shall be fastened to the body of the valve in such a way that there is no possibility of the seat lifting.

72.6 Springs used in safety valves shall not show a permanent set exceeding 1 percent of their free length 10 min after being released from a cold compression test closing the spring solid.

72.7.1 The spring in a safety valve or safety relief valve in service for pressures up to and including 250 psi shall not be used for any pressure more than

10 percent above or 10 percent below that for which the safety valve or safety relief valve is marked. For higher pressures the spring shall not be reset for any pressure more than 5 percent above or 5 percent below that for which the safety valve or safety relief valve is marked.

72.7.2 If the operating conditions of a valve are changed so as to require a new spring under PG-72.7.1 for a different pressure, the valve shall be adjusted by the manufacturer or his authorized representative who shall furnish and install a new nameplate as required under Par. PG-110.

PRESSURE RELIEF DEVICES

UG-125 General [1]

(*a*) All pressure vessels within the scope of this Division of Section VIII, irrespective of size or pressure, shall be provided[2] with protective devices in accordance with the requirements of UG-125 through UG-134.

(*b*) An unfired steam boiler as defined in U-1(e) shall be equipped with protective devices required by Section 1 of the Code in so far as they are applicable to the service of the particular installation.

(*c*) All pressure vessels other than unfired steam boilers shall be protected by pressure-relieving devices that will prevent the pressure from rising more than 10 percent above the maximum allowable working pressure, except when the excess pressure is caused by exposure to fire or other unexpected source of heat.

(*d*) Where an additional hazard can be created by exposure of a pressure vessel to fire or other unexpected sources of external heat (for example, vessels used to store liquefied flammable gases), supplemental pressure relieving devices shall be installed to protect against excessive pressure. Such supplemental pressure relieving devices shall be capable of preventing the pressure from rising more than 20 percent above the maximum allowable working pressure of the vessel. A single pressure

relieving device may be used to satisfy the requirements of this paragraph and (c), provided it meets the requirements of both paragraphs.

(*e*) Pressure relieving devices shall be constructed, located, and installed so that they are readily accessible for inspection and repair and so that they cannot be readily rendered inoperative (see Appendix M), and should be selected on the basis of their intended service.

(*f*) Pressure indicating gages, if used, shall preferably be graduated to approximately double the operating pressure but in no case to less than 1.2 times the pressure at which the relieving device is set to function.

(*g*) Rupture disks may be used in lieu of safety valves on vessels containing substances that may render a safety valve inoperative, or where a loss of valuable material by leakage should be avoided, or contamination of the atmosphere by leakage of noxious gases must be avoided (see UG-127).

(*h*) Vessels that are to operate completely filled with liquid shall be equipped with liquid relief valves, unless otherwise protected against overpressure.

(*i*) The protective devices required in (a) need not be installed directly on a pressure vessel when the source of pressure is external to the vessel and is under such positive control that the pressure in the vessel cannot exceed the maximum allowable working pressure at the operating temperature except as permitted in (c) and (d) (see UG-98).

NOTE: Pressure reducing valves and similar mechanical or electrical control instruments, except for pilot operated valves as permitted in UG-126(b), are not considered as sufficiently positive in action to prevent excess pressures from being developed.

(*j*) Pressure relieving devices shall be constructed of materials suitable for the pressure, temperature, and other conditions of the service intended.

UG-126 Safety and Relief Valves

(*a*) Safety and relief valves shall be of the direct spring loaded type, except as permitted in (b).

(*b*) Pilot valve control or other indirect operation of safety valves is not permitted unless the design is such that the main unloading valve will open automatically at not over the set pressure and will discharge its full rated capacity if some essential part of the pilot or auxiliary device should fail.

(*c*) Safety and relief valves for steam or air service shall be provided with a substantial lifting device so that the disk can be lifted from its seat when the pressure in the vessel is 75 percent of that at which the valve is set to blow.

---

[1]Safety valve—An automatic pressure relieving device actuated by the static pressure upstream of the valve and characterized by full opening pop action. It is used for gas or vapor service.

Relief valve—An automatic pressure relieving device actuated by the static pressure upstream of the valve which opens further with the increase in pressure over the opening pressure. It is used primarily for liquid service.

Safety relief valve—An automatic pressure actuated relieving device suitable for use either as a safety valve or relief valve, depending on application.

[2]Safety devices need not be provided by the vessel manufacturer, but overpressure protection shall be provided prior to placing the vessel in service.

(*d*) Safety and relief valves for service other than steam and air need not be provided with a lifting device, although a lifting device is desirable if the vapors are such that their release will not create a hazard.

(*e*) If the design of a safety or relief valve is such that liquid can collect on the discharge side of the disk, the valve shall be equipped with a drain at the lowest point where liquid can collect (for installation, see UG-134).

(*f*) Seats or disks of cast iron shall not be used.

(*g*) The spring in a safety or relief valve in service for pressures up to and including 250 psi shall not be reset for any pressure more than 10 percent above or 10 percent below that for which the valve is marked. For higher pressures, the spring shall not be reset for any pressure more than 5 percent above or 5 percent below that for which the safety or relief valve is marked.

UG-127 Rupture Disks

(*a*) The cross-sectional area of the connection to a vessel shall be not less than the required relief area of the rupture disk.

(*b*) Every rupture disk shall have a specified bursting pressure at a specified temperature, shall be marked with a lot number, and shall be guaranteed by its manufacturer to burst within 5 percent (plus or minus) of its specified bursting pressure.

(*c*) The specified bursting pressure at the coincident operating temperature shall be determined by bursting two or more specimens from a lot of the same material and of the same size as those to be used. The tests shall be made in a holder of the same form and pressure area dimensions as that with which the disk is to be used.

(*d*) A rupture disk may be installed between a spring loaded safety or relief valve and the vessel provided:

(*1*) The valve is ample in capacity to meet the requirements of UG-132(a) and (b);

(*2*) The maximum pressure of the range for which the disk is designed to rupture does not exceed the maximum allowable working pressure of the vessel;

(*3*) The opening provided through the rupture disk, after breakage, is sufficient to permit a flow equal to the capacity of the attached valve and there is no chance of interference with the proper functioning of the valve; but in no case shall this area be less than the inlet area of the valve; and

(*4*) The space between a rupture disk and the valve should be provided with a pressure gage, try cock, free vent, or a suitable telltale indicator. This arrangement permits the detection of disk rupture or leakage.[1]

(*e*) A rupture disk may be installed on the outlet side[2] of a spring loaded safety or relief valve which is opened by a direct action of the pressure in the vessel provided:

(*1*) The valve is so constructed that it will not fail to open at its proper pressure setting regardless of any back pressure that can accumulate between the valve disk and the rupture disk;[3]

(*2*) The valve is ample in capacity to meet the requirements of UG-132(a) and (b);

(*3*) The disk is designed to rupture at not more than the maximum allowable working pressure of the vessel;

(*4*) The opening provided through the rupture disk, after breakage, is sufficient to permit a flow equal to the rated capacity of the attached safety or relief valve; but in no case shall this area be less than the inlet area of the safety or relief valve;

(*5*) Any piping beyond the rupture disk cannot be obstructed by the rupture disk or fragments;

(*6*) All valve parts and joints subject to stress due to the pressure from the vessel and all fittings up to the rupture disk are designed for not less than the maximum allowable working pressure of the vessel;

(*7*) Any small leakage or a larger flow through a break in the operating mechanism that may result in back pressure accumulation within enclosed spaces of the valve housing other than between the rupture disk and the discharge side of the safety or relief valve, so as to hinder the safety or relief valve from opening at its set pressure, will be relieved adequately and safely to atmosphere through telltale vent openings;

---

[1]Users are warned that a rupture disk will not burst at its design pressure if back pressure builds up in the space between the disk and the safety or relief valve which will occur should leakage develop in the rupture disk due to corrosion or other cause.

[2]This use of a rupture disk in series with the safety or relief valve is permitted to minimize the loss by leakage through the valve of valuable, or of noxious or otherwise hazardous, materials, and where a rupture disk alone or disk located on the inlet side of the safety valve is impractical.

[3]Users are warned that an ordinary spring loaded safety or relief valve will not open at its set pressure if back pressure builds up in the space between the valve and rupture disk. A specially designed safety or relief valve is required, such as a diaphragm valve or a valve equipped with a bellows above the disk.

(8) The contents of the vessel are clean fluids, free from gumming or clogging matter, so that accumulation in the space between the valve inlet and the rupture disk (or in any other outlet that may be provided) will not clog the outlet; and

(9) The services are such that atmospheric temperatures are not exceeded.

NOTE: When a rupture disk is installed on the outlet of a safety or relief valve fitted with a lifting device, a valved vent shall be located between the valve disk and the rupture disk to permit the checking of the operative conditions of the valve. The distance between the valve and the rupture disk shall be a practical minimum.

Users are warned that replacing a ruptured disk on the outlet of a safety or relief valve may be attended by some danger if done without first reducing the pressure in the vessel, particularly when noxious or otherwise hazardous contents might be discharged.

UG-128 Liquid Relief Valves

Any liquid relief valve used shall be at least $^1/_2$ in.-iron pipe size.

UG-129 Marking

(a) Safety and Relief Valves. Each safety and relief valve $^1/_2$-in. pipe size and larger shall be plainly marked by the manufacturer with the required data in such a way that the marking will not be obliterated in service. Smaller valves may be exempted from these marking requirements. The marking may be placed on the valve or on a plate or plates securely

 FIG. UG-129 OFFICIAL SYMBOL FOR STAMP TO DE-NOTE THE AMERICAN SOCIETY OF ME-CHANICAL ENGI-NEERS' STANDARD

fastened to the valve. The Code symbol shall be stamped on the valve or nameplate, but the other required data may be stamped, etched, impressed, or cast on the valve or nameplate. The marking shall include the following:

(1) The name or identifying trademark of the manufacture,

(2) Manufacturer's design or type number,

(3) Size_____in.,

(The pipe size of the valve inlet)

(4) Set Pressure_____psi,

(5) Capacity_____lb of saturated steam per hour, or_____cu ft of air (60 F and 14.7 psia) per min.

NOTE: In addition, the manufacturer may indicate the capacity in other fluids (see Appendix J).

(6) ASME symbol as shown in Fig. UG-129.

(b) Safety and relief valves certified for a steam

discharging capacity under the provisions of Section I of the Code and bearing the official Code symbol stamp of that Section of the Code for safety valves, may be used on pressure vessels. The rated capacity in terms of other fluids shall be determined by the method of conversion given in Appendix J. (See UG-131(h).)

(c) Liquid Relief Valves. Each liquid relief valve shall be marked with the following data:

(1) Name or identifying trademark of the manufacturer,

(2) Manufacturer's design or type number,

(3) Size_____in.,

(Pipe size of inlet)

(4) Set Pressure_____psi,

(5) Relieving capacity_____gal of water per min at 70 F.

(d) Rupture Disks. Every rupture disk shall be plainly marked by the manufacturer in such a way that the marking will not be obliterated in service. The marking may be placed on the flange of the disk or on a metal tab permanently attached thereto.[1] The marking shall include the following:

(1) The name or identifying trademark of the manufacturer,

(2) Manufacturer's design or type number and lot number,

(3) Size_____in.,

(4) Maximum bursting pressure_____psi,

(5) Minimum bursting pressure_____psi,

(6) Coincident disk temperature_____F,

(7) Capacity_____lb of saturated steam per hour, or_____cu ft of air (60 F and 14.7 psia) per min.

NOTE: In addition, the manufacturer may indicate the capacity in other fluids (see Appendix J).

UG-130 Use of Code Symbol Stamp

Each safety and relief valve to which the Code symbol is to be applied shall be fabricated by a manufacturer who is in possession of a Code symbol stamp (see Fig. UG-129) and a valid certificate of authorization obtainable under the following conditions:

(1) Permission to use the symbol referred to in UG-129(a)(6) will be granted by the Society pursuant to the provisions of this paragraph.

(2) Any manufacturer may apply to the Boiler and Pressure Vessel Committee of the Society upon forms issued by the Society for permission to use the

[1] The marking may be coded and identified with the marking given on the certificate supplied with each rupture disk.

stamp. Each applicant must agree that if permission to use the stamp is granted, it will be used according to the rules and regulations of this Division of Section VIII of the Code that any safety or relief valve to which the symbol is applied will have the certified capacity stamped upon the valve after tests as required in UG-131, and that any stamp will be promptly returned to the Society upon demand, or in case the applicant discontinues the manufacture of safety and relief valves to which the Code symbol is applied, or in case the certificate of authorization issued to such applicant has expired and no new certificate has been issued. The holder of any such stamp shall not permit any other manufacturer to use his stamp.

(3) Permission to use the stamp may be granted or withheld by the Society in its absolute discretion. If permission is given, and the proper administrative fee paid, a certificate of authorization evidencing permission to use such symbol, expiring on the triennial anniversary date thereafter, will be forwarded to the applicant. Each such certificate will be signed by the Chairman and Secretary, or other duly authorized officer or officers, of the Boiler and Pressure Vessel Committee. Six (6) months prior to the date of expiration of any such certificate, the applicant must apply for a renewal of such permission and the issuance of a new certificate.

(4) The Society reserves the absolute right to cancel or refuse to renew such permission, returning fees paid for the prorated unexpired term.

(5) The Boiler and Pressure Vessel Committee may at any time and from time to time make such regulations concerning the issuance and use of such stamps as it deems appropriate, and all such regulations shall become binding upon the holders of any valid certificates of authorization.

(6) All steel stamps for applying the symbol shall be obtained from the Society.

UG-131 Certification of Capacity of Safety and Relief Valves

(a) Before the symbol is applied to any safety or relief valve, the valve manufacturer shall have the capacity of his valves certified as prescribed in (d) or (e).

(b) Capacity certification tests shall be conducted on saturated steam, or air, or natural gas. When saturated steam is used, corrections for moisture content of the steam shall be made.

(c)(1) Capacity certification tests shall be conducted at a pressure not to exceed 110 percent of the pressure for which the safety or relief valve is set to operate. The reseating pressure shall be noted and recorded.

(2) Capacity certification of pilot operated safety and relief valves may be based on tests without the pilot valves installed, provided prior to capacity tests it has been demonstrated by test to the satisfaction of the official observer that the pilot valve will cause the main valve to fully open within 110 percent of the set pressure of the main valve and that the pilot valve in combination with the main valve will meet all the requirements of this Division.

(d)(1) A capacity certification test is required on a set of three valves for each combination of size, design, and pressure setting. The stamped capacity rating for each combination of design, size and test pressure shall be 90 percent of the average capacity of the three valves tested.

NOTE: The capacity of a set of three valves shall fall within a range of plus or minus 5 percent of the average capacity. Failure to meet this requirement shall be cause to refuse certification of that particular safety valve design.

(2) If a manufacturer wishes to apply the Code symbol to a design of safety or relief valves, four valves of each combination of pipe size and orifice size shall be tested. These four valves shall be set at pressures which will cover the approximate range of pressures for which the valves will be used. The capacities, as determined by these four tests, shall be plotted against the absolute flow test pressure and a curve drawn through these four points. If the four points do not establish a reasonable curve, the authorized observer shall require that additional valves be tested. From this curve, relieving capacities shall be obtained. The stamped capacity shall be 90 percent of the capacity taken from the curve.

(e) Instead of individual capacity certification as provided in (d), a coefficient of discharge K may be established for a specific safety valve design according to the following procedure:

(1) For each design the safety valve manufacturer shall submit for test at least three valves for each of three different sizes (a total of nine valves) together with detailed drawings showing the valve construction. Each valve of a given size shall be set at a different pressure.

(2) Tests shall be made on each safety or relief valve to determine its capacity-lift, popping and blow-down pressures, and actual capacity in terms of the fluid used in the test. Valves having an adjustable blowdown construction shall be adjusted prior to testing so that the blowdown does not

exceed 5 percent of the set pressure. A coefficient K_D shall be established for each test run as follows:

$$K_D = \frac{\text{Actual flow}}{\text{Theoretical flow}} = \text{coefficient of discharge}$$

where actual flow is determined quantitatively by test, and theoretical flow is calculated by the appropriate formula which follows:

For test with dry saturated steam

$$W_T = 51.5\ AP$$

For test with air

$$W_T = 356\ AP\sqrt{\frac{M}{T}}$$

For test with natural gas

$$W_T = CAP\sqrt{\frac{M}{ZT}}$$

where

W_T = theoretical flow, pounds per hour
A = actual discharge area through the valve at developed lift, square inches
P = (set pressure × 1.10) plus atmospheric pressure, pounds per square inch absolute
M = molecular weight
T = absolute temperature at inlet (degrees Fahrenheit plus 460)
C = constant for gas or vapor based on the ratio of specific heats
$= \frac{C_p}{C_v}$ (see Fig. UA-230)
Z = compressibility factor corresponding to P and T

The average of the coefficients K_D of the nine tests required shall be multiplied by 0.90, and this product shall be taken as the coefficient K of that design.

NOTE: All experimentally determined coefficients, K_D, shall fall within a range of plus or minus 5 percent of the average K_D found. Failure to meet this requirement shall be cause to refuse certification of that particular valve design.

(3) The official relieving capacity of all sizes and pressures of a given design, for which K has been established under the provisions of (2), that are manufactured subsequently shall then be calculated by the appropriate formula in (2) multiplied by the coefficient K (see Appendix J).

(f) Tests shall be conducted at a place where approved equipment and personnel are available to conduct pressure relieving capacity tests. Tests shall be made in the presence of and certified by an authorized observer. The place, personnel, equipment, and the authorized observer shall be subject to approval by the Boiler and Pressure Vessel Committee. Laboratory approval is subject to review within each five year period.

(g) A data sheet for each safety valve tested shall be filled out and signed by the manufacturer and by the authorized observer witnessing the tests. Such data sheet will be the manufacturer's authority to stamp valves of corresponding design, size, and pressure range. When changes are made in the design, the certification test must be repeated.

(h) It shall be permissible to rate safety valves under PG-69.1.2 of Section I with capacity ratings at a flow pressure of 103 percent of the set pressure, for use on pressure vessels, without further test. In such instances, the capacity rating of the valve may be increased to allow for the flow pressure permitted in (c), namely, 110 percent of the set pressure, by the multiplier.

$$\frac{1.10\,p + 14.7}{1.03\,p + 14.7}$$

where p = set pressure, pounds per square inch, gage. Such valves shall be marked in accordance with UG-129. This multiplier shall not be used as a divisor to transform test ratings from a higher to a lower flow.

UG-132 Determination of Pressure Relieving Requirements

(a) Except as permitted in (b), the aggregate capacity of the pressure relieving devices connected to any vessel or system of vessels for the release of a liquid, air, steam or other vapor shall be sufficient to carry off the maximum quantity that can be generated or supplied to the attached equipment without permitting a rise in pressure within the vessel of more than 10 percent above the maximum allowable working pressure when the pressure relieving devices are blowing.

(b) Protective devices as permitted in UG-125(d), as protection against excessive pressure caused by exposure to fire or other sources of external heat, shall have a relieving capacity sufficient to prevent the pressure from rising more than 20 percent above the maximum allowable working pressure of the vessel when all pressure relieving devices are blowing.

(c) Vessels connected together by a system of adequate piping not containing valves which can isolate any vessel may be considered as one unit in figuring the required relieving capacity of pressure relieving safety devices to be furnished.

(*d*) Heat exchangers and similar vessels shall be protected with a relieving device of sufficient capacity to avoid overpressure in case of an internal failure.

(*e*) The official rated capacity of a pressure relieving safety device shall be that which is stamped on the device and guaranteed by the manufacturer.

(*f*) The rated pressure relieving capacity of a safety valve for other than steam or air may be determined by the method of conversion given in Appendix J.

UG-133 Pressure Setting of Safety Devices

(*a*) When safety or relief valves are provided, they shall be set to blow at a pressure not exceeding the maximum allowable working pressure of the vessel at the operating temperature, except as permitted in (b). If the capacity is supplied in more than one safety or relief valve, only one valve need be set to open at a pressure not exceeding the maximum allowable working pressure of the vessel; the additional valves may be set to open at a higher pressure, but not to exceed 105 percent of the maximum allowable working pressure of the vessel (see UG-125(c)).

(*b*) Protective devices permitted in UG-125(d) as protection against excessive pressure caused by exposure to fire or other sources of external heat shall be set to operate at a pressure not in excess of 110 percent of the maximum allowable working pressure of the vessel. If such a device is used to meet the requirements of both UG-125(c) and UG-125(d), it shall be set to operate at not over the maximum allowable working pressure.

(*c*) If the operating conditions of a valve are changed so as to require another spring rated for a different pressure, the relief setting shall be adjusted by the manufacturer or by an individual certified by the manufacturer of that safety valve; the valve shall be remarked by either of them in conformance with UG-129.

(*d*) Rupture disks when installed in place of, or in series with, a spring loaded safety valve, shall be rated to rupture at a pressure not to exceed the maximum allowable working pressure of the vessel at the operating temperature.

NOTE: It is recommended that the design pressure of the vessel be sufficiently above the intended operating pressure to provide sufficient margin between operating pressure and rupture disk bursting pressure to prevent premature failure of the rupture disk due to fatigue or creep.

(*e*) The pressure at which any device is set to operate shall include the effects of static head and constant back pressure.

(*f*) The set pressure tolerances, plus or minus, of safety or relief valves, shall not exceed 2 psi for pressures up to and including 70 psi; and 3 percent for pressures above 70 psi.

UG-134 Installation

(*a*) Safety and relief valves and rupture disks shall be connected to the vessel in the vapor space above any contained liquid, or to piping connected to the vapor space in the vessel which is to be protected.

(*b*) The opening through all pipe and fittings between a pressure vessel and its pressure relieving device shall have at least the area of the pressure relieving device inlet, and in all cases shall have sufficient area so as not to unduly restrict the flow to the pressure relieving device. The opening in the vessel wall shall be designed to provide direct and unobstructed flow between the vessel and its pressure relieving device.

(*c*) When two or more required pressure relieving devices are placed on one connection, the inlet internal cross-sectional area of this connection shall be at least equal to the combined inlet areas of the safety devices connected to it, and in all cases shall be sufficient so as not to restrict the combined flow of the attached devices.

(*d*) Liquid relief valves shall be connected below the normal liquid level.

(*e*) There shall be no intervening stop valves between the vessel and its protective device or devices, or between the protective device or devices and the point of discharge, except:

(*1*) When these stop valves are so constructed or positively controlled that the closing of the maximum number of block valves possible at one time will not reduce the pressure relieving capacity provided by the unaffected relieving devices below the required relieving capacity; or

(*2*) **Under the conditions set forth in Appendix M.**

(*f*) The safety devices on all vessels shall be so installed that their proper functioning will not be hindered by the nature of the vessel's contents.

(*g*) Discharge lines from pressure relieving safety devices shall be designed to facilitate drainage or shall be fitted with an open drain to prevent liquid from lodging in the discharge side of the safety device, and such lines shall lead to a safe place of discharge. The size of the discharge lines shall be such that any pressure that may exist or develop will not reduce the relieving capacity of the relieving devices below that required to properly protect the vessel. (See UG-126(e) and Appendix M.)

APPENDIX J*
CAPACITY CONVERSIONS FOR SAFETY VALVES

UA-230

The capacity of a safety or relief valve in terms of a gas or vapor other than the medium for which the valve was officially rated may be determined by application of the following formulas: [1]

For steam:

$$W_s = 51.5 KAP$$

For Air:

$$W_a = CKAP \sqrt{\frac{M}{T}}$$

$$C = 356$$
$$M = 28.97$$
$$T = 520 \text{ when } W_a \text{ is the rated capacity}$$

For any gas or vapor:

$$W = CKAP \sqrt{\frac{M}{T}}$$

where

$W_s =$ rated capacity, pounds of steam per hour

$W_a =$ rated capacity, converted to pounds of air per hour at 60 degrees Fahrenheit, inlet temperature

$W =$ flow of any gas or vapor, pounds per hour

$C =$ constant for gas or vapor which is a function of the ratio of specific heats, $k = c_p/c_v$ (see Fig. UA-230)

$K =$ coefficient of discharge (see UG-131(d) and (e))

$A =$ actual discharge area of the safety valve, square inches

---

*Appendix J of the ASME Boiler and Pressure Vessel Code.

[1] Knowing the official rating capacity of a safety valve which is stamped on the valve, it is possible to determine the overall value of KA in either of the following formulas in cases where the value of these individual terms is not known:

Official Rating in Steam	Official Rating in Air
$KA = \dfrac{W_s}{51.5P}$	$KA = \dfrac{W_a}{CP} \sqrt{\dfrac{T}{M}}$

This value for KA is then substituted in the above formulas to determine the capacity of the safety valve in terms of the new gas or vapor.

$P =$ (set pressure $\times 1.10$) plus atmospheric pressure, pounds per square inch absolute

$M =$ molecular weight

$T =$ absolute temperature at inlet (degrees Fahrenheit plus 460)

These formulas may also be used when the required flow of any gas or vapor is known and it is necessary to compute the rated capacity of steam or air.

Molecular weights of some of the common gases and vapors are given in Table UA-230.

For hydrocarbon vapors, where the actual value of k is not known, the conservative value, $k = 1.001$ has been commonly used and the formula becomes,

$$W = 315 \, KAP \sqrt{\frac{M}{T}}$$

When desired, as in the case of light hydrocarbons, the compressibility factor, Z, may be included in the formulas for gases and vapors as follows:

$$W = CKAP \sqrt{\frac{M}{ZT}}$$

Example 1:

Given: A safety valve bears a certified capacity rating of 3020 lb of steam per hour for a pressure setting of 200 psi.

Problem: What is the relieving capacity of that valve in terms of air at 100 F for the same pressure setting?

Solution:
For steam:

$$W_s = 51.5 \, KAP$$
$$3020 = 51.5 \, KAP$$
$$KAP = \frac{3020}{51.5} = 58.5$$

For air:

$$W_a = CKAP \sqrt{\frac{M}{T}}$$

$$= 356 \, KAP \sqrt{\frac{28.97}{460 + 100}}$$

$$= (356)(58.5) \sqrt{\frac{28.97}{560}}$$

$$= 4750 \text{ lb per hr}$$

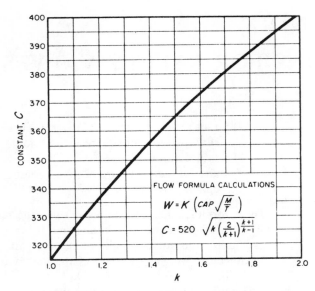

k	Con-stant C	k	Con-stant C	k	Con-stant C
1.00	315	1.26	343	1.52	366
1.02	318	1.28	345	1.54	368
1.04	320	1.30	347	1.56	369
1.06	322	1.32	349	1.58	371
1.08	324	1.34	351	1.60	372
1.10	327	1.36	352	1.62	374
1.12	329	1.38	354	1.64	376
1.14	331	1.40	356	1.66	377
1.16	333	1.42	358	1.68	379
1.18	335	1.44	359	1.70	380
1.20	337	1.46	361	2.00	400
1.22	339	1.48	363	2.20	412
1.24	341	1.50	364		

Flow formula calculations:

$$W = K \left(CAP \sqrt{\tfrac{M}{T}} \right)$$

$$C = 520 \sqrt{k \left(\tfrac{2}{k+1}\right)^{\frac{k+1}{k-1}}}$$

FIG. UA-230 CONSTANT C FOR GAS OR VAPOR RELATED TO RATIO OF SPECIFIC HEATS ($k = c_p/c_v$)

Example 2:

Given: It is required to relieve 5000 lb of propane per hour from a pressure vessel through a safety valve set to relieve at a pressure of P_s, pounds per square inch, and with an inlet temperature of 125 F.

Problem: What total capacity in pounds of steam per hour in safety valves must be furnished?

Solution:
For propane:

$$W = CKAP \sqrt{\frac{M}{T}}$$

value of C is not definitely known. Use the conservative value, $C = 315$

$$5000 = 315 \, KAP \sqrt{\frac{44.09}{460 + 125}}$$

$$KAP = 57.7$$

For steam:

$$W_s = 51.5 \, KAP = (51.5)(57.7)$$
$$= 2790 \text{ pounds per hour set to relieve at } P_s, \text{ pounds per square inch}$$

Example 3:

Given: It is required to relieve 1000 lb of ammonia per hour from a pressure vessel at 150 F.

Problem: What is the required total capacity in pounds of steam per hour at the same pressure setting?

Solution:
For ammonia:

$$W = CKAP \sqrt{\frac{M}{T}}$$

Manufacturer and user agree to use $k = 1.33$

From Fig. UA-230, $C = 350$

$$1000 = 350 \, KAP \sqrt{\frac{17.03}{460 + 150}}$$

$$KAP = 17.10$$

For steam:

$$W_s = 51.5 \, KAP = 51.5 \times 17.10$$
$$= 880 \text{ pounds per hour}$$

Example 4:

Given: A safety valve bearing a certified rating of 10,000 cu ft per minute of air at 60 F and 14.7 psia (atmospheric pressure).

Problem: What is the flow capacity of this safety valve in pounds of saturated steam per hour for the same pressure setting?

Solution:
For air:

Weight of dry air at 60 F and 14.7 psia is 0.0766 per cu ft.

$$W_a = 10,000 \times 0.0766 \times 60 = 45,960 \text{ lb per hour}$$

$$45,960 = 356 \, KAP \sqrt{\frac{28.97}{460 + 60}}$$

$$KAP = 546$$

For steam:

$$W_s = 51.5 \, KAP = (51.5)(546)$$
$$= 28200 \text{ pounds per hour}$$

APPENDIX M*
INSTALLATION AND OPERATION

UA-350 Introduction

(a) The rules in this Appendix are for general information only, because they pertain to the installation and operation of pressure vessels, which are the prerogative and responsibility of the law enforcement authorities in those states and municipalities which have made provision for the enforcement of Section VIII of the Boiler and Pressure Vessel Code.

(b) It is permissible to use any departures suggested herein from provisions in the mandatory parts of this Division of Section VIII when granted by the authority having legal jurisdiction over the installation of unfired pressure vessels.

UA-351 Corrosion

(a) Vessels subject to external corrosion shall be so installed that there is sufficient access to all parts of the exterior to permit proper inspection of the exterior, unless adequate protection against corrosion is provided or unless the vessel is of such size and is so connected that it may readily be removed from its permanent location for inspection.

(b) Vessels having manholes, handholes, or cover plates to permit inspection of the interior shall be so installed that these openings are accessible.

(c) In vertical cylindrical vessels subject to corrosion, to insure complete drainage, the bottom head, if dished, should preferably be concave to pressure.

UA-352 Marking on the Vessel

The marking required by this Division shall be so located that it will be accessible after installation and when installed shall not be covered with insulation or other material that is not readily removable (see UG-116(i)).

UA-353 Pressure-Relieving Safety Devices

The general provisions for the installation of pressure relieving devices are fully covered in UG-

134. The following paragraphs contain details in arrangement of stop valves for shutoff control of safety pressure relief devices which are sometimes necessary to the continuous operation of processing equipment of such a complex nature that the shutdown of any part of it is not feasible. There are also rules in regard to the design of discharge piping from safety and relief valves, which can only be general in nature because the design engineer must fit the arrangement and proportions of such a system to the particular requirements in the operation of the equipment involved.

UA-354 Stop Valves Between Pressure Relieving Device and Vessel

(a) A vessel, in which pressure can be generated because of service conditions, may have a full-area stop valve between it and its pressure relieving device for inspection and repair purposes only. When such a stop valve is provided, it shall be so arranged that it can be locked or sealed open, and it shall not be closed except by an authorized person who shall remain stationed there during that period of the vessel's operation within which the valve remains closed, and who shall again lock or seal the stop valve in the open position before leaving the station.

(b) A vessel or system (see UG-132(c)) for which the pressure orginates from an outside source exclusively may have individual pressure relieving devices on each vessel, or connected to any point on the connecting piping, or on any one of the vessels to be protected. Under such an arrangement, there may be a stop valve between any vessel and the pressure relieving devices, and this stop valve need not be locked open, provided it also closes off that vessel from the source of pressure.

UA-355 Stop Valves on the Discharge Side of a Pressure Relieving Device (See UG-134(e).)

A full-area stop valve may be placed on the discharge side of a pressure relieving device when its discharge is connected to a common header with other discharge lines from other pressure relieving devices on nearby vessels that are in operation, so that this stop valve when closed will prevent a discharge from any connected operating vessels from backing up beyond the valve so closed. Such a stop

---

*Appendix M of the ASME Boiler and Pressure Vessel Code.

valve shall be so arranged that it can be locked or sealed in either the open or closed position, and it shall be locked or sealed in either position only by an authorized person. When it is to be closed while the vessel is in operation, an authorized person shall be present, and he shall remain stationed there; he shall again lock or seal the stop valve in the open position before leaving the station. Under no condition should this valve be closed while the vessel is in operation except when a stop valve on the inlet side of the safety relieving device is installed and is first closed.

UA-356 Discharge Lines From Safety Devices

(a) Where it is feasible, the use of a short discharge pipe or vertical riser, connected through long-radius elbows from each individual device, blowing directly to the atmosphere, is recommended. Such discharge pipes shall be at least of the same size as the valve outlet. Where the nature of the discharge permits, telescopic (sometimes called "broken") discharge lines, whereby condensed vapor in the discharge line, or rain, is collected in a drip pan and piped to a drain, are recommended.[1]

(b) When discharge lines are long, or where outlets of two or more valves having set pressures within a comparable range are connected into a common line, the effect of the back pressure that may be developed therein when certain valves operate must be considered (see UG-134(g)). The sizing of any section of a common-discharge header downstream from each of the two or more pressure relieving devices that may reasonably be expected to discharge

simultaneously shall be based on the total of their outlet areas, with due allowance for the pressure drop in all downstream sections. Use of specially designed valves suitable for use on high or variable back pressure service should be considered.

(c) All discharge lines shall be run as direct as is practicable to the point of final release for disposal. For the longer lines, due consideration shall be given to the advantage of long-radius elbows, avoidance of closeup fittings, and the minimizing of excessive line strains by expansion joints and well-known means of support to minimize line-sway and vibration under operating conditons.

NOTE: It is recognized that no simple rule can be applied generally to fit the many installation requirements, which vary from simple short lines that discharge directly to the atmosphere to the extensive manifold discharge piping systems where the quantity and rate of the product to be disposed of require piping to a distant safe place.

UA-357 General Advisory Information on the Characteristics of Safety Relief Valves Discharging Into a Common Header

Because of the wide variety of types and kinds of safety relief valves, it is not considered advisable to attempt a description in this Appendix of the effects produced by discharging them into a common header. Several different types of valves may conceivably be connected into the same discharge header and the effect of back pressure on each type may be radically different. Data compiled by the manufacturers of each type of valve used should be consulted for information relative to its performance under the conditions anticipated.

---

[1]This construction has the further advantage of not transmitting discharge-pipe strains to the valve. In these types of installation, the back pressure effect will be negligible, and no undue influence upon normal valve operation can result.

7 Application Guidelines for Analytical Systems

Roy L. McCullough, Laurence C. Hoffman

The development and use of process analyzers for onstream applications has been a contributing factor to the growth of the chemical, petrochemical and refining industries in recent years. With proper design, installation and use, process analyzers have proven to be invaluable by providing better quality products, increased yields and solutions to many process problems. The transition from laboratory use to process analysis and control, however, has not been without difficulty. Early model analyzers were merely adaptations of laboratory models, unsuitable for many of their assigned tasks and often poorly engineered from their conception. While today's process analyzers perform capably and have been accepted by operating and management personnel, an improperly engineered installation can still lead to wasted money, mistrust among operators and a maintenance headache.

Factors in Successful Applications

Analyzer installations require a great amount of study and appraisal if they are to be properly engineered. There should be close cooperation between the instrument engineer and other personnel of the laboratory, production and process engineering departments for proper application and design.

Some analyzer applications are straightforward and require practically no research in order to determine what type analyzer should be used and under what conditions they will operate. Frequently, however, the application is not so obvious and an investigative program must be undertaken to determine the best analyzer system to be used. Such a program is outlined below listing the essential steps in the selection of compatible instruments. The work assignment of the particular functions depends on the cir-

cumstances and the plant departmental structure. However, the final responsibility of acquiring correct data and the proper application of the analyzer usually rests upon the instrument engineer.

Factors in the investigative program include definition, study, experimentation, design, checkout and follow-up.

Definition of the Process Problem
1. Define the problem and list probable and possible solutions.
2. If closed loop control is proposed, what equipment constitutes the rest of the loop?
3. Must the analytical method be selective or is a non-selective method permissible?
4. Will the transmission, preparation or analysis of the sample require a time lag such as to invalidate its value as a control signal?

Process and Sample Survey
1. Determine the process conditions, including flow, temperature, pressure, viscosity, stream composition and stream contaminates such as solids, oils and water.
2. If the sample phase is liquid, can the liquid be vaporized without fractionization or polymerization?

Feasibility Study
1. Get true analysis of the stream; include, if possible, analysis during periods of process upsets and startups.
2. Determine the range of the desired component measurement so that an analyzer may be selected to fit the requirements.
3. Make a cost estimate which includes the analyzer, sample system, miscellaneous hardware, standard

samples and installation labor. This estimate, together with savings data furnished by the process engineering group, is used to determine the economic feasibility. (Many analyzers today are justified on the basis of safety or pollution monitoring rather than economics, however.)

Experimental Analyzer Test

If doubts exist regarding either the analysis or the process control function, a pilot study may be run, using a surplus analyzer or possibly a rental unit. The process data accumulated during such a pilot run helps determine the feasibility of the application and enables the engineer to place the analyzer on-stream with a minimum of confusion and experimentation. The study enables the engineer to actually monitor the process under varying conditions and compile information such as accuracy, reproducibility and reliability.

Final Design

At this stage of the program, the engineer should have enough information for final design if the project is to proceed. The analyzer loop has been operated under actual process conditions, statistical data on the analyzer and the process have been gathered, and there has been sufficient operating experience to dictate the design of a sample system and provide information on calibration samples. All items may be specified for purchase and drawings prepared and issued.

Preinstallation Check-out

Ideally, a complex analyzer should be checked out at the manufacturer's plant before delivery. This should include a complete operating test with standard samples to check all functions, including output, switching, auto-zero, etc. After delivery of the analyzer, these same tests should be duplicated in the plant instrument shop as a check against shipment damage. The shop tests afford an excellent time to train instrument technicians on the operation, calibration and maintenance of the analyzer. Instruction manuals, schematics and spare parts lists should be properly filed at this time.

Installation and Check-out

Field installation is normally made by construction personnel. However, all work should be followed by maintenance personnel in case field changes must be made. When installation is complete, the analyzer may be placed in service, but under no circumstances should it be turned over to operating personnel until it has been thoroughly tested. This may take from one to several weeks, depending on the complexity of the analyzer and the problems encountered.

Project Follow-up

A conference with operating supervisory personnel is in order after one to three months of service by the analyzer. The purpose of the meeting is to ascertain that the initial objectives were achieved and if any additional benefits were obtained.

Sample Systems

The design of the sample system for an analyzer is equally as important as the application of the analyzer itself. The sample system may vary in complexity from a simple shutoff valve and single tubing connection to a complicated system consisting of solenoid valves, filters, vaporizers, regulators and other items costing several thousand dollars.

Basic Principles

The basic principles of sampling are as follows:

1. Extraction of a representative sample from the stream
2. Cleaning the sample
3. Minimizing the time lag in getting the sample to the analyzer
4. Meeting the pressure, temperature and sample flow rate requirements of the analyzer
5. Meeting other special requirements of the analysis such as vaporizing a liquid sample, removing water drops, etc.
6. Disposition of the sample.

The application of these basic principles is discussed in the following paragraphs.

Sample Location

Extracting a representative sample is sometimes difficult. The sample point must be in an active stream to assure that it provides the information required by the analysis.

Sample points on process lines should always be installed on the top or side of the pipe, never on the bottom. This prevents the flow of condensed liquids in vapor streams from entering the sample valve and also eliminates the entrance of fine solids which may be swept along the pipe. The side tap is preferred on liquid streams to minimize the chances of getting entrained vapors. The probe shown in Figure 7.1 is commonly used to extract a sample from the center of the stream, thereby avoiding condensed liquids in a vapor stream and solids in a liquid stream. The end is usually tapered with the taper facing away from the direction of flow, except when isokinetic sampling methods are needed because of particulate matter in the stream.

Isokinetic sampling obtains the most representative sample of particulate matter or entrained liquids by extracting the sample at the same velocity as the flowing stream. Samples extracted at higher velocities than the stream velocity are usually low in particulates while samples extracted at lower velocities than the stream velocity are high in particulates due to the disturbance of the flow pattern around the sample probe.

The sample tube is usually made of ¼-inch OD stainless steel tubing but can be as small as ⅛-inch OD on clean streams. The packing gland assembly is available from

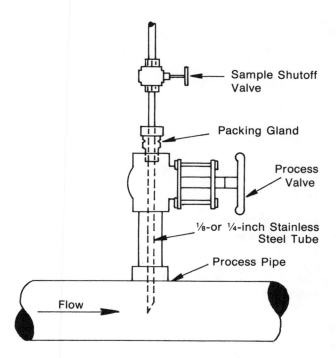

Figure 7.1. The sample probe assembly shown illustrates the technique of sampling from the center of the stream to avoid condensates in vapor streams and solid particles in liquid streams.

Labels in figure:
- Sample Shutoff Valve
- Packing Gland
- Process Valve
- ⅛-or ¼-inch Stainless Steel Tube
- Process Pipe
- Flow

several suppliers and allows the probe to be removed for cleaning without upsetting the process. The block valve must meet piping requirements for the process lines. All components in the probe assembly should meet the material and pressure requirements of the process line also.

A clean stream is important to successful low maintenance operation of most analyzers.

Filters

It should be assumed that all sample streams contain solid contaminates such as rust particles, polymers, etc.; therefore, minimum requirements call for a fine mesh wire filter. Extremely fine solids sometimes require the use of a porous metallic or ceramic filter. The gaskets and diaphragms of all pressure regulators, flow controllers, filters, etc., should be examined to determine that they will not deteriorate or absorb components from the stream. The filter might be considered as one of the most critical components in the sample conditioning system simply because so many other functions are performed after the stream has been cleaned.

Because of the inherent maintenance problems associated with filtration, only the minimum volume of sample necessary to purge the sample valve or cell should be filtered. A bypass filter (Figure 7.2) performs the dual function of filtering the required amount of sample while using the bypass stream to sweep the filter clean. A similar unit

(Figure 7.3) offers the advantages of high pressure rating (1,500 psig), high temperature rating (500°F), low differential pressure (less than 1 psi) and pore sizes from 0.3 to 20 microns.

Some vapor samples contain so much entrained solids that filtration is not practical until the vapor is scrubbed either by bubbler devices or by small spray scrubbers. The use of scrubbers may cool samples to the extent of altering component composition, so care must be exercised to assure that nothing is removed which needs analysis. The scrubber is usually followed by a filter. Its volume must be kept to a minimum since added volume introduces more time lag into the sample system.

Sampling Time Lags

Sampling time lags delay the analysis and may reduce the usefulness of the measurement. The analyzer needs to be located close to the sample point, but this is not always feasible. A bypass loop such as that shown in Figure 7.4 is often used to reduce time lags in long sample runs where the analyzer will handle only a small flow rate. This system can be used on liquids or gases whenever there is a point of lower pressure for sample return or when the sample can be vented or run to a sewer. However, the bypass system is not as effective for liquid as for vapor systems since it is difficult to reach velocities over 3 to 5 fps without encountering high pressure drops. Velocities can be increased slightly for the same pressure drop by using larger diameter tubing for the bypass. The bypass system is very effective for vapor flows where normal velocities range from 25 to 40 fps with maximum velocities around 100 fps. Pressure regulators are generally located at the sample tap for vapor systems.

Transport tubing should be protected by installing it in raceway or thin-wall pipe. The tube should be sloped downward toward the analyzer to drain liquids, and no pockets should be left in the line without a condensate trap. Tube size is determined by flow conditions. Flow volume should be as low as possible and velocity should be as high as is practical to minimize lag time.

Sample Conditioning

An effort should be made to match the sample requirements of the analyzer to that of the stream, thereby minimizing extra system components such as vaporizers, regulators, sample coolers, etc. If a liquid sample pressure is to be reduced, care should be exercised to assure that flashing does not occur. However, it is not always possible to match the analyzer to the process conditions, and the sample must necessarily be conditioned to meet the analyzer requirements.

Temperature

Many manufacturers make small heat exchangers to cool the sample. Cooling water may be allowed to flow unregulated through the exchanger, leaving the sample temperature uncontrolled. Some analyzers are temperature sensitive, however, and the measurement can be improved

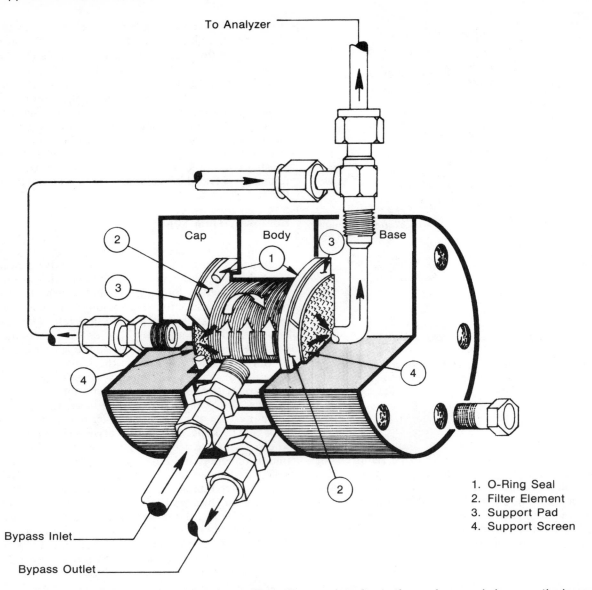

Figure 7.2. A bypass filter performs a dual function. It filters the sample going to the analyzer and also uses the bypass stream to sweep the filter clean.

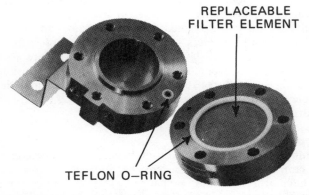

Figure 7.3. High pressure bypass filters are available for services to 1,500 psig and for temperatures to 500°F. Pore sizes range from 0.3 to 20 microns.

by controlling the sample temperature. In some cases, close temperature control is essential to the separation and measurement of components.

Pressure

Pressure is also a common mismatch. Regulators are used to reduce high pressures and to regulate varying low pressures. To prevent maintenance problems and increase performance, the regulator should be preceded by a fine mesh strainer. Materials should be noncorrosive.

Single stage regulation is usually sufficient for sample pressure reduction. Figure 7.5 shows a small regulator with inlet pressure ranges to 3,000 psig and controlled outlet ranges to 500 psig. Temperature range is −40 to 500°F. Wettable parts are stainless steel and Teflon with other machinable alloys available. Figure 7.6 shows both electric

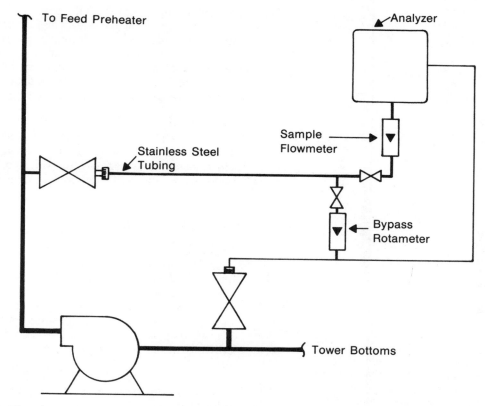

To Feed Preheater

Analyzer

Stainless Steel Tubing

Sample Flowmeter

Bypass Rotameter

Tower Bottoms

Figure 7.4. Time lags are reduced by using high fluid velocities to bypass loops located near the analyzer.

Figure 7.5. Pressure regulator used for high pressure sample streams. Inlet pressures to 3,000 psig and outlet pressures up to 500 psig can be accommodated at temperatures from −40 to 500°F.

and steam-heated combination vaporizer-regulators for inlet pressures to 5,000 psig and for temperatures from 32 to 450°F. Controlled outlet pressure ranges are 3 to 30 and 10 to 100 psig. On vapor systems the regulator is usually located at the sample point to keep a high sample velocity with a low flow rate.

In most other cases the pressure regulator can be located near the analyzer. When reducing the pressure of hot liquid samples, the regulator is usually located downstream of the sample cooler to avoid flashing since most liquid analyzers cannot tolerate vapor bubbles in the sample stream. When time lag and sample disposal do not present a problem, a constant level overflow tank can be used to maintain constant pressure on the analyzer while unused sample overflows to the sewer. A needle valve can be used to regulate the sample flow to the overflow tank, and the system will not require a pressure regulator.

When pressure increase rather than pressure reduction is necessary (for example, using a sample from an open ditch), pumps can be used to raise the sample pressure. Strict attention must be given to the pump system design to ensure an adequate NPSH (net positive suction head). Positive displacement pumps are normally used since flow rates are small. They may require a relief valve or rupture disc for overpressure protection.

Flow

Many analyzers are sensitive to velocity changes, and the sample flow rate must be monitored and/or controlled. Where sample conditions are constant, a simple rotameter suffices.

When sample conditions are not constant, a rotameter-constant differential regulator combination should be used. The manufacturer's literature usually gives maximum

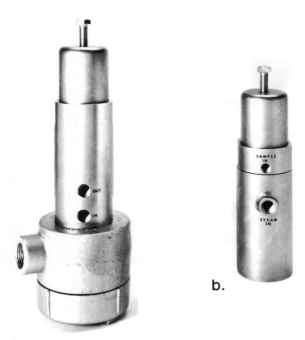

b.

a.

Figure 7.6a and b. These combination sample vaporizer-regulators can be used on inlet pressures to 5,000 psig and has outlet ranges of 3 to 30 and 10 to 100 psig. Figure 7.6a is an electrically heated type, and Figure 7.6b is a steam heated type. (Courtesy of The Arcas Company.)

recommended velocity and points out sensitivity to flow rate (but not always).

Vaporizers

If a liquid sample must be vaporized, it can be done with a small heat exchanger or with a combination vaporizer-regulator.

A vaporizer-regulator is often used when converting a liquid sample to vapor analysis. Figure 7.6 shows both electric and steam heated regulators, and Figure 7.7 shows how they are constructed internally. Stainless steel is standard with Monel, Hastelloy and other materials furnished on request. Inlet pressure ratings to 5,000 psig are available with outlet pressures from 3 to 100 psig.

The dual-service vaporizer-regulator shown in Figure 7.7 will vaporize liquid samples with boiling points up to 400°F and regulate the vaporized samples within close tolerances for stable flow rates.

The heat source may be steam or electrical. Figure 7.8 shows a modular concept of a vaporizer and regulator. The control element is a stainless steel bellows with integral relief valve, compared to conventional diaphragm elements.

In dead-end sample systems (where the unused sample is not returned to the process), it is usually advisable to vaporize and/or reduce in pressure at the sample point in order to reduce lag time.

Usually analyzers which require vapor samples can tolerate no entrained liquid. The vaporized samples may still require additional conditioning to ensure against the entrained liquids. Components such as separators, traps, demistors or centrifuges are used for this purpose. When vaporizing a sample, precautions must be taken to prevent fractionation, chemical reaction or solids drop-out.

Coalescers

The presence of free water in a sample can result in false analysis, requiring the use of a coalescing unit. Figure 7.9 shows a coalescing bypass-filter which coalesces water from the stream, filters the required sample and bypasses the remainder.

Other Conditioning Items

There are many other sample conditioning items such as filters, scrubbers, oil separators, driers, etc., for sample conditioning that may be required to meet the analyzer requirements. Since there are many types of analyzers, each with individual requirements, no attempt is made to discuss all possible treatments. It is important when applying any analyzer to know the composition of the stream being measured, particularly impurities and solids, and to look critically at the application to see if any of these will affect the analysis and, if so, minimize their effects.

In specifying the additional items for specific applications, these general requirements should be considered.

1. Body material—should be compatible with stream components and rated for process conditions
2. Element material—should be noncorrosive metals or ceramic or synthetics (such as Teflon) to minimize corrosion and/or deterioration which would contaminate the stream
3. Minimum volume—each added component adds volume and therefore time lag to the system
4. Pressure drop—in low pressure streams this may become critical

Sample Disposal

Sample disposal sometimes presents significant problems. Safety has always been a consideration since many samples are either toxic or flammable or both. In some applications the sample may simply be dumped or vented to the atmosphere. In others, where venting becomes dangerous, other solutions are required.

Liquids

When disposing analyzed samples of water or other inexpensive, nonpolluting , nonhazardous liquids, the most common method is to run it to the sewer. This method may also be used on hazardous or polluting samples if an appropriate chemical sewer is located nearby. If the sample is expensive or cannot be conveniently run to the sewer, it is commonly returned to the process system at a point where the pressure is lower than the sample pressure and as steady as possible.

A typical example is a situation in which the sample is taken from a pump discharge and returned to the pump suction. The drop across a heat exchanger or some other piece of equipment or even across a valve may be used, but these pressure drops usually vary and are therefore less desirable. Any time that the sample is being returned to a closed system and there is a possibility that the closed system pressure will build up equal to the sample pressure, a back pressure regulator should be installed in the vent line. This allows the sample pressure to be raised above the closed system pressure, and sample flow will not be stopped by higher downstream pressures. Extreme cases may require the use of a pump to return the sample to the system.

Vapors

Vapor samples present slightly different problems. Nonhazardous gases are normally vented to the atmosphere while hazardous gases are often vented to a flare header or returned to the system, whichever is more convenient. Gases which cannot be vented to a flare header or returned to the system may be scrubbed of the corrosive or toxic components and then vented to the atmosphere or flare header.

Vent flow can be very critical to the successful performance of an analyzer. In a return vent system, the vent pressure and flow must be sufficent to prevent back flow and back diffusion. In a free vent system, continuous flow is equally important to prevent back diffusion of air into the analyzer. In addition, the installation of free vent discharge tubing must be high enough to take advantage of prevailing wind dispersal for safety reasons.

Typical Sample Systems

Basic principles of sample systems have been discussed in the previous paragraphs. The application of these principles is illustrated by describing specific examples for various gas chromatograph analyses.

Since there is no universal sample conditioning system, each application requires specific engineering design for a particular set of process conditions. Just as process streams can be divided into general categories, however, sample conditioning systems may also be categorized into basic systems with similar functions. The four systems discussed here (single and multistream vapor, and single and multistream liquid) cover many of the problems encountered in analyzer applications.

Single Stream Vapor

A single stream vapor conditioning system with alternate sets of sample conditions is shown in Figure 7.10. The system consists of a sample flowmeter with needle valve, a bypass flowmeter with needle valve, a micron filter, a three-way valve, a pressure gauge and optional components for varying stream conditions. Sample flow is metered to the analyzer and vented in many cases. Bypass flow may be vented, returned to the process or piped to the flare header, as the situation requires. The calibration gas pressure should equal the sample pressure to avoid the need for reset-ting the flow to the analyzer. If the required analyzer flow is high enough, the bypass flow may be eliminated.

Condition *b* of Figure 7.10 uses a pressure regulator to reduce high pressures and also has a relief valve to protect the system in the event of regulator failure. Condition *c* uses a vaporizer-regulator to vaporize entrained or pure liquid streams. Most liquid streams should be vaporized at the sample point or beyond the bypass takeoff point, as shown, to minimize lag time or the amount of sample to be vaporized. The manual three-way calibration valve may be replaced by a remotely actuated valve to permit calibration checks by operating personnel.

Single Stream Liquid

Figure 7.11 depicts a single stream liquid conditioning system. The filter is a coalescing type to remove entrained water from the sample. Sample flow is measured downstream of the analyzer to decrease sample volume between the necessary hardware and the sensing unit. Note the use of a back pressure regulator to prevent vaporization of the liquid sample in the conditioner or in the analyzer.

Three-Stream Vapor

A three-stream vapor conditioning system is shown in Figure 7.12. Inlet conditions and components are the same as previously presented. Primary changes are made in the flow scheme by the addition of solenoid valves and changes in the bypass and vent streams. The safety relief valves are piped to a vent to prevent their discharge in the instrument and housing area. The solenoids are normally actuated in sequence, changing the flow from vent to analyzer. Double block and bleed solenoid valves may be used as a precaution against leaking valves. This system uses a differential back pressure regulator in the bypass flow stream to control bypass flow and maintain pressure on the sample stream to the analyzer. It also maintains the sampled stream at a higher pressure than the bypassed streams. In the event of a leaking solenoid, the leak is from the active stream to the bypassed stream, eliminating the need for the second block and bleed solenoids.

Three-Stream Liquid

Figure 7.13 shows a three-stream liquid conditioning sample system. A feature of this system is the addition of bleed orifices across the output ports of each solenoid. The sample flow pressure is higher than the vent pressure due to the action of the differential back pressure regulator; thus, there is a small flow from the on-line stream through each of the three orifices. In this instance, the orifice is built into the solenoid valve and serves to sweep into the vent any sample that remains in the port of a deactivated valve and prevents contamination of one stream by another. This system also is equipped with a back pressure regulator on the vent outlet to prevent flashing in the system. The sample pressure of all three streams must be higher than that which is expected at any time in the vent return header.

The conditioning systems discussed above are available as packaged units from practically all analyzer manufacturers. In addition, many other miscellaneous components are

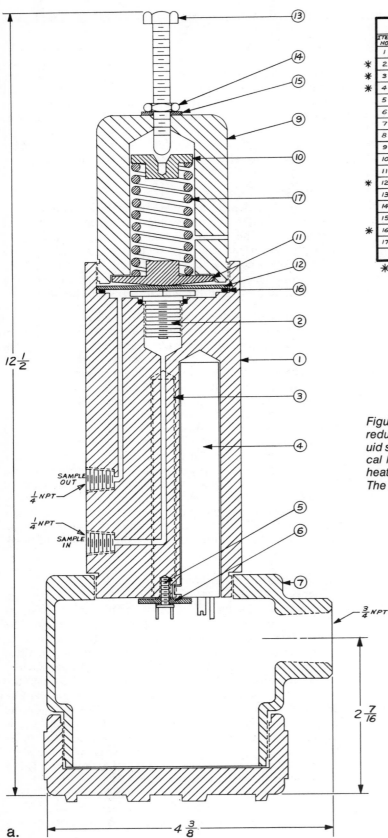

ITEM NO.	NO. REQD	PART NO.	DESCRIPTION
1	1	7-4634	REGULATOR FLASH CHAMBER
2	1	7-4642	POPPET ASS'Y INCLUDES O-RING
3	2	7-4480	HEATER CARTRIDGE, 150W-120V.
4	1	2-5830	THERMOSWITCH, 100°-400°F
5	1		SCREW, #6-32 X 1/2 S.S. PAN HD. MACH.
6	1	7-4635	RETAINER PLATE
7	1	7-4636	VAPORIZING REG. CONDULET
8	1	7-4670	LOW PRESS. REG. KIT (ITEMS 9-17)
9	1	7-4644	REG. SPRING ENCLOSURE
10	1	7-4637	SPRING COMPRESSION PLATE
11	1	7-4650	DIAPHRAM COMPRESSION PLATE
12	1	7-4646	REGULATOR DIAPHRAGM
13	1	7-4651	COMPRESSION BOLT
14	1		HEX NUT 5/16-18 (S.S.)
15	1		WASHER, FLAT - 5/16 (S.S.)
16	1	7-4653	O-RING TFE. #132
17		7-4469	AVAILABLE SPRING - 35 PSI

✳ RECOMMENDED SPARE PARTS

Figure 7.7a and b. Dual service vaporizer-regulators for reducing and regulating pressures and for vaporizing liquid samples may use either low pressure steam or electrical heaters as a heat source. Figure 7.7a is electrically heated while Figure 7.7b is steam heated. (Courtesy of The Arcas Company)

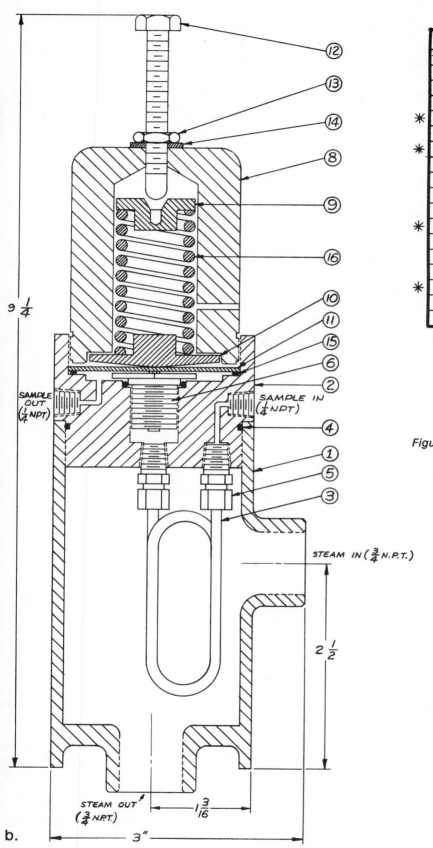

ITEM NO.	NO. REQ'D	PART NO.	DESCRIPTION
1	1	7-4649	STEAM CHAMBER
2	1	7-4645	REGULATOR BODY
3	1	7-4652	SAMPLE COIL
✳ 4	1	7-4654	O-RING VITON #033
5	2	4-11045	FITTING - 1/8" PIPE TO 1/8" TUBE (S.S.)
✳ 6	1	7-4642	POPPET ASSEMBLY INCLUDES O-RING
7	1	7-4670	LOW PRES. REG. KIT (ITEMS 8-16)
8	1	7-4644	REGULATOR SPRING ENCLOSURE
9	1	7-4637	SPRING COMPRESSION PLATE
10	1	7-4650	DIAPHRAGM COMPRESSION PLATE
✳ 11	1	7-4646	REGULATOR DIAPHRAGM
12	1	7-4651	COMPRESSION BOLT
13	1		.HEX NUT 5/16 -18 (S.S.)
14	1		WASHER, FLAT - 5/16 (S.S.)
✳ 15	1	7-4653	O-RING TFE. #132
16		7-4469	AVAILABLE SPRING - 35 PSI

PARTS LIST

✳ RECOMMENDED SPARE PARTS

Figure 7.7 continued

Figure 7.8. In the modular concept of vaporizer-regulators, separate units are furnished and simply piped together with a short piece of tubing. (Courtesy of Beckman Instruments)

Figure 7.9. A coalescer is a device that collects a particular fluid component by causing droplets to be formed large enough to be removed from the stream. The bypass-filter coalescer shown allows the desired sample components to pass for analysis while collecting and bypassing the coalsced components. (Courtesy of Beckman Instruments, Inc.)

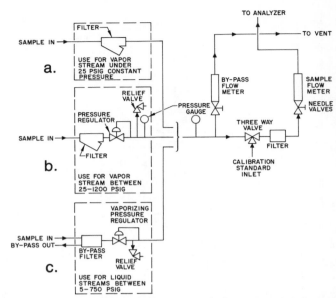

Figure 7.10. Three different configurations of piping and accessories are shown for sample conditioning assemblies for single stream vapor applications. (Courtesy of Beckman Instruments, Inc.)

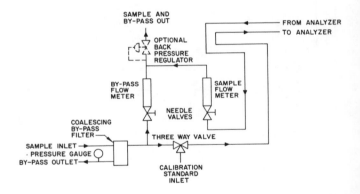

Figure 7.11. This single stream liquid sample conditioning assembly uses a coalescing bypass filter and places the sample flowmeter downstream of the analyzer to minimize sample lag time. (Courtesy of Beckman Instruments, Inc.)

available from these sources either in modular form or for remote mounting. Extreme care should be exercised when specifying a system with an analyzer since the vendor has no way of anticipating individual needs. A complete specification of the process conditions to be met should be included. Items inadvertently omitted in the specifications are (a) contaminants in the system and (b) back pressure on the sample return line. Familiarity with the most commonly used components of sample conditioning systems is essential to the successful design and operation of analyzer systems.

Evaluating Analyzer Components

In the following paragraphs, components of several types of analyzers are discussed to aid in making the proper selection for various applications. Components are examined and advantages and disadvantages of each are listed.

Chromatographs

Chromatographs comprise a relatively high percentage of analyzer applications. Applications are usually complex in comparison to other types of analyzers. An understanding of the various components pertinent to chromatography is essential to the design of good systems. The most widely used detectors are the thermal conductivity (TC) and the ionization types. Of the ionization detectors, the hydrogen flame is most often used while use of the TC type is about evenly divided between the two-thermistor type and the two- or four-filament hot wire.

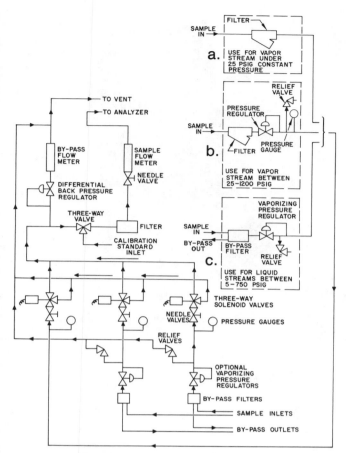

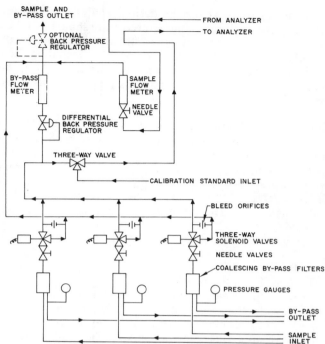

Figure 7.12. The sample conditioning assembly for multipoint vapor system analysis is very similar to single stream systems except for the additional solenoid valves to control stream flows. (Courtesy of Beckman Instruments, Inc.)

Figure 7.13. A notable feature of this sample conditioning assembly for multipoint liquid systems is the addition of bleed orifices at each stream inlet solenoid valve. They prevent stream-to-stream contamination by sweeping samples away from deactivated valves. (Courtesy of Beckman Instruments, Inc.)

The thermistor detector responds faster than the hot-wire filament and is therefore useful for fast eluting peaks. Signal response is essentially linear for both types.

Sensitivities of the thermistor and the two hot-wire filaments are roughly the same while the four hot-wire filaments are considerably more sensitive. General recommendations of minimum ranges is 0 to 1% for the thermistor and two hot-wire filament detectors and 0 to 250 ppm for the hydrogen flame detector, although lower ranges are possible under ideal conditions.

Thermistor detectors are used in low to moderate temperatures while hot-wire filaments may be used in high temperature applications.

The electronic bridge circuitry for all TC detectors is essentially the same; it is relatively simple and easy to repair. Due to high impedance and low signal current, the hydrogen flame electronic circuitry is much more sophisticated and requires expensive test instruments for repair and calibration.

In summation, thermistor detectors have the advantage of fast response but are limited to low and moderate temperature service. They will decompose in a hydrogen carrier stream because the small hydrogen atom tends to penetrate its shell.

The hot-wire filament is suitable for high temperature service but has a slower response time than the thermistor.

The hydrogen flame ionization detector has fast and essentially linear response and is very sensitive. Calibration is more complicated and less stable than TC units. Three carrier and combustion gases are required for hydrogen flame ionization detectors compared to one for TC detectors.

Columns

The separation column is the heart of a chromatographic analyzer because it is here that the components of a sample are separated from each other for measurement by the detector. Reference may be made to Chapter 9 of Volume 1 for a more thorough discussion of various parameters of column design which is an art within itself. Major manufacturers of chromatographic analyzers maintain files on column specifications for hundreds of successful applications in every industry. If specified by the purchaser, a column guaranteed for the desired analysis will be furnished by the analyzer manufacturer. Care should be taken, however, in transmitting the stream analysis to the manufacturer so that all components, including contaminants, are listed, as the manufacturer can warrant his column to perform only on the basis of the information furnished.

Sample and Column Valves

The sample and column valves are critical items in a process analyzer since they must operate thousands of times without failure, frequently in high pressure and high temperature services. Because of the extreme service conditions, valve materials should be 316 stainless steel or better, and most valves require close dimensional tolerance control.

For maximum operating efficiency, a valve should:

1. Be leak tight
2. Have fast actuation time
3. Have high reproducibility of delivered volume
4. Have low dead volume
5. Produce very small sample volumes where required
6. Have no surfaces in contact with the sample that require lubrication

Most manufacturers equip their analyzers with one of four types of sample and column valves. Each has certain advantages for particular applications.

Sliding Plate Valve. Figure 7.14 shows a sliding plate sample valve with an external sample volume loop used in vapor applications. The sliding plate, located between two rectangular plates, is made of Teflon, eliminating the need for a lubricant. Lubricants absorb some components from the sample and cause tailing of peaks by releasing them at a slow rate into the carrier stream. For liquid samples, the external loop is eliminated, and the hole through the slider serves as the sample volume. This is a fast action valve with excellent volume reproducibility and has a maximum pressure rating of 150 psig. The Teflon slider limits the temperature rating to 200°F due to softening of the Teflon. In some high temperature applications, the valve may be located in a cool zone with the column located in the hot zone. The sliding plate valve is also used in various column configurations as a column switching valve.

Rotary Valve. Figure 7.15 shows a rotary sample valve (available with 6 to 10 ports) which uses an etched flat Teflon ring to switch flow between ports. Maximum pressure rating is 300 psig. A model is available with metal-to-metal rotary surfaces to permit higher pressure and temperature service. However, these surfaces require lubrication, resulting in tailing of peaks.

Diaphragm Valve. Figure 7.16 depicts a six-port diaphragm sample valve. Air pressure through three alternate ports of the upper body forces a thin diaphragm to

Figure 7.14. The sliding plate sample valve is fast acting and has excellent volume reproducibility. (Courtesy of Beckman Instruments, Inc.)

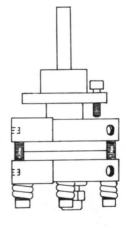

Figure 7.15. Rotary sample valves are available with 6 to 10 ports. Metal-to-metal surfaces are available to permit their use in high temperature service.

close, matching ports on the lower valve body (Figure 7.17). One set of three ports is always closed. The valve is fast acting with a minimum of dead volume. Only external sample volume loops may be used.

Plunger-Diaphragm Valve. The plunger-diaphragm valve is similar in construction and operation to the diaphragm valve. Figure 7.18 shows a cutaway operational schematic illustrating that three alternate plungers are piston driven at any one time, either blocking or connecting adjacent ports. Figure 7.19 presents both liquid and vapor sample configurations. This type valve is fast and has good reproducibility. Pressure rating is 300 psig, optional to 1,-500 psig. Temperature rating is 300°F, optional to 450°F.

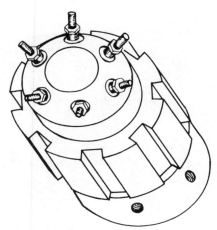

Figure 7.16. The sixport diaphragm valve is fast acting and is characterized by a small dead volume.

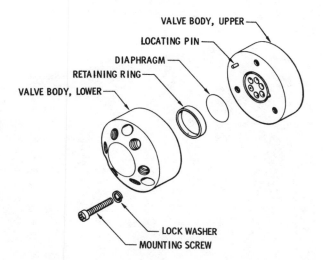

Figure 7.17. This is an exploded view of the body of the diaphragm valve shown in Figure 7.16.

VALVE BODY, UPPER
LOCATING PIN
DIAPHRAGM
RETAINING RING
VALVE BODY, LOWER
LOCK WASHER
MOUNTING SCREW

Liquid sample volumes are 0.5 and 2.0 microliters standard, and vapor sample volumes are as low as 25 microliters.

Programmer

The programmer is the control unit of a process chromatograph. It is housed in a sheet metal enclosure, is not explosionproof (although it usually may be purged to meet electrical classification requirements), and is constructed for panel mounting. Programmer electrical circuitry may be divided into four systems:

1. Power supplies—low voltage DC and 115v AC
2. Logic and attenuation—measuring bridge and attenuating resistors
3. Distribution—output signal to recorders, controllers, etc.
4. Sample valve and stream switching elements

GAS VALVE
VERTICAL CUT AWAY

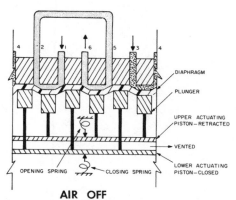

DIAPHRAGM
PLUNGER
UPPER ACTUATING PISTON—RETRACTED
VENTED
LOWER ACTUATING PISTON—CLOSED
OPENING SPRING   CLOSING SPRING

AIR OFF

Figure 7.18. This cutaway schematic of the plunger-diaphragm sampling valve shows how the piston driven plungers block or connect adjacent ports. (Courtesy of Seiscor, Div. of Seismograph Service Corp.)

A chromatographic programmer should be capable of the following functions:

1. Auto-zero—automatically zeroing the bridge measuring circuit at least once per analysis
2. Auto (trend)—bargraph or trend presentation
3. Calibrate (standard)
4. Time—to check component gate timing
5. Manual (chromatogram)

The key to programmer performance lies in the sequence timer mechanism since it controls all analyzer functions. Most programmers use a cam timer which is made up of a group of cams on a shaft driven by a synchronous motor. Each cam actuates a mercury switch or a microswitch. The microswitch is usually preferred because it can be set more easily and more precisely than the mercury switch, and because mercury switches are less dependable because the mercury oxidizes. Figure 7.20 shows a typical multistream, multicomponent programmer with a cam-actuated microswitch timer. The switches actuate relays which, in turn, switch component potentiometers and other bridge elements. Switches and relays should be hermetically sealed to keep out dust and moisture.

Another approach to programmer timing is illustrated by the unit in Figure 7.21 which uses a digital timer in place of the cam timer and switches. The digital timer has 0.1% adjustability and repeatability compared to 2.0% for the cam timer. The program cycle length and function start and stop times are quickly programmed, utilizing push-button switches with digital time indicators. This is fast and accurate, compared with changing motors and gears as well as adjusting cams on the cam timer. The entire program is displayed at all times in three-digit numbers. The program

LIQUID VALVE
HORIZONTAL CUT AWAY

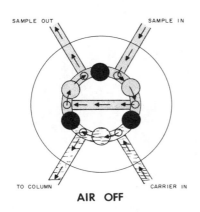

GAS VALVE
HORIZONTAL CUT AWAY

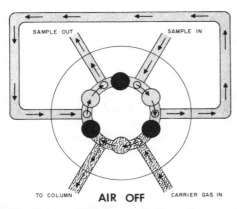

Figure 7.19. Schematic views of the plunger diaphragm sampling valve shown in Figure 7.18 show flow lines for liquid or vapor samples. (Courtesy of Seiscor, Div. of Seismograph Service Corp.)

Figure 7.20. A typical multistream, multicomponent chromatograph programmer with a cam actuated microswitch timer. (Courtesy of The Arcas Company)

DIGITAL TIMER WITH
DIRECT DIGITAL DATA ENTRY
FOR ARCAS SERIES 405
PROCESS GAS CHROMATOGRAPH

Figure 7.21. This chromatographic programmer uses a plastic disc in place of the cam timer and switches to control analyzer functions. (Courtesy of The Arcas Company)

functions are displayed by light emitting diodes (LED), and the program cycle is indicated by a three-digit LED readout. The digital timer has 11 channels, four with adjustable start and fixed duration time and seven with adjustable start and adjustable stop time. Each function, such as sample inject, column switch, auto-zero and attenuator switching, has its own channel with the times displayed in three digits. By pushing the "reset" switch, the programmer will reset to zero time and start the program over. The programmer can be stopped at any point in the program for an indefinite period by putting the stop switch in the off position. The program is not destroyed by a power failure or by turning the power off.

The length of the program cycle can be up to 1,000 counts and can be set with the "end of cycle" switches anywhere between 001 and 999. The duration of the counts are selected to be 2, 1, ½, ¼, ⅛, 1/16, or 1/32 seconds. For 2-second counts, the timer has a resolution of 2 seconds with a maximum cycle time of 33 minutes and 20 seconds. The count time selection is binary until at 1/32 of a second resolution the maximum cycle time is 31.25 seconds. This wide selection of times will cover all types of chromatography applications, from long difficult analysis to the super high speed units. The program switches are push-button 10-position switches with digital position indicators. This allows the entire program to be presented continuously in digital form. Any one of the function's start time or stop time can be changed one or more time counts without changing any other part of the program, and the change can be made while the programmer is running.

The programmer outputs are 5v DC—either IC or dry contact. The cycle indicator is three-digit BCD. The unit can be monitored by, substituted for or operated in parallel with a digital computer.

Listed below are several features that a good programmer should have.

1. Plug-in module construction, especially on critical services, as a maintenance aid
2. Manual or auto stream selection for multiple stream programmers
3. Stream identification lights on the programmer (or stream selector unit) and stream identification marks on the recorder chart

Analyzers normally are available with standard signal ranges to 10v DC and frequently are purchased with several different signal levels. In many instances, where a plant computer is in use or planned, it may be wise to require provision for future takeoff of the signal for this purpose.

Although most chromatographs are still used for monitoring purposes only, a variety of auxiliary electronic circuits are needed in conjunction with the basic programmer and recorder in many instances. The trend toward smaller panel instruments has caused many of these auxiliary units to be housed separately from and yet still controlled by the programmer. Some of these are

1. Stream selector—for auto or manual sequences
2. Amplifier—for low level signals such as those from ionization detectors
3. Analog units—for signal preparation such as memory boards, retransmitting slidewires, computing relays, etc.

Infrared Analyzers

There is very little choice in selecting an optical system for infrared analyzers because the nondispersive infrared analyzer (NDIR) is manufactured nearly exclusively for process applications. The NDIR system achieves waveband isolation by techniques which either filter or mask out unwanted light wavelengths and is a compromise which offers fewer maintenance problems and greater sensitivity even though it sacrifices some selectivity.

Care should be exercised in the design of sample conditioning systems for IR analyzers. Infrared requires a clean sample under carefully controlled conditions. Vapor sample should be flow controlled with the downstream pressure maintained above atmospheric. Maximum accuracy is obtained with a temperature controlled environment.

Reference gases should be specified and purchased with close tolerances. As a rule, calibration requires a zero gas and one or more span gases. The zero gas is usually a pure gas which exhibits no adsorption at the same wavelengths as the component to be analyzed and is used to zero automatically the analyzer at predetermined intervals. The span gases should be similar to the process stream with varying amounts of the components to be analyzed. If a

third component absorbs at the same wavelength as the desired component, this compensation is made by filling the filter cell with the interfering gas. Severe overlapping of spectra cannot be eliminated satisfactorily, and a substantial increase of the interfering gas in the process will be erroneously read as the principal variant. Flow and pressure conditions should be the same for both the reference and sample streams.

The programmer for IR analyzers may or may not be separately housed from the sensing unit. In any event, both units are usually located together with a recorder, if required, in the control room. Housing would be compatible with the area electrical requirements. If the housing is explosionproof, then electronic control knobs should protrude through the front panel to permit calibration without opening the housing.

Conductivity Measurement

Electrolytic conductivity is a measure of the ability of a solution to carry electrical current. The measurement is nonspecific, and to be useful, there must be some knowledge of the ions that are in solution. Conductivity measurements are most useful in well-defined solutions where an indication of a change in ion concentration is desired. The measurements are widely used as a measure of water purity. A typical application is the monitoring of boiler feedwater to determine deviations from normal operation. The conductivity of boiler feedwater is normally very low except when impurities are entering the returned condensate from sources such as a leak in a heat exchanger. When leaks occur, the conductivity increases greatly. This knowledge allows the operator to cut off the contamination source until the leak is found and corrected, thereby averting damage to the boiler.

When conductivity measurements are used primarily as alarm functions where the alarm condition is set at a much higher value than the normal condition, they operate very satisfactorily even when there is little knowledge of the solution. When more detailed information is desired from the measurement, more information on the ions in solution must be available. Most ions exhibit a near-linear curve at low concentrations, and as concentration increases, the curve tends to flatten with most solutions, exhibiting a maximum conductivity or peak at about 20%.

A few solutions exhibit more than one peak. In such cases there must be enough knowledge of the conductive properties of the solution for the area of operation on the curve to be known. For example, at 25°C the conductive peak for hydrochloric acid occurs at 19%. A specific conductance of 0.75 mhos occurs at 12% concentration and another at 28%. If the measurement range were 0 to 40% concentration, the particular conductance value would be obtained at 12 and 28%, but there would be no way of knowing which concentration was being measured. In such cases conductivity measurements alone do not reveal sufficient information. Another analytical technique is needed.

Another major problem with conductivity measurement is the large variation in conductivity with temperature. Most solutions normally increase in conductivity from ½ to 3% per degree centigrade change. Ultra pure water conductivity, however, changes over 10% per °C with a solution temperature around 50°C. These changes indicate that temperature compensation is a very important part of precise conductivity measurements.

When a conductive solution has a constant temperature coefficient, the temperature compensator can correct for temperature variations quite easily, but if several different ions affect the measurement and the concentration of these ions change independently, some form of temperature regulation is necessary. Direct temperature control may be justified if no simpler solution exists and the potential results justify the cost.

Perhaps the simplest method of temperature control, and one that works well for some cases, is the cooling of a sample by overdesign of the cooler so that sample temperature closely approaches the cooling water temperature. It usually changes slowly enough (as the seasons change) to allow conductivity readings to be corrected, based on laboratory tests for temperature effects.

The accuracy of some measurements can be improved greatly by heating the sample. One example of this is sodium hydroxide which in the 10 to 25% range cannot be measured at 25°C but can be measured quite handily at 100°C.

pH Measurement

One of the most common analytical measurements used today, pH is often used for automatic control on many processes. There are several precautions, however, that must be observed in their application.

The glass electrode used for pH analysis has many limitations including:

1. A high electrical resistance
2. A high temperature coefficient of resistance
3. Absorbs preferentially a variety of ions other than hydrogen
4. The necessity of the glass surface to maintain an equilibrium with water

Because of its low voltage output signal, cable shielding is important and ground loops in the cable or conduit must be avoided.

The measurement of pH is subject to errors due to absorption of other ions such as sodium, lithium and potassium. The sodium ion error is the most common problem, occurring mostly at higher pH levels (i.e., lower hydrogen ion activity).

Temperature compensation is often required. The output of the glass electrode varies with temperature. Compensation normally supplied with a pH system corrects for these changes but not for hydrogen ion activity which also changes with temperature. Hydrogen ion activity changes cannot be easily compensated since the temperature coefficient varies with the type of solution. The error thus introduced can be large enough to make the pH measurement

useless. When this occurs, the sample must be temperature controlled or the method of temperature compensation must be "matched" to the process stream condition.

The measurement of pH in hydrocarbon streams is questionable because of the necessity of maintaining equilibrium between the glass electrode and water.

Reference electrodes must be selected to provide stable reference voltages. The silver-silver chloride and calomel electrodes have been used for years as reference electrodes for pH and ORP (Oxidation Reduction Potential) measurements. The electrode consists of an internal half cell, the electrolyte solution (KCl) and the liquid junction. The internal half cell maintains contact with the solution being measured through a flowing electrolyte junction. This junction is usually fiber, paladium, ceramic or a ground glass sleeve. A ground glass sleeve maintains the surest contact with the solution being measured since it is the least susceptible to plugging, but it requires that the KCl electrolyte be refilled more frequently.

Pressure does not affect the output of the reference electrode, but a constant outflow of electrolyte must be maintained through the junction; therefore, the electrolyte pressure must be greater than the solution pressure. In atmospheric installations, the electrolyte level itself provides the head. In submerged assemblies, a KCl head tank is used in conjunction with a sidearm electrode to provide the required head. Self-pressurizing electrodes can be used in closed flow cells up to 15 psig. Higher pressures require use of an air pad on the electrode and a differential pressure regulator to maintain the air pad about 5 to 15 psi above the solution pressure.

Beckman's new Lazaran reference electrode does not have the flowing electrolyte junction and appears to have many advantages over previous electrodes. It will work at pressures up to 150 psig without external pressurization. The measurement of pH at high temperatures (200°F and above) tends to shorten the electrode life.

Analyzer Housings

The increase in both the number and types of analyzers installed in the modern process plant has created a need for an evaluation of housing requirements for these instruments. Since the installation of an analyzer house is often a sizeable investment, the question of its justification often arises and justly so.

Purpose

Most process analyzers can be purchased in weatherproof and explosion-proof housings. Under normal operating conditions the analyzer can operate in the open (or with a minimum of protection) without being severely affected by the weather. However, during periods of calibration or repair when the instrument housing must be open, problems arise. Any analyzer, optical or otherwise, is adversely affected by blowing rain or dust. In addition, almost all precision analyzers require temperature control, and this is accomplished more easily in a temperature controlled en-

Figure 7.22. Where close temperature control is not required, analyzers may be installed satisfactorially in housings such as this. Entrance for maintenance purposes may be from the front or rear. (Courtesy of Winston Mfg. Co.)

vironment. Analyzer repairs, frequently requiring several hours of maintenance time, are more satisfactorily performed in a controlled atmosphere housing. A third justification for housing lies in the elimination of explosive and corrosive gases in the area of an instrument that must be opened for maintenance.

Construction

A small analyzer, not requiring close temperature control or sample conditioning equipment, may be installed in a meter type enclosure as shown in Figure 7.22. Entrance for maintenance purposes is restricted to two sides through the front and rear doors. Built of interlocking panels, the house is sturdy, weatherproof and may be conveniently mounted in crowded areas.

To afford better accessibility to an analyzer, a "shed" cover (Figure 7.23) may be used—sometimes with one wall added to protect from blowing rain.

Greater protection and installation flexibility are achieved by using closed analyzer houses (Figure 7.24). Constructed of interlocking panels, versatility is added by the optional selection of panels containing vents, windows, doors, fans, etc. Fireproof models are available to comply with individual company specifications. The house may be installed on a raised slab 2 to 4 inches thick to keep out water and other liquids.

The removal of vapors and the prevention of their entrance from outside sources should be a major consideration in the design of instrument housings. Ventilation requirements are determined by individual application factors, such as the presence of explosive gases, the velocity of prevailing breezes through the area, whether the analyzers have spark producing devices or whether they require environmental temperature control. Minimum requirements for any installation should include vents located near the floor in the walls of the upwind and downwind sides of the building. When vapors are present in the area, the upwind vent should be replaced with a forced-draft fan located high on the wall or on the roof. Volume displacement of the fan should be sufficient to displace the volume of the building once each 3 minutes. Air conditioned houses should be maintained at about 75°F throughout the year. Single pass, explosionproof units should be used with the intake at elevated level, if necessary, to prevent the entrance of hazardous gases. The cracks in the building should be sealed and one floor level vent installed. The vent should be small enough to provide a positive pressure in the house. For winter temperatures a steam radiator with thermostat may be used to maintain the temperature at a constant level.

Lighting in the analyzer house should be conducive to precision repair work. A general rule of thumb is one strategically located light over each analyzer in the building.

Utilities Required

Because of vapor potential, each analyzer should be wired for a Class I, Division I, Electrical Classification with each

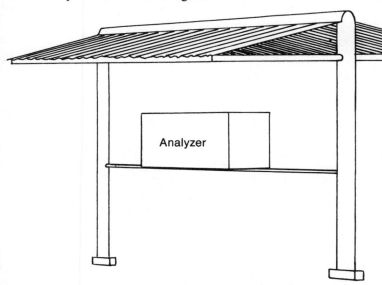

Figure 7.23. A "shed type" analyzer house allows more working room for maintenance purposes. A wall may be added to partially shut out blowing rain.

Figure 7.24. Closed houses afford the greatest protection for analyzers. They are frequently air conditioned for summer and winter to provide close temperature control of the analyzer units. (Courtesy of Winston Mfg. Co.)

electrical conduit sealed. Separate conduits should be installed for signal and power leads. Many plants require analyzers to be powered from a separate power panel with emergency backup rather than from a lighting panel. Each analyzer should have its own power switch located at the unit. An explosionproof receptacle should be provided in the house for small hand tools and other uses.

Both air and water are often used in analyzers for valve actuation, purging, cooling, etc. If they are not required for operation, they are useful for cleaning and unplugging sample lines and their components. A floor drain may be permissible, provided consideration is given to destination of the drain—the pad outside the analyzer house, the chemical sewer, etc. Toxicity and explosivity of the vented material must be considered.

Steam is often used to vaporize or prevent condensing of a sample in the analyzer. In chromatographic analyzers using temperature programming, heaters are programmed to give a rate of rise throughout the detection cycle (see Chapter 9, Applied Instrumentation, Volume 1). Electric heating can also perform this function.

Other connections external to the analyzer house may include sample in, calibration gases, vents, air conditioner ductwork. The minimum elevation of air conditioner inlets must be such that Electrical Classifications and personnel safety are preserved. Air conditioner inlet stacks usually have a minimum elevation of 25 feet above existing terrain.

8 Control Panels

William G. Andrew, Gordon C. Tucker,
John G. Royle

Control panel designs, like the hardware used in them, have changed appreciably in the past few years and, without doubt, will continue to do so. The reasons for these changes have resulted primarily from

1. The reduction in size of individual instruments
2. The greatly increased quantity of instruments for which a single operator is responsible
3. The availability of new hardware

Prior to the 1950s, large case instruments then in use permitted only a few instruments in a relatively large area (Figure 8.1). A typical 8-foot wide by 7-foot high panel (56 square feet) might contain 12 controllers and recorders plus a few indicators, alarms and miscellaneous controls.

With the advent of miniature control stations, panel designs in the mid 1950s contained many more instruments. (Figure 8.2). The same typical panel (8 feet by 7 feet) contained over twice as many controllers and recorders with more indicators, alarms and miscellaneous controls than its counterpart a few years earlier.

Even greater changes in panel design occurred in the 1960s as the high density layout concept (see Figure 8.3) was introduced along with further miniaturization of panel hardware. This great concentration of instruments allowed the same size panel (8 feet by 7 feet) to contain more than 100 controllers and recorders plus additional indicators, alarms and miscellaneous devices. Concepts involving the three distinct eras of design are discussed more fully in subsequent paragraphs.

Industry is now entering another era in which data acquisition and display systems and computer-controlled processes have produced layouts and special designs

somewhat different from the three shown above. This, too, is covered in greater detail later.

Before design details of panels and control room layouts are discussed, however, it is well to recall the basic purpose and function of the control panel, which is to provide a means of communication between the process and the process operators. It contains the instrumentation network which gathers, processes, controls and displays technical data necessary for efficient, safe plant operation. The control panel serves as the nerve center for the reception and dispatching of information relative to plant operation.

Control Room Layout

Control rooms for centralized plant control have changed and will continue to change as instrumentaton and automatic control concepts evolve into more sophisticated systems. The first control rooms were small and crude by today's standards, often being no more than partitioned areas in process structures. The modern control room is a quiet, air conditioned, well-lighted work center with the decor and furniture to emphasize its importance to the plant operation.

The design of control rooms and control panels that comprise the control center varies a great deal from one company to another; it varies even within the same company from one plant to another. The reasons include (a) changes in instrument hardware, (b) improved control concepts, (c) varying philosophies of plant operations and (d) personal preferences of the people who lay out and design the panels and control centers.

While the economics involved are always pertinent, intangible considerations in control room design are often

197

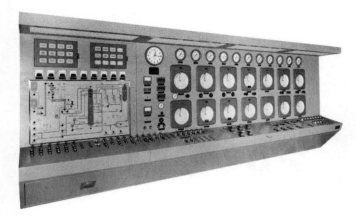

Figure 8.1. Prior to the early 1950s, large case instruments permitted only a few instruments in a relatively large panel area. (Courtesy of The Foxboro Co.)

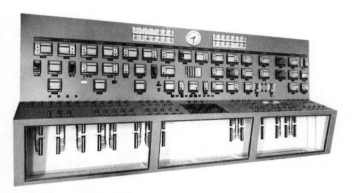

Figure 8.2. The introduction of miniature case instruments by the mid-1950s permitted more than twice as many instruments to be used in a given amount of panel space as large scale instruments. (Courtesy of The Foxboro Co.)

slanted toward personal preferences. However, the basic purpose of the control center must be maintained; that is, to provide (a) an area of personal comfort, (b) an environment compatible with low instrument maintenance costs and (c) limited traffic to personnel not directly concerned with the controls or their maintenance.

The final design and layout of the control center usually results from efforts of several groups, and the instrument engineer usually has the responsibility for determining that all areas of design meet his requirements.

Space requirements for panels, consoles and accessory cabinets should be carefully considered. The planning should include future expansions as well as current requirements. When future expansion is anticipated, the building should be designed for expansion at a minimum cost with few interruptions of existing operations.

Control panels and consoles should be arranged so that the operator has adequate visibility of the entire panel area for which he is responsible. Rectangular layouts similar to that shown in Figure 8.4 are usually preferable to long straight line layouts that remove operators from their equip-

Figure 8.3. Even greater miniaturization, together with the high density concept introduced in the 1960s, allowed much more information and control ability in a given amount of space. (Courtesy of The Foxboro Co.)

ment. Half of the layout shown might be covered by one operator. U-type layouts are also convenient for operator coverage. A scale layout of the control room floor and wall areas is very useful in obtaining an optimum layout. Plastic or wooden blocks may be used to show the locations of panels, racks, consoles and other equipment. They need not be fully detailed.

Sufficient space must be allowed between the rear of the control panels and room walls or accessory cabinets for maintenance accessibility. A clearance of 5 feet is desirable. Emergency egress from the panel rear should be considered. Rear door size and hinge location are important factors. With rear panel doors open and test equipment carts in place, quick exits might be hampered. Egress around either panel end is desirable.

Provision must also be made for interconnecting tubing and wiring between cabinets and from the control room to the field areas. The most flexible approach is the use of computer flooring to provide bottom entrances to panels, cabinets and consoles. This type of floor allows flexibility in the initial layout and also greatly reduces expansion costs when expansions occur.

Other possibilities include the formation of trenches in the floor under panels, the installation of individual conduits in the floor, the use of the ceiling area for conduit or tray, or tray and conduit in back of the panels.

Door openings in the control room must be laid out to avoid congestion in passageways. Doors should be large enough to permit entry of any size panel, cabinet or console that might be added to the building.

Within panel areas, accessory devices should be mounted so as not to interfere with future cutouts and expansions. This can be prevented by including instructions to that effect in the specifications sent to panel fabricators.

Electric Power Systems

As control systems become more complex, the need for stable power sources increases. Brief discussions follow on instrument power requirements, instrument power distribution, control room lighting and communications systems.

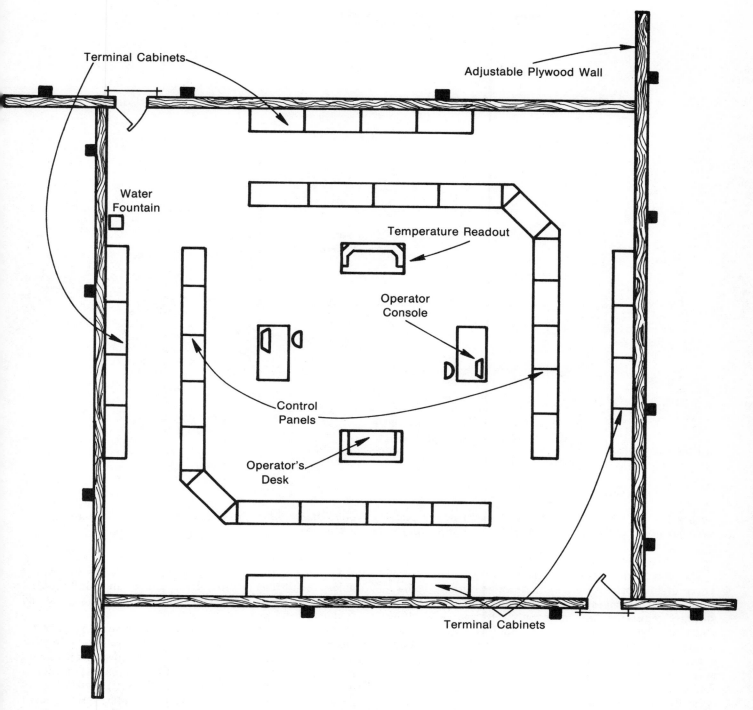

Figure 8.4. Control room layouts should be made to provide convenient operator surveillance for many instruments.

Instrument Power Requirements

The nature of electric generating and distribution equipment is such that it is subject to interruptions of varying lengths. Interruptions can be in the form of voltage dips due to motors starting, surges and/or flickers due to lighting, or relatively long term outages due to equipment or line failures. The requirements for instrument power systems depend primarily on (a) the nature of the process itself—how it responds to outages, (b) the characteristics of the instrument control systems and (c) the history of electrical power failures in the region of the plant.

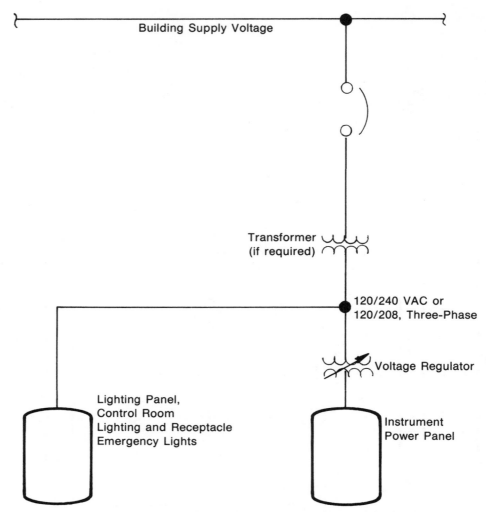

Building Supply Voltage

Transformer (if required)

120/240 VAC or 120/208, Three-Phase

Voltage Regulator

Lighting Panel, Control Room Lighting and Receptacle Emergency Lights

Instrument Power Panel

Figure 8.5. As a minimum, most instrument power systems require stabilization by constant voltage transformers which typically hold outputs to ±1% variation for ±15% input voltage changes.

The complexity of these systems varies considerably. Three or four basic systems are described briefly below.

Most instrument systems require at the very least a voltage regulated power supply. Figure 8.5 shows schematically a hookup used to hold voltage variations within the tolerances acceptable for most instrument systems. A typical transformer for a system of this type would hold voltage output to ±1% with an input variation of ±15% of nominal voltage. The total harmonic distortion typically would be less than 3%.

Most plants and processes require some kind of electrical power backup system in addition to voltage regulation. If the instrument control systems are designed so that a lag of a few cycles or even a few seconds can be tolerated without a major shutdown, a backup system similar to Figure 8.6 may be used.

A standby generator driven by a steam turbine or a gasoline or diesel engine is provided. A transfer switch is necessary in this case to start the generator and connect it to the power system upon loss of normal power.

Systems of this type may be on line in a few cycles, but unless the turbine or other driver is kept up to full speed, a few seconds delay is necessary for it to reach full speed and frequency and provide full output.

Many modern plants with complex control systems require more sophisticated power backup systems. When the instrument systems are designed to shut down the processes safely after a power failure, some kind of battery backup is necessary. Interlock relays, solenoids, etc., used as safety devices will drop out even on short interruptions, so it becomes necessary to provide uninterruptible power systems to prevent process upsets and delays.

The simplest method of achieving this type of backup is to use DC operated relays, solenoids and other interlocking devices. A battery and battery charger can then be used to meet the DC requirement. Figure 8.7 shows a typical system of this nature.

When AC power is required for emergency operation, an inverter must be provided along with the battery and battery charger. A static transfer switch is a desirable feature to in-

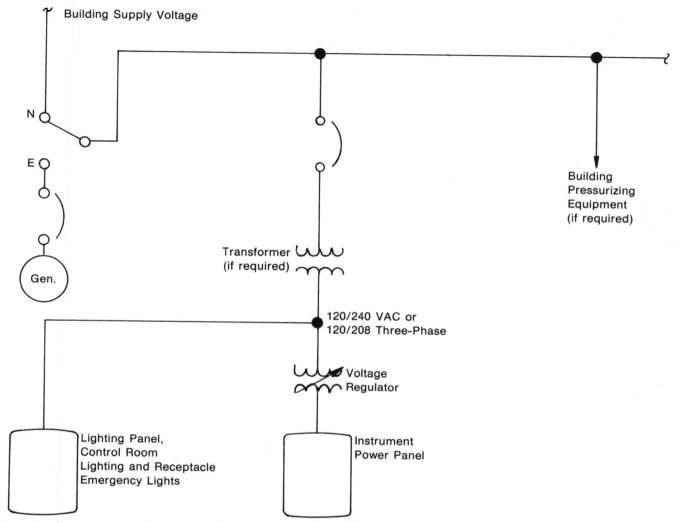

Figure 8.6. *The use of emergency generators for instrument power backup in the event of primary power failures is a common provision in processing plants.*

clude in inverter systems, especially for large systems. It connects the load to the AC line when the inverter fails. Figure 8.8 depicts this type system which is designed to switch over without interruption of power.

When battery systems are used, the ampere-hour requirements can be determined by using one or more of the following to develop a design criteria:

1. 1½ x the average power interruption interval
2. 2 x the required time interval to switch to an alternate power supply such as an engine driven generator
3. 1½ x the time interval the process could be operated after loss of primary power
4. Time interval of instrument operation required for orderly process shutdown

The battery charger should be sized to furnish power to all the instruments while at the same time having sufficient power to recharge the battery bank after discharge. An 8-hour recharge time may be considered reasonable.

The electrical equipment required for instrument backup power should be located in a separate room near the central control room. The emergency generator (when used) is usually located near the control center. The room used for electrical backup power may also be used for mechanical equipment required for the building—heating, ventilating and air conditioning.

Instrument Power Distribution

The method of distributing instrument power to individual users is a major factor in design reliability. Since all electrical devices can fail, due consideration must be given to the consequences of these failures.

When control instruments are powered by 115 volt AC sources, separate fuses or circuit breakers should be

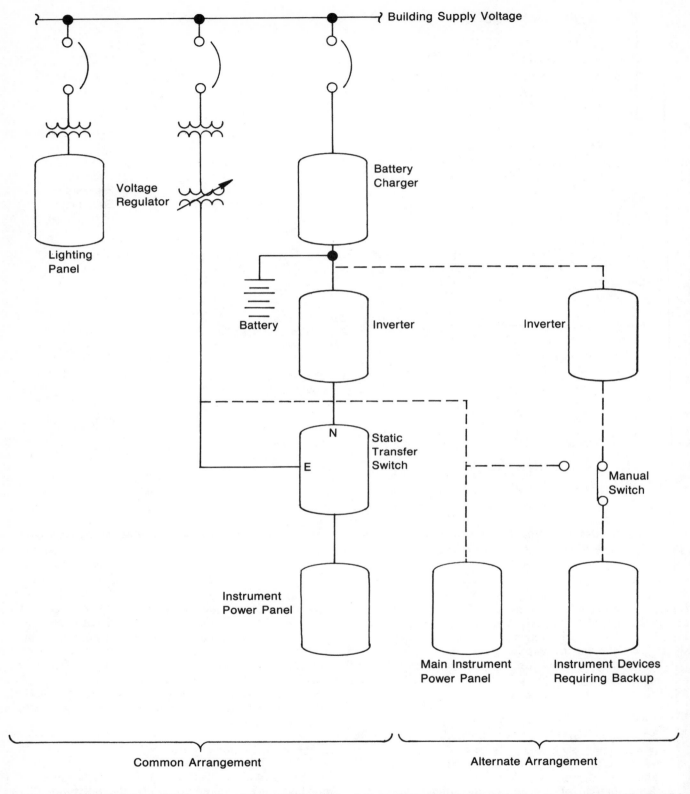

Figure 8.7. A typical, more sophisticated, uninterruptable power system that utilizes battery backup, an inverter and fast, solid-state switching to assure quick switchover to the backup power system. DC relays, solenoids and other interlocking devices are used in this system.

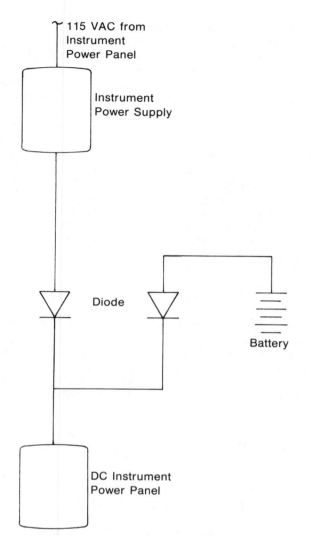

115 VAC from
Instrument
Power Panel

Instrument
Power Supply

Diode

Battery

DC Instrument
Power Panel

Figure 8.8. When AC power is used as backup, a static switch is used to connect instrument power to the load when the inverter fails.

supplied for each control device. This isolates the initial failure and prevents the shutdown of others.

When the instruments use an external DC supply, the system should be designed so that failure of one supply does not shut down the others. The system should also be designed so that wiring failures at any point should not shut down a large part of the power distribution system.

When inverters are not used for instrument power supply, circuit breakers are commonly used for circuit protection. When inverters are used, all circuit protective devices between the inverter and the instruments should be high-speed (current limiting) fuses. Otherwise, the time required to open a branch circuit breaker might be longer than the time required to blow a fuse in the inverter, causing the entire system to fail.

Control Room Lighting

Control room lighting systems should be designed to provide comfortable conditions for the operator so that operating fatigue is minimized. Current practice is to provide an illumination of 50 foot-candles for desks and consoles at the desk top. Illumination of 30 to 40 foot-candles should be provided on the vertical surface of control panels between approximately 45 and 60 inches above the floor.

The area behind the panels should be lighted to about 25 foot-candles. Some panel areas may require a lower lighting intensity to improve readability or visibility of equipment such as back-lighted displays or alarms, in which case dimmers should be considered. Rear-of-panel lighting should be purposely designed into the panel when it is known that the panel will be in a low light area, for troubleshooting and repair ease. Other areas may require higher than ordinary light for close work.

Emergency lights of some type should be provided for the control room. In situations where no other emergency backup power is used for instrumentation or other power requirements, battery-operated lights should be used for emergency use.

When backup systems include emergency generators, steam or gas driven, emergency lighting would come from those sources.

Communication Systems

Modern control centers require good communications between control room operators and field operators because of increased operator responsibility in the control room and because of increased interconnections of processes within plants. Communications between control rooms of the various plants should be maintained. For instance, a process unit needs to communicate with the utilities department(s) and with other units that may furnish raw materials or receive finished products.

The systems that provide these services (with the necessary redundancy) may include a mixture of the following approaches:

1. Plant telephone system
2. Telephone bellboys
3. Two-way radio
4. Hard-wired point-to-point communications
5. Paging systems

Electrical Classifications

A significant factor in the design of control panels is the electrical classification of the control room. It dictates the type of equipment and electrical hardware used and the methods of wiring. Chapter 2 provides guidelines on electrical area classification and describes the methods that are used to lower the hazard ratings.

Because of the higher cost of panel equipment and hardware and the increased difficulty of their maintenance, the trend is toward pressurization of control rooms so that they meet General Purpose requirements. Division II classifications are still sometimes used, but Division I classifications are now rare except for old, existing control rooms.

Figure 8.9 illustrates the contrast of wiring methods used for general purpose and Division I classifications and suggests the economy of the former.

Control Panel Types

There are three basic panel types—flatface, breakfront and console. These may be subdivided into additional groups determined by the method of construction used.

Flatface Panels

For many years flatface panels have been the most popular design. They probably will remain popular for a long time. Flatface designs are subdivided into three types—the turnback, angle frame and angle frame modular designs.

Turnback Design

The flatface panel of 10-gauge steel (a common thickness) with 2-inch turnback construction is a simple, economical design. It is usually fabricated in sizes from 2 to 4 feet wide and heights from 6 to 8 feet. Depths range from 18 to 36 inches. They are often mounted on a 6-inch concrete curb or steel channel base with independent bracing at the top of the panel. The bracing members serve an additional function as conduit and tubing supports. This type of panel design is used for simple gauge panels, for direct connected large case recorders and controllers and for small groups of local instruments which require minimum rear-of-panel support and only a few panel mounted auxiliaries. The elimination of support framing behind the panel facilitates installation and maintenance of the instruments and associated hardware (see Figure 8.10).

Open Angle Frame Design

The second type of flatface panel is the open frame design (Figure 8.11). The framework is self-supporting, and the panels are usually made of 3/16- or 1/4-inch steel plate with a 2-inch turnback and an open angle iron frame extending from 24 to 36 inches to the rear of the panel face. The panel face with its 2-inch turnback serves as the front frame and 1½- x 2- x 1/4-inch angle iron is used for the open framing on the rear of the panel. The width of this panel varies considerably. Four-foot sections are common, but longer widths may be furnished. The panel height is usually 7 feet-6 inches, and it may be mounted directly to the floor or placed on a curb or channel iron. This type of panel construction is generally devoted to small units where there are no requirements for top or end closing.

Tops may be added to these open frames, and several units can be placed together to form a line of panels. Sides are placed on the outside ends of the end panels, giving a single unit appearance. This is a popular approach which leaves open ends for ease of maintenance, repair and troubleshooting.

Angle Frame Enclosed Design

A third type of flatface panel design uses an angle frame enclosed construction (Figure 8.12). The essential difference from the open frame type just described is the attachment of enclosing sections on the sides, top and rear. Full length doors are usually provided for access to instruments and accessories.

Angle frame modules are built in several sizes. Heights range usually from 7 to 8 feet; 30-inch depths are common; and 4-foot modules are typical widths. A 1/16-inch panel overhang on each side of the 4-foot frame is usually allowed to assure tight fits at seams between panel sections. Sections are bolted together to provide the necessary overall length.

The complete group of panels is usually mounted on studs with panel bottoms at or near floor level. This type of panel construction is probably the most commonly used at the present time. It is adaptable to all types of instruments, electronic or pneumatic, single or multicase miniatures with their varying space requirements. Provision can easily be made for rear-of-panel mounted components, piping and wiring.

The simplicity of the panel and frame construction makes panel additions and extensions economical and panel revisions practical. Panel fronts can be made removable so that revisions and modifications can be accomplished more easily. This is not a common practice, however. Removable doors allow easy access for maintenance, repairs, additions and changes.

Breakfront Panels

Another type of control panel design is the breakfront configuration, sometimes referred to as the "stand-up console" (see Figure 8.13). The panel front is formed or broken into several planes to effectively increase the usable area of the panel front without a corresponding increase in overall height.

The general construction of the breakfront panel is a basic 1½- x 1½ x 1/4-inch angle iron frame with a formed panel front. The lower "bench" section of the panel should be 3/16- or 1/4-inch plate to support the weight of the mounted instruments properly. The angle or center section may be 10-gauge steel with internal cross members to support the instrument cases, front and rear, with limited weight stress on the panel face. The top vertical portion of the panel may be 10-gauge steel, or 1/4-inch phenolic may be used. The overall height of this type panel varies from 6 to 7 feet and panel width from 4 to 6 feet.

For ease of maintenance, the section of panel front beneath the panel bench break should be removable, and two full doors should be provided at the rear. Doors should have rubber gasketing, three-point latches and automotive locking handles. Removable panels should be used for end closures.

It may be noted that the additional front-of-panel utilization of this type configuration is obtained at the sacrifice of internal and rear mounting space. This may require external rack mounting or cabinet mounting of components often mounted in the rear sections of other panels.

a.

b.

Open Wiring Specifications

General

All wiring is bundled and run open or enclosed in vented plastic wireway.

All conductors run open are bundled and bound at regular intervals, not to exceed 12 inches, with nylon cable ties, or equal.

All wires within a bundle are run parallel to one another and are not twisted.

Bundles have a uniform appearance, a circular cross section, and are securely fastened to the panel (cabinet) framework.

Enclosures

Where the use of open bundles is not practical, conductors are run in vented plastic wireway.

When it becomes necessary to enclose equipment for protection or safety, such equipment is enclosed in sheet metal housings. Where conductors are run to a metal enclosure from either an open bundle or from plastic wireway, the enclosure is fitted with a suitable insulating bushing.

All wireways that are requested to be constructed of conduit or electrical metallic tubing (E.M.T.) terminate at a sheet metal pull box, for customer convenience, within the panel or cabinet. All pull boxes are provided with pilot holes and plugs for standard conduit fittings.

Conductors carrying alternating current and conductors carrying direct current, or carrying different voltages that are from the same source, may occupy the same wireway provided all are insulated from the maximum voltage of any conductor in the wireway.

Wiring carrying voltages that originate at a different source are not run in the same wireway.

Figure 8.9. The cost of material hardware and the ease of maintenance are contrasted in these views of general purpose (a) and explosionproof (b) wiring methods used for connecting instrument systems. (a. Courtesy of The Foxboro Co.)

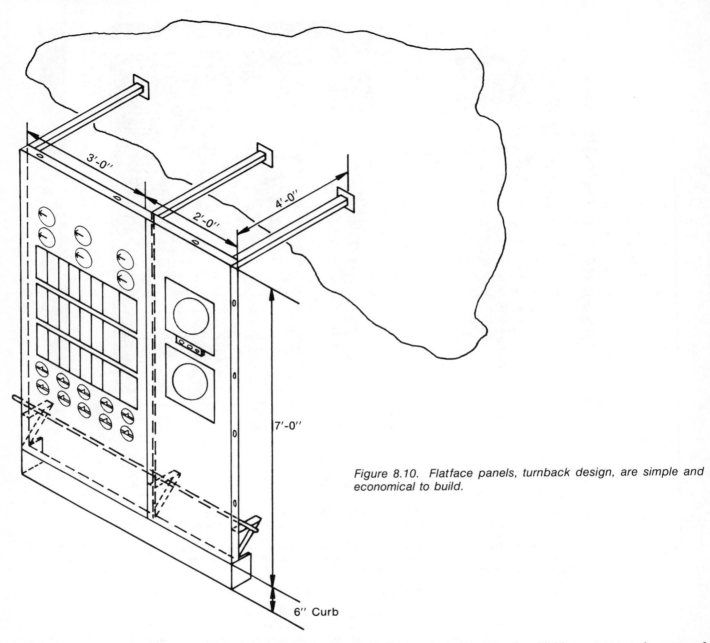

3'-0"

4'-0"

2'-0"

7'-0"

6" Curb

Figure 8.10. Flatface panels, turnback design, are simple and economical to build.

The breakfront panel design lends itself to high density instrumentation which utilizes miniature or multicased instruments and requires many push buttons and pilot lights that must be observed by operators. This panel configuration is perhaps best suited to electronic instrumentation, although either pneumatic or electronic may be used. The instrument internals of either type must be removable from the front of the panel.

Consoles

A console by definition is "a panel or desk on which are mounted dials, switches and other apparatus used in centrally controlling electrical and mechanical devices." This definition is indicative of a highly concentrated center of control instrumentation. Two typical designs and configurations are discussed and illustrated below.

Desk Type

An economical console may be made by modifying a commercially available office desk. The addition of a small panel for attachment to the desk top often meets the requirements of a console application. The addition may be formed from 10-gauge metal, fabricated from phenolic plastic or from a combination of metal and plastic.

The configuration of the section can be "tailored" to the requirements of the instrumentation and the desk selected.

Figure 8.11. *The open angle design is used when panel sizes become large enough to require supporting structures for support and for ease of handling; installation sizes may vary from 3'-0" to 7'-0" or 8'-0" in width.*

Figure 8.13. *The breakfront design or standup console increases the effective panel area without increasing height and/or width of panels. (Courtesy of The Foxboro Co.)*

Figure 8.12. *The angle frame enclosed design provides the same stable framework of the open design and further encloses and protects instruments, tubing and wiring. It is particularly useful where many wiring terminals are required.*

Two considerations of design are (a) a low console profile and (b) a concealed wire or cable duct for electrical leads.

This type of assembly and fabrication is used for multipoint readout systems for temperature, level or other functions that may be monitored through a single unit. It utilizes a switching matrix for individual point selection. It also provides a convenient location for communication equipment, and the desk area allows working room for desk type duties.

Custom Type

Figure 8.14 shows a custom-built console and vertical full-graphic panel in use in a chemical processing plant. The operators have key data digitally displayed on CRTs on the console and are able to observe the status of key valves and circuit breakers on the active graphic panel. Analog recorders, controllers, and indicators for important process loops are located on adjacent auxiliary consoles and vertical panels.

The construction of this type of panel is usually of specially formed steel sections assembled with welded seams. Entry into these units is normally from under the floor. Special under-floor ducts or computer flooring is normally used for connecting cables.

Modular 19-inch Rack-Type

Many manufacturers now furnish instrument cases and mounting plates to allow installation of their instruments in 19-inch (nominal 21 inches overall width) racks. A number

Figure 8.14. This custom-built console allows easy operator access to critical control loops while standing panels back of the consoles offer easy surveillance to many other less critical systems. (Courtesy of Shell Oil Company)

of manufacturers now furnish desks and consoles which will accommodate the 21 ± inch depth of control instruments, and which are a standard product line rather than being custom-built in each case (Figure 8.15).

Flat blank 19-inch plates can be purchased in several heights. Special cutouts are sometimes obtainable through the rack manufacturer, but are often done by the purchaser or at a panel shop.

Easily replaceable rack plates can be custom-painted for color coding by process unit. Since their cost is very nominal, obsoleted plates are often thrown away when replaced by new ones.

Comparison of Panel Types

Each panel type has its advantages and disadvantages. These advantages and disadvantages should be considered carefully before selecting the type of panel to be used.

Flatface or vertical panels are the most economical. The turnback variety is used primarily for local control, and its equipment is readily accessible.

The open angle frame type is only slightly more expensive. Its equipment is accessible and it is useful not only for local control applications but also for central control room requirements. Its appearance may not be particularly pleasing, but it is functional and economical.

The angle frame enclosed type is esthetically more acceptable for modern control centers. Although more costly than the open type, the enclosure feature allows economical wiring practices, protects equipment from dust and offers increased protection from tampering by unauthorized personnel.

Breakfront panels are more costly to build compared to the flatface variety. Their equipment is more difficult to maintain, and the lack of panel space may require the allotment of other space for mounting accessory equipment. However, they offer more space for a given width.

Consoles are used for special applications. Because most are custom-made, their cost is relatively high. Because their height is kept low, the number of instruments per panel foot is also low. They are well-suited for concentrations of critical loops that need close monitoring. Seldom will they be used exclusively for a particular control area.

Panel Layout

The panel layout has an appreciable effect on process operation. If the operator can easily watch the key loops and keep them in close control of the process, the operation goes much more smoothly and probably is more profitable.

The layout also affects instrument maintenance. The units that require much attention should be located for easy, convenient access.

Face Layout

The instruments and controls in the panel should be arranged to display information from the process to the operator in a logical manner. Usually this means grouping all instruments associated with a particular process section and arranging these groupings in the panel in the order of process flow.

To allow for process and/or control changes, space should be allowed in the panels for future additions so they will fit into the process scheme in a logical sequence.

The three general sizes of instruments most commonly used are

1. Large case conventional, nominally 14 inches wide x 18 inches high
2. Miniature, 6 inches wide x 6 inches high
3. High density multicase, 3 inches (or less) x 6 inches high

Large case instruments are usually mounted three rows high with the bottom row at a relatively low level. They are rarely used in modern control centers because of space limitations.

Miniature instruments are usually mounted four rows high with instruments on 10-inch centers vertically and on 9-inch centers horizontally. The use of only four rows reduces space congestion at the rear of the panels and provides accessibility to instruments for maintenance.

The mounting configuration for high density instruments varies with the manufacturer. A typical vertical spacing places the line of instruments on 12-inch centers. A common characteristic of high density instruments is that they mount in multiunit cases which may be installed and wired prior to installing the units—indicators, recorders and/or controllers. Single unit cases are also available. When several units are installed side by side, bracing of the instruments is almost always required.

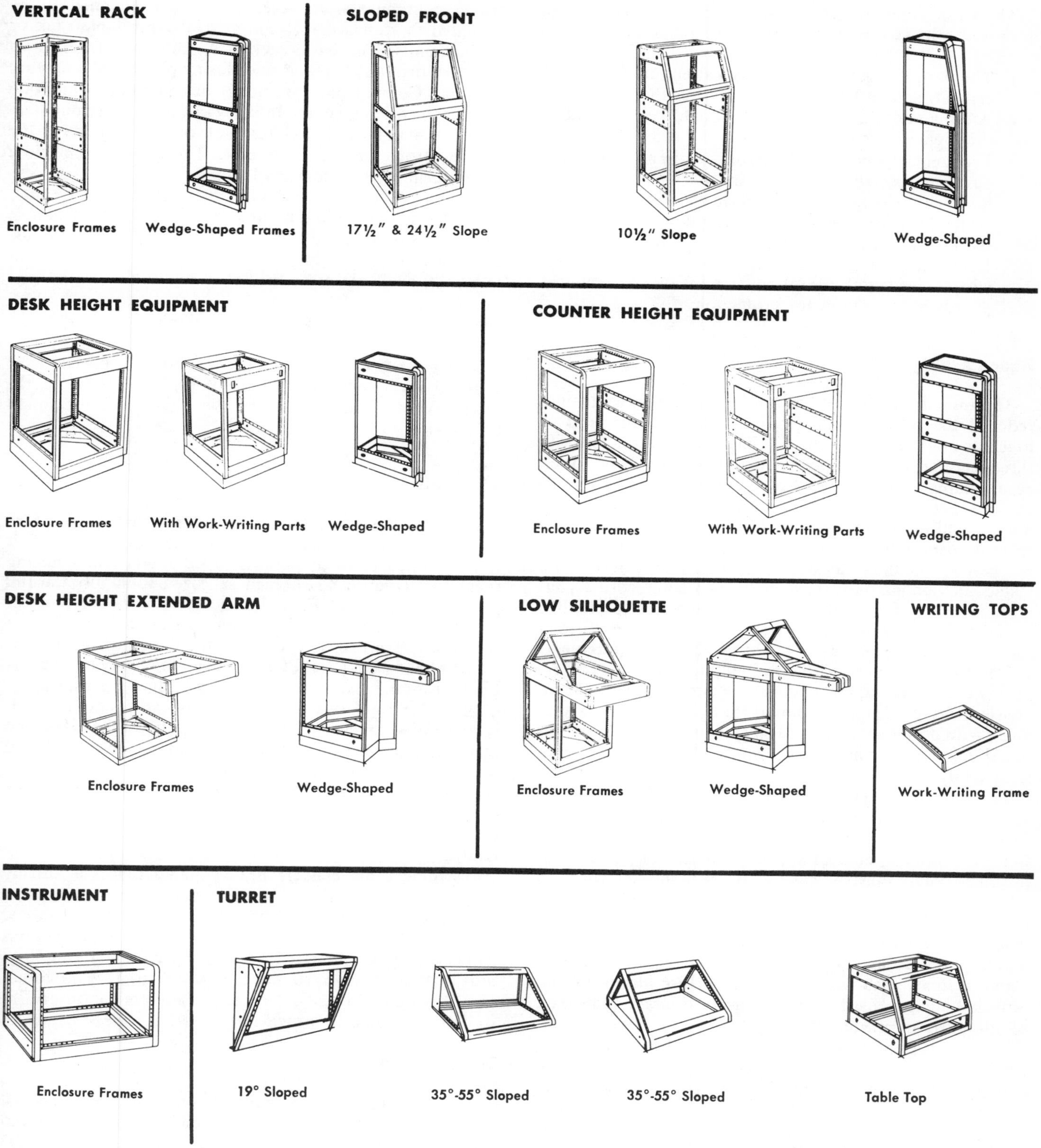

VERTICAL RACK

Enclosure Frames

Wedge-Shaped Frames

SLOPED FRONT

17½" & 24½" Slope

10½" Slope

Wedge-Shaped

DESK HEIGHT EQUIPMENT

Enclosure Frames

With Work-Writing Parts

Wedge-Shaped

COUNTER HEIGHT EQUIPMENT

Enclosure Frames

With Work-Writing Parts

Wedge-Shaped

DESK HEIGHT EXTENDED ARM

Enclosure Frames

Wedge-Shaped

LOW SILHOUETTE

Enclosure Frames

Wedge-Shaped

WRITING TOPS

Work-Writing Frame

INSTRUMENT

Enclosure Frames

TURRET

19° Sloped

35°-55° Sloped

35°-55° Sloped

Table Top

Figure 8.15. Rack-type panels are now manufactured in many standard configurations. (Courtesy of Emcor Products)

The arrangement of instruments in any of the panel types discussed follows essentially the same guidelines.

1. The loops requiring the most attention (the most critical) should be mounted close together.
2. Alarms and indicating devices should be mounted at the highest levels.
3. Frequently operated controllers, hand switches, pushbuttons, etc., should be located at a convenient level within easy reach of tall or short operators.
4. Instruments that require little attention should occupy the lowest panel areas.
5. Emergency switches should be located in positions where accidental operation is unlikely. Mechanical guards may be used for protection.

Rear Layout

The miniaturization of control instrument faces has not reduced the total volume requirements of their cases as much as might be expected. The large case instrument with three recorder pens and one control function utilizes a total case volume of approximately 1,140 cubic inches or 285 cubic inches per point. The 6- x 6-inch "miniature" instrument requires about 1,575 cubic inches to house three recorder pens with one controller added or 394 cubic inches per point. The "high density" 3- x 6-inch unit, normally a single function instrument except for two-pen recorders, has a requirement of 740 cubic inches for each unit case or 370 cubic inches per point. The face size reduction loses much of its significance when viewed from the volume requirements behind the panel face. With essentially the same volume requirement for instruments behind the panel and with the increased wiring or piping required for a high density layout, along with power supplies, relays, alarms, bulkhead and terminal box requirements, the ingenuity of the panel designer is certainly taxed.

The rear panel layout must be arranged to maintain accessibility of all components for maintenance without removal of other components, wiring or piping. The location of terminals, control adjustments and calibration adjustment points must be easy to reach, adjust or check.

Accessory equipment can often be mounted in back of the panel face in the lower and upper regions of the panel where few, if any, instruments extend through the panel. There is usually more accessory equipment space required for electronic systems than for pneumatic because alarm and interlock functions used in electronics are larger than their counterparts in pneumatic systems.

When sufficient space is not available at the panel rear, auxiliary racks or cabinets are necessary. These should be mounted along walls in back of the panels as shown in Figure 8.4.

Auxiliary Racks and Cabinets

Auxiliary racks for accessory equipment and termination points may be custom-built (a common practice) to blend in or conform more closely with the custom-built panels. Electronic components, however, are often designed to be mounted in standardized, commercially available rack or cabinet units. The basic rack is a preformed frame with standard drilling. Top and side closing plates and doors are optional. Components specifically designed for installation in this type of rack are mounted by the instrument manufacturer on standard size plates that become a part of the cabinet front in the final unit. These units are more economical than custom-made units.

For pneumatic accessory equipment or electronic equipment that is not made for standard mounting racks, simple open racks may be built in the field or shop-fabricated, using welded or bolted members of angle iron or other commercially available preformed steel material.

Panel Piping and Tubing

Piping and tubing methods for control panels have changed slowly over the years. For practical purposes, the discussion is divided into air header piping and instrument tubing runs.

Air Headers

Galvanized pipe is normally used to supply instrument air to control panels from the main instrument air supply. Air is usually piped to dual, parallel piped filters and reducing regulators (Figure 8.16), each set sized to supply the full panel requirements. Each set is valved for isolation for repair and replacement. The reducing station should have a pressure gauge and a relief valve for the panel air header. The air header and the filter-reducing units may be sized on the basis of 0.5 standard cubic feet per minute per air user.

From the regulating station, the panel air header is usually made of seamless red brass pipe although galvanized iron and aluminum are also used. The header should be sloped from the dual filter regulator station to a ½-inch blowdown valve located on the bottom of the header pipe at the low end of the header. A header sized gate valve should be placed at the end for future use if the header needs to be extended.

Instrument air headers are available that are already tapped for individual air users to reduce labor cost in panel fabrication (Figure 8.17). The individual air takeoffs should be supplied with ¼-inch brass valves, needle or packless diaphragm type, mounted on ¼-inch nipples or couplings brazed or silver soldered into the main header. A minimum of 10% (preferably 20%) spare air takeoffs for future installations should be provided.

Tubing Runs

All tubing runs should be in horizontal or vertical planes, carefully grouped for accessibility and ease in tracing and clamped for rigidity. Panel tubing is usually ¼-inch OD copper, aluminum or plastic. When plastic tubing is used, plastic trays with slotted entry openings are available for routing tubing behind panels between instruments and terminal points.

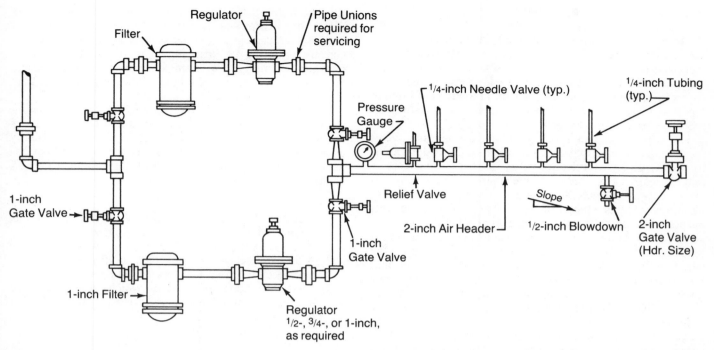

Figure 8.16. Panel air supply systems should have parallel filters and regulators sized so that either line can supply sufficient panel air while the other filter or regulator is being serviced.

Incoming and outgoing transmission lines should be terminated with bulkhead fittings in a bulkhead bar at the rear of the panel. These points should be labeled for field connections with at least 10% spares provided for future use.

Each auxiliary pressure switch or similar device which requires testing may be provided with (a) a valve and plugged tee and (b) a tube fitting or pipe union for ease of removal.

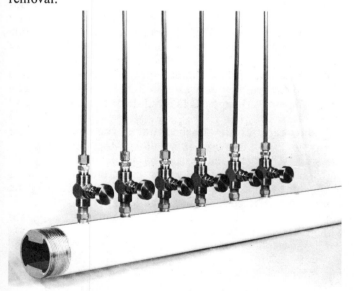

Figure 8.17. Instrument air headers are available which are already tapped for individual air users, reducing fabrication costs. (Courtesy of James C. White Co.)

Plastic tubing has been available since the early 1950s. Its use is increasing for pneumatic installations because of the reduced labor cost compared to metal tubing. Its appearance may not be as neat as formed copper and aluminum runs, but the use of slotted plastic or metal ducts to contain the tubes results in a satisfactory appearance.

Plastic tubing is manufactured in several colors, and installations may be color coded to conform to the Instrument Society of America RP7.2. Different colors are used for air supply, transmitter, valve and feedback lines.

Plastic tubing fittings are normally furnished in brass, are easily installed and can be tightened by hand.

When copper tubing is used, the copper should be sufficiently ductile to be easily formed, yet rigid enough to require minimum support when short runs are used. A commonly used type is ASTM-B68, ¼-inch OD with 0.030-inch wall thickness. Compression fittings are used to connect the tubing.

Panels piped with copper tubing are beautiful pieces of art if the craftsmen are good at tube bending. As previously noted, however, labor costs for piping rise significantly.

Aluminum tubing is sometimes used in panels installed in atmospheres containing contaminants that attack copper. It is soft enough to be formed easily but tends to work harden. It should not be used where vibration is a problem. Compression aluminum fittings are used in preference to flared or soldered types, just as for copper tubing fittings. They are much more economical from a labor standpoint.

Stainless steel, rubber, nylon, polyvinyl or glass tubing materials are available when needed for special panel piping requirements.

Panel Wiring

Most control rooms of plants built in the past few years are located in general purpose areas, and electrical equipment and wiring methods conform only to General Purpose requirements. This has not been true in the past, and many control rooms are located in Division II areas (and some in Division I areas) where wiring practices must conform to the requirements for hazardous atmospheres.

The use of air conditioning and pressurization allows most control rooms that are located within hazardous areas to be classified General Purpose by using a clean source of makeup air. The reduced mechanical protection requirements of this classification permits wiring to be run in wireways or ducts. One of the best types of wireway for this purpose is a slotted plastic channel that allows wires to be laid in the wireway with entrances and exits at the points required. Alternate wiring methods are (a) wrapping the wire bundles in spiral-cut plastic and (b) using cable ties that bolt or glue to the surface at short intervals—approximately 6 inches to 1 foot. Slotted wireways should be sized so that no more than 40% of the cross-sectional area of the wireway is filled.

All AC wiring in a panel should be routed in steel wireway to reduce noise in instrument signal wiring. If this is not feasible, it should be physically separated from the DC signal wiring as much as possible. Parallel runs of AC and DC wiring closer than 1 foot should be avoided.

Instruments requiring AC power should have a disconnect of some kind in the power leads. The most economical method is to provide a receptacle for each instrument. Instrument power leads are normally #14 AWG while single conductor signal and alarm wiring is usually #16 AWG. Multiwire cables are usually #18 to #22 gauge. One criterion that must be considered for instrument signal wiring is the mechanical strength of the conductors.

Terminal blocks should be provided for all external wiring to the panel, except for thermocouple wiring which should be direct. In open panels, terminal blocks must be placed in General Purpose enclosures or provided with some other insulated isolation. Snap-on or screw mounted covers for this purpose are provided by many terminal block manufacturers. Terminal blocks in enclosed panels require no cover strips or other enclosures. Terminal blocks should be of the mechanical clamp connection type. Wire termination at these blocks should be by bare conductor. Terminations at items not provided with clamp connectors should be made with self-insulating lugs.

Instrument terminals are generally contained in the instrument case or provided with insulated cover strips by the manufacturer, so they are suited to open or enclosed panels.

Circuits of 440 v are not permitted in central control room panels by many companies because of hazards to people who normally do not work with voltages higher than 110 v. If 440 v circuits are used, separate covers marked "440 Volts" should be provided for easy identification by maintenance personnel.

Wire termination lugs are of two basic forms—forked tongue or ring. These two types may be specified as nonin-sulated or self-insulating. Wire lugs are sized by the wire gauge they accept and the screw terminal size they fit. The lugs of sizes used in panels are usually attached to the wire end by soldering or crimping. The forked tongue, self-insulating crimp type is the most popular for panel wiring. Heat shrink tubing is useful in insulating lugs of other than the self-insulating type.

To assist in installation and troubleshooting, all panel instruments, back of panel components and terminal blocks should be identified by labels. Embossed tape is usually adequate for this. All wiring should be identified by permanent, nonsmearing labels at each end. Labeling methods include:

1. Adhesive backed vinyl or linen strips. Each strip is preprinted with a number (0 through 9), and multiple strips are used to make the required wire number.
2. Adhesive backed labels of linen or vinyl on which the numbers are typed before application to the wire.
3. Heat shrinkable spaghetti on which the desired wire number is printed with a verityper before placing on the wire.

Nameplates and Tags

Engraved nameplates should be provided for each instrument on both the front and rear of the control panel. All the components mounted behind the panel including air supply valves, air bulkhead terminals, electrical switches and circuit breakers, pressure switches, relays, transducers and other auxiliary components should be supplied with metal or plastic tags or nameplates.

Engraved nameplates mounted on the panel front normally display an abbreviated service legend and an instrument number. Chart factors and multipliers may be included on the front of panel nameplates or placed on separate plates mounted inside the instrument door. Rear of panel tags or nameplates normally include only instrument numbers.

Panel nameplates are manufactured of $1/16$-inch laminated plastic engraved through to a different color core. The engraving stock is supplied in a number of background and core colors.

The nameplates are attached to the panel face and other components by the use of small self-tapping screws or pressure sensitive, double coated tape. Front of panel nameplates should be fixed to the panel face. Instruments may then be moved and exchanged without disturbing the nameplates. A third less commonly used method of mounting nameplates is the use of extruded mounting strips into which the plates may be slipped.

The nameplate should be of a length appropriate to the instrument case width and high enough to provide a minumum of three lines of $1/8$-inch engraved characters. Engraving blanks $2\frac{3}{4}$ inches long x $\frac{7}{8}$ inch high will meet most requirements for front of panel mounting. Rear of panel plates may be $1\frac{1}{4}$ x $\frac{1}{2}$ inch.

Painting

The proper paint selection and steel preparation before painting determines the life expectancy and overall appearance of the panel's final finish. The control panels, racks, consoles or cabinets should be sandblasted to white metal or chemically cleaned and phosphatized before painting. The processes remove oil, rust and mill scale and provide an etched surface to assure good finish bonding to the metal surfaces. After cleaning, a minimum of two coats of primer surfacer should be applied and sanded to provide a smooth surface for the final finish coats which should be about 1 mil thick to provide a blemish-free automotive type finish.

The finish of the final color coats should be a suede or semigloss texture to minimize light reflection. Color selection for the exterior finish should be made from factory mixed colors. Local custom-mix colors should be avoided. Factory-mix colors tend to fade less, and repairs to mars and scratches on the panel finish can be made more satisfactorily. The interior or back of panel finish may be white or the same color as the exterior surfaces. White finishes enhance any light available for maintenance.

Modern paint technology has developed many special purpose finishes, but enamel and lacquer remain the most popular panel and enclosure finishes. Baked rather than "air-dried" finishes have a slightly longer appearance life.

Panels installed in operating areas may require the extra protection of galvanizing or special coating. This special protection is applied after the steel fabrication is complete and is usually restricted to the supporting frame and brackets. Epoxy based paint or other special environmental finish may be applied to the front panel surfaces.

Graphic Displays

Graphic displays are a special panel concept used to display process and control information as a simplified process flow diagram.

The flow diagram is manufactured from (a) laminated plastic or phenolic engraving stock, (b) engraved and back painted translucent vinyl or similar plastic, (c) translucent symbols sandwiched between translucent plastic pieces, (d) snap-in plastic squares with painted symbols, (e) formed metal equipment symbols with extruded metal lines, (f) symbols painted directly on the face of the panel, or (g) symbols generated by a computer and displayed on a CRT.

There are two general forms of process flow diagrams—"full graphic" and "semigraphic" displays. The full graphic concept is manufactured on a scale large enough to locate miniature case instruments in the equipment outlines (Figure 8.18).

This type of presentation of the process diagram results in low density arrangement of instrumentation on the face of the panel because of the size of the equipment symbols. The economics of the increased overall panel length requirements and poor area utilization, together with the cost and difficulty of flow diagram modification, limit the use and popularity of this concept. At one time, the full graphic

Figure 8.18. Full graphic panels provide graphic representations of equipment and lines large enough to mount miniature instruments in positions portraying their use. (Courtesy of The Foxboro Co.)

display was fairly popular but is seldom used now except for special systems such as conveying and storage systems. In these instances, instruments are usually limited to lights, pushbuttons, alarms and similar small devices that require little mounting space.

The semigraphic process diagram presentation is of a scale size that may be attached to a 24- to 30-inch section of the panel area of a flatfaced or breakfront panel configuration. The most desirable panel area employed to mount the semigraphic display is the top of the panel face. The bench section of the breakfront panel or the entire face of a pushbutton and pilot light console may be used.

The semigraphic display may be reproduced in several different forms utilizing different materials. The simplest form is a blueprint reproduction of the mechanical flow sheets mounted in a plastic or glass-front frame on the panel face (see Figure 8.19). The information displayed may be easily changed and updated by replacing the reproduction as drawings change. The usefulness of the display is limited almost entirely to operator training and education. One manufacturer has carried this concept further by producing translucent lines and symbols with adhesive backing. Carefully placed between plastic sheets, this quickly produced, graphic flow diagram can be just as easily changed, using standard symbols available from the manufacturer. Symbols are available in several sizes and many colors. Magnetically mounted lights can be placed behind the flow symbols, creating an alarm-active graphic (Figure 8.20).

The most popular form of semigraphic display is a plastic or phenolic engraving stock diagram attached to a phenolic panel section similar to that shown in Figure 8.19. Metal backgrounds are not recommended for use with plastic symbols and lines because their different temperature expansion

Figure 8.19. A popular form of semigraphic representation is the use of phenolic engraving stock attached to the steel panel face. (Courtesy of The Foxboro Co.)

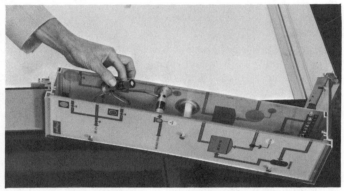

Figure 8.20. Magnetically mounted lights are placed behind flow symbols, creating an alarm-active graphic. (Courtesy of Fitzgerald Engineering Co., Inc.)

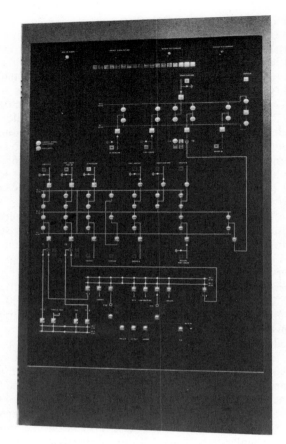

Figure 8.21. Snap-together square tiles can create an easily-changed graphic. (Courtesy of Entrelec Division of Cogenel, Inc.)

coefficients tend to break the bonding of the adhesives used to attach the symbols and lines to the metal.

The graphic display and the instruments on the panel face should be arranged so that instrument symbols on the flow diagram are located above the instruments indicated by the symbols. The instrument symbol and the instrument nameplates may be color-coordinated for rapid identification and association by the operator.

The semigraphic may also contain an annunciator system using remote relays and "bullseye" lights located in the graphic display to identify points of malfunction. Color coded pilot lights are used to indicate on-off status of pumps, open-close action of valves and other remote equipment conditions or functions shown on the display.

Recent innovative graphic designs enable the user to correct flow symbology with ease. Several manufacturers now furnish snap-together square tiles. Symbols on these tiles can be backlighted for active graphic operation. To change

the graphic, affected tiles are removed and replaced with those having the correct symbols.

The tiles are mechanically supported at regular intervals using adjustable metal standoffs. Since a slight amount of flexibility can be provided in the tiles and their supports, large graphic assemblies can even be curved, within limits (Figure 8.21).

The use of metal symbols and lines mechanically attached to the panel face or other background panel retains the same basic design and function as the phenolic graphic display, except that color coding of the symbols and lines is accomplished by painting or anodizing. The metal graphic display has a much longer mechanical life than other commonly used manufacturing materials but is more difficult to modify and update. These two features somewhat restrict its use.

The painting of the flow diagram directly on the panel is another method of presenting process information. The cost is relatively low, but changes are not easily made and the method is not a popular one.

Backlighted graphic panels of small dimensions may be developed using back engraved, color-filled vinyl plastic sheets sandwiched between glass or clear plastic sheets.

These are mounted in a frame with an enclosure on the back to contain the light source. This type of display is usually furnished by vendors of annunciator and supervisory systems and purchased as part of an equipment or systems package.

Panel Bid Specifications

The preparation of good control panel specifications is not an easy job. Drawings, sketches and diagrams are prepared which furnish an overall view of the work to be done. These must be supplemented by written specifications which provide general conditions as well as detailed information on the panel fabrication requirements.

The combination of drawings, sketches and written specifications should convey to the panel fabricator the exact amount of work to be done. The job will be done more economically if instructions are clear and no doubt exists on the job scope. The length and detail of these written specifications depend on the complexity of the panel and equipment to be installed and the engineering details to be completed by the panel fabricator.

The following topics should be included and discussed in the specifications.

Scope

The scope should define the extent of the design and construction required. It should list the number and size of panel sections to be fabricated and define the engineering design responsibility of the fabricator. It also should designate who furnishes instruments and equipment. Typically the contractor or user furnishes instrument items such as controllers, recorders, indicators, annunciators, converters and similar items, while the panel fabricator furnishes pilot lights, push-buttons, selector switches, electrical relays, pressure switches and other similar hardware.

The responsibility for furnishing instruments and materials is clarified simply by providing a list of user-furnished items with a statement that the fabricator furnishes anything else indicated by the drawings and specifications. If it is desired that fabricator-furnished items, such as pressure switches, be compatible with existing spares, then a list of acceptable vendors should be included in the specifications, or complete pressure switch specifications should be included for the fabricator to use when bidding.

Bids

Bidders for panel fabrication should be instructed on the content of their proposals. It should include such information as:

1. Total price
2. Price breakdown if units can be broken down
3. Delivery schedule
4. Delivery cost and method of delivery
5. Schedule for drawing preparation
6. Unit rates for additions and deletions to contract

Drawings

A list of drawings with identifying numbers and titles supplied by the user should be furnished as part of the specifications. The requirements for drawings and engineering details to be supplied by the fabricator should be described. The prescribed manner for customer approval of fabrication drawings should be listed.

Typically, the contractor or user furnishes only drawings, schematics and diagrams that develop the physical design of the panels and present key control concepts that need to be followed. The fabricator is usually responsible for shop and fabrication details, physical wiring diagrams, and tubing and piping layouts. It is sometimes advantageous to have the design done by the contractor or user rather than by the fabricator. The fabricator can do the constructing directly from furnished drawings. This will generally result in a lower price, less hours spent reviewing the job (approval prints), and perhaps a quicker delivery. These things should be reviewed in the light of man-hours to be expended to complete the preliminary design already done to permit bidding.

Codes and Standards

Applicable codes covering installation and work to be performed by the fabricator should be listed. Deviations from work methods of such codes called for in drawings or written specifications should be made clear.

Applicable codes and standards may include the National Electrical Code, Section 12 (Instrument Panels); Part 1 of API RP 550, the American Petroleum Institute Manual on process instrumentation; and pertinent ISA Standards and recommended practices (ISA RP7.2, Color Coding for Panel Tubing, for example).

Panel Construction

The type of panels should be described. Specific details should be given on mechanical construction such as material types and thicknesses, joints, finishes, support requirements, mounting requirements, type of door hardware, accessory supports needed, etc.

Painting

Describe the method of preparing the steel before painting. Specify the type paint and finish desired, describe the paint by manufacturer, color and application methods. Indicate whether the panel interiors are to be of the same or a different color.

Graphic Section (If Required)

Specify the materials to be used, the size of equipment and line representations, the color preferences for equipment and lines and the method of attaching lines and symbols to the background surface. If the methods of graphic fabrication are not well known, bidders may be asked to describe materials and application methods for evaluation during the bid analysis period.

Nameplates and Tags

List the proper legends, specify materials, the size of plates and engraving, and indicate locations and mounting requirements. Specify the type tags and mounting methods for equipment in back of the panel. Accessory equipment needs to be identified, and panel mounted equipment with nameplates in front need to be identified at the panel rear also.

Piping

Specify the tubing materials to be used, the support methods preferred and the bulkhead locations desired. Give the air header size, its location, the type of materials used and the individual takeoff connections, valves, air sets, etc.

Electrical Systems

Provide details on materials desired, wiring methods, power distribution requirements, termination locations and termination methods. Be sure to give the electrical classification of the panel and panel area.

Instrument Equipment

List and specify by manufacturer and type all equipment to be supplied by the purchaser. List and designate acceptable manufacturers of items to be supplied by the fabricator. Give information on wiring, disconnecting, switching and fusing. Provide mounting preferences and locations for auxiliary instrument items.

Testing and Inspection

Describe the testing and inspection to be performed by the fabricator. List the equipment and fabricator personnel required to assist in customer inspection.

Delivery

Establish the desired schedule for:

1. Approval drawings to be furnished by the fabricator
2. Panel completion
3. Inspection period
4. Arrival at the job site

Shipping

Specify the crating requirements, type of conveyance, F.O.B. point and insurance requirements.

Guarantees

Set conditions under which a panel or equipment may be rejected and establish the length of warranty time for fabricator-furnished materials.

A typical set of instrument specifications is given below.

Scope. This specification covers the design and construction of five (5) prefabricated panels, 4 feet wide x 8 feet high x 30 inches deep, which shall be furnished complete with all instruments mounted thereon and all piping and wiring installed to terminal points. Any joints in structure, piping or tubing or discontinuity in wiring, required for shipping sections, should be so designed to require minimum work in the field to reassemble the sections. All such field connections should be clearly marked or tagged.

The user will furnish only the following drawings:

1. Front of panel layout with approximate location of equipment
2. Dimension drawings of user furnished equipment
3. Electrical schematic drawings
4. Schematic loop drawings

The fabricator shall furnish the following detailed drawings:

1. Steel fabrication drawings
2. Dimensioned drawings of the graphic section drawn to approximate scale of 3 inches per foot
3. Tubing and piping drawings including bulkhead layout
4. Electrical wiring and terminal board drawings
5. Any other drawings required for fabrication and assembly

The user will furnish all items shown in an appendix. (This would be attached to and become a part of the specifications.) All other equipment and material will be furnished by the panel fabricator.

Bids. The bid proposal shall be based on these specifications and the drawings and applicable codes listed herein. They shall include furnishing a sepia of dimensional and arrangement drawings for approval and a sepia of all final drawings required in the scope of this specification.

Six sets of parts lists shall be furnished for all fabricator furnished instruments.

The bid proposal shall include:

1. The total lump sum price
2. Estimated delivery price and proposed method of delivery
3. Schedule for: (a) preparation of drawings for approval; (b) receipt for user furnished items; (c) completion date
4. Unit rates for additions and deletions to all major equipment items

Drawings, Codes and Standards. The panels shall conform to the drawings listed in an appendix (This becomes part of the specifications).

Applicable codes and standards include:

1. The National Electric Code
2. ISA RP 7.2—Color Coding for panel tubing
3. API RP 500, Part I—Process Instrumentation and Control (Section 12—Instrument Panels)

Methods specifically mentioned in these specifications and drawings shall take preference to the codes listed.

Panel Construction

1. The panel shall be a free-standing, open type design, 20 feet long, 8 feet high and 30 inches deep. The panel shall be fabricated in five separate sections as detailed on a drawing. The channel base shall not be furnished.
2. The panel face shall be fabricated from 3/16-inch steel plate. The plate shall be cut to size on special squaring shears to ensure tight flush joints when butted together.
3. Materials shall be selected for levelness and smoothness and shall be free of tool or clamp marks.
4. All edges, including cutouts for the installation of equipment, shall be ground smooth.
5. The panels shall be provided with sufficient stiffener brackets welded to the rear of the panels to provide rigidity. The number and location of these brackets shall be indicated on the fabricator's drawing and shall be subject to approval by the buyer.
6. All racks, raceways and supporting steel members shall be arranged not to interfere with maintenance of the equipment.

Painting. The method of preparing the panel steel and painting shall be as follows:

1. The steel shall be sandblasted or chemically treated to remove all rust, mill scale, oil and foreign matter.
2. A suitable filler shall be applied to all pits and blemishes in the steel.
3. The frame and the back surface of the panels and partition sections shall be painted with two coats of sealing-primer and surfacer and two coats of light green lacquer.
4. The front surface of the panels shall be painted with three coats of sealing-primer and surfacer. The entire face of the board and partitions is to be sanded between coats. The finish shall be a minimum of two coats of high grade light green automotive lacquer.

Piping

1. All piping and tubing shall be arranged neatly and suitably supported and installed to permit easy access to the rear of the instruments for adjustments, repair and/or removal without the necessity of removing any other piece of equipment or piping. Tubing shall not be supported from instruments. Piping and tubing shall not run indiscriminately from instrument to instrument.
2. The air supply header at the base of the panel shall be fed by a filter and reducing valve assembly consisting of a set of two filters (with 40 micron elements) and two reducing valves. The filter-reducing valve assemblies shall be piped and valved in parallel so that either one set or both can be used simultaneously. Provide a 0-60 psig pressure gauge downstream of the regulators. Air filters shall be furnished with replaceable filter elements capable of removing solids 40 microns and larger.
3. The air supply header shall be 1½-inch brass pipe extending the full length of the panelboard with ¼-inch brased takeoff connections on a minimum of 1½-inch centers. One takeoff shall be provided for each air user or for each blanked cutout for future air users. A minimum of 10% valved and plugged spare takeoffs shall be provided. Takeoffs shall have ¼-inch brass nipples and ¼-inch needle valves. A ½-inch brass blow-down valve shall be provided at the end of the air supply header. A relief valve shall be installed on the air header (set at 25 psig) to protect the instruments. The end of the header shall be valved.
4. Pneumatic transmission, controlling and supply lines shall be ¼-inch OD polyethylene tubing. Tubing fittings shall be forged brass for use with plastic tubing. Lines shall be color coded in accordance with ISA RP 7.2.
5. Incoming and outgoing air lines shall terminate at bulkhead fittings mounted in bulkhead bars near the top of the panel. The field connections of the bulkhead fittings shall be for ¼-inch polyethylene tubing. At least 10% spare bulkhead fittings shall be provided.
6. Each incoming and outgoing instrument connection shall be properly identified with its correct function, e.g., transmitter (T), control valve (V), etc., in addition to the instrument loop symbol and number.

Electrical

1. The wiring is to be of General Purpose classification for nonhazardous areas as defined in the National Electrical Code. The wiring shall be neatly cabled and supported in nonflammable plastic wiring raceways.
2. The 120 volt AC branch circuits to the instruments shall serve no other load. Each frame section shall be provided with a ¼-inch x 1-inch (minimum) bare soft copper solid bar grounding bus. The panel shall be bonded to the bus. The ends of the bus shall be so prepared that proper splice plates or bonding jumpers may be attached. The bus shall be arranged so that when all sections of future panel boards are in place, the grounding bus shall be one continuous system.
3. Power wiring shall be type THW, stranded, with 600 volt insulation. The minimum wire size is to be #14 AWG. All wiring is to be connected to screw type terminals. No solder connections to terminals will be made. All wiring will be arranged in such a manner so as not to obstruct access to any instrument or any future instruments.
4. Terminal strips, housed in General Purpose enclosures, are to be provided for all supply and control circuits entering the panel.
5. Each instrument requiring electrical power is to be provided with a circuit breaker or other protective means customary to vendor's standard practice.

Breakers shall be mounted in a metal enclosure for personnel protection.

6. The panel fabricator shall be responsible for a complete circuitry check of the wiring prior to shipment.

Equipment

1. All instruments, instrument cases and equipment furnished by the user are listed in an appendix. Instrument cases shall be installed in the panels and shipped with the panel. Instruments which are easily removable from their cases shall be shipped separate from the panel.
2. Pressure switches and solenoid valves along with all piping, tubing and electrical materials will be furnished by the panel fabricator. Specifications are attached for pressure switches and solenoid valves.

Nameplates and Tags

1. Nameplates are to be provided for each instrument on both the front and rear of the control panel.
2. Nameplates and tags are to be of laminated plastic $\frac{3}{32}$ inch thick, with black surface and white interior. Characters shall be $\frac{3}{16}$ inch high x 0.030 inch stroke. Plates shall be $\frac{1}{8}$ inch high x $3\frac{1}{2}$ inches long. Nameplates and tags shall be fastened with self-adhesive backing.
3. All components in the rear of the panel, including air supply valves, air bulkhead terminals, electrical switches and circuit breakers, pressure switches, relays and other auxiliary devices shall be identified. All wires shall be tagged at both ends with EZ Code or equal marking. All terminal strips shall be identified.
4. The tubing bulkhead fitting nameplate will have the instrument item number and correct function.
5. The instrument air "takeoff" nameplate will have the instrument item number.

Testing and Inspection

1. The panel fabricator shall be responsible for checking each air supply and transmission or control line to assure continuity to each terminal point.
2. A functional check shall be performed for each indicator and recorder by hooking up a regulated supply to the input terminals in the presence of the user inspector. The controller output will be checked by using a pressure gauge on the output terminal.
3. Electrical leads shall be checked with a buzzer to verify continuity.
4. The alarm system shall receive a functional check, but calibration of switches will not be necessary.

Delivery Requirements. The panel fabricator's quotation shall state the following:

1. Time required to submit fabrication drawings after receipt of order

2. Time required to complete steel fabrication and painting after receipt of approved fabrication drawings
3. Time required to mount, pipe and wire instruments after completion of fabrication and painting

(The user will inspect the panel after fabrication and painting before piping and wiring begins.)

Shipping Requirements. The fabricator shall either:

1. Have the panel crated for shipment in such a manner as to protect the panels and instruments from damage,
2. Or ship in a dedicated enclosed van. Submit costs for each method. The choice of method will be determined 10 days prior to shipment.

Guarantee Required. Fabricator agrees that equipment furnished under this order will perform in accordance with the specifications. If test runs indicate that equipment will not perform in accordance with the specifications, fabricator shall, at his expense, make such revisions, changes or additions as may be necessary to fulfill the requirements of the specifications. Fabricator guarantees all material and workmanship for a period of 12 months after equipment is placed in service or 18 months after shipment.

Panel Inspection

There are several schools of thought relative to how extensive panel inspection and check-out should be followed. Practices vary from a somewhat casual inspection by a customer representative to a very comprehensive functional check-out by the customer engineering personnel at the fabricators plant. There are many occasions when a casual inspection is sufficient, particularly if the panel size is small. The high cost of travel and living expenses plus the time consumed for functional check-out may be a waste of money.

On other occasions when the construction schedule is tight or if the control wiring and/or tubing is complex, a complete functional check-out may be needed and may be an economical approach. The particular circumstances or condition should dictate the approach to take.

A casual inspection would be a visual inspection as to appearance and workmanship, completeness of installation and equipment and adherence to general specifications.

A comprehensive visual and functional check-out would include a thorough check of panel dimensions, materials of construction, panel finish, compliance with tubing and wiring specifications, quality of workmanship, proper tagging and proper location of instruments. It would include a complete point-to-point check of the pneumatic runs for accuracy and agreement with loop and installation drawings. The wiring check would be a complete point-to-point "ring out" and check for agreement with installation drawings and equipment vendor prints of the complete system and an inspection of all field connection terminals and labeling.

Power should be applied to the panel and all switches. Pilot lights, annunciators and similar equipment should be checked. Functional tests with simulated signals should be made where practicable.

These checks would be performed by the customer's engineering personnel and the fabricator's wiring and piping mechanics. The inspection would develop a "punch list" of any discrepancies and corrections required.

The performance of the above inspection does not preclude the requirement of a system functional check and calibration inspection being performed at the installation site. The check outlined covers primarily the actual installation work performed by the fabricator and of the hardware supplied by him.

Following is an extensive check list that may be used for check-out.

1. Panel construction features should agree with drawings and specifications.
2. Are panel dimensions within tolerance ($\pm\frac{1}{8}$ inch is common)?
3. Location of lifting hooks if provided.
4. Panel color and finish should be as specified.
5. Location of instruments on panel. As each instrument is checked, note nameplate for service information as well as ranges of dials, scales and charts for compliance to specifications.
6. If the panel is graphic or semigraphic, determine that it agrees with drawings in every respect—sizing and color of lines and symbols.
7. Are the instruments properly supported?
8. Can instruments be removed for maintenance without removing wiring, piping or other equipment?
9. Are the pressure switches accessible for adjusting?
10. Check accessibility of piping and wiring terminal points for field connections.
11. Air header compliance with respect to size and required number of takeoffs. Are there sufficient spares?
12. Do tubing and tubing fittings comply with specifications?
13. Are the tubing ducts properly supported?
14. Does the electrical wiring hardware meet specifications—quality, size, and type of wire, terminals and other fittings?
15. Are there spare electrical terminals?
16. Does the alarm system check out okay?
17. Does the wiring for the electronic signal system meet specifications relative to wire size, shielding, grounding and separation from power wiring?
18. Is the wiring properly labeled and coded?
19. Does the wiring meet the electrical area classification?

As the panel is inspected, a list should be made of all items that need correcting. Copies of this list should be distributed to the panel vendor, the purchasing department and the project engineer. Corrective action should be confirmed prior to shipment of the panel.

9 Instrument Air Systems

LeVern Ashley, William G. Andrew

Instrument air systems must be as reliable as any other utility serving the plant system. The quality of instrument air is so vital to modern process plants that it is difficult to overemphasize its importance. This becomes evident when one considers the small nozzles and passages through which instrument air passes in the functioning of pneumatic control devices. Even in plants that are primarily electronic, pneumatic power is still used for most of the final control elements. Also, local control loops are still likely to be of the pneumatic type.

The air used must be clean, dry and oil-free to ensure that small lines, restrictions and nozzles will not be fouled or plugged by scales, dirt, oil or water.

Water in the lines can cause:

1. Corrosion and rusting of air systems and instrument devices
2. Scaling that may cause damage to delicate instrument devices and plug lines and nozzles
3. Removal of lubricants from controllers
4. Blocked or ruptured instruments or air lines when freezing temperatures occur

In view of the continued importance of dependable, high quality instrument air systems, a review is presented of some of the factors that should be considered in designing instrument air systems. These factors include:

1. Sizing criteria
2. Pressure levels
3. Source criteria
4. Compressor selection
5. Dryer selection
6. Distribution systems

Sizing Criteria

The capacity requirement of an instrument air system is determined by listing all the consumers which are to operate simultaneously. This is not an easy task, for there is no way of knowing exactly how many instruments will be operating at a given time. Many instruments are of a constant bleed type—there is a constant use of some air regardless of the state of the measurement or control function of the instrument. More air is used when upsets or wide variations in measurement or control occur than under stable process conditions. Some rules-of-thumb are used, however, to help overcome the lack of knowledge of actual conditions.

A listing is given in Table 9.1 of several air users and their rule-of-thumb capacity requirements.

When manufacturer's data are available on the consumption of these users, that data should be used. However, in the absence of reliable data, the values in Table 9.1 should be helpful. The values given are in excess of the steady-state air consumption requirements, but a large safety factor is needed in any case.

Spare capacity for future additions should be included. The amount should vary depending on several factors. It

Table 9.1. Air Users and Their Capacity Requirements	
Instrument Air User	Capacity Requirement, scfm
Each instrument air pilot	0.50
Diaphragm valve positioner	0.75
Piston positioner	3.00
Purge or blowback	10.00

220

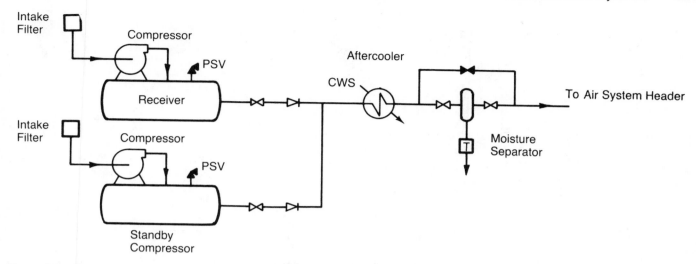

Figure 9.1. When air requirements are low, a typical system may consist of an air-cooled compressor (plus spare) with an air receiver complete with cooling, lubricating, regulating and starting system integrally mounted on the air receiver. (Courtesy of Hydrocarbon Processing)

should be at least 10% for instrument additions and changes even when no physical enlargement of the process is anticipated. If processes are to be enlarged or added, spare capacity should be in proportion to the anticipated additions.

At least 10% extra capacity should be allowed for air dryer losses and leaks in the distribution system. In determining compressor requirements another 20% of the maximum system demand should be added to avoid overloading the compressor.

Pressure Level

The pressure levels used for instrument air systems vary from about 40 psig to as high as 120 psig. The most common ranges are between 80 and 100 psig.

The increasing use of piston and cylinder operators for high torque requirements have added to the need for higher pressure source levels. Seldom is the 40 psig pressure level deemed sufficient in today's processing plant.

When a decision is reached on the pressure necessary in the main distribution header, additional allowances need to be made for pressure drops in the drying and accessory equipment. Ten to 15 psi should be allowed for the pressure drop for the entire cleaning and drying system, which might consist of an aftercooler, water separator, pre- and after-filters and air dryer.

Air Supply Source

The preferred instrument air system is one which is separated from any other air system. Usually there are two distinct systems used in a process plant: (a) an instrument air system and (b) a plant air system furnishing air for operating and maintenance purposes such as for cleaning uses and power requirements for pneumatic tools.

Occasionally there is a need for a third system to meet a process requirement.

Because plant air demands (and process also) are unpredictable, causing wide variations in header pressure, the instrument air system is preferably an independent system or is appropriately isolated from disturbances in the plant or process air systems.

When air capacity requirements are small, self-contained systems such as shown in Figure 9.1 may be used. It consists of an air-cooled compressor (plus spare), an air receiver (complete with cooling, lubricating, regulating and starting system integrally mounted on the compressor), an aftercooler, filters and dryer. Instead of the spare or standby compressor, a tie-in might be made to a plant air system for makeup.

Figure 9.2 shows a typical instrument air system that may be used for large or small capacity requirements. It consists of essentially the same equipment (no integral mounting, of course) but also includes a tie-in from the plant air system to meet emergency requirements such as compressor failure or for other unusual demands on the system.

An alarm is advisable any time makeup air is required from the plant air system. A standard pressure control system such as shown in Figure 9.2 may be used, or an alternate scheme using a self-contained regulator (Figure 9.3) may suffice. In either case, the control valve or regulator should be single seated for tight shutoff.

When plant air is available in such a quantity as to make a separate system unjustifiable, an instrument air system as shown in Figure 9.4 should be used. The equipment and accessories are essentially the same as used in the previously mentioned systems.

Additional insurance for the instrument air system is provided if the instrument air connection is made at the plant air receiver and a regulator valve shuts off plant air

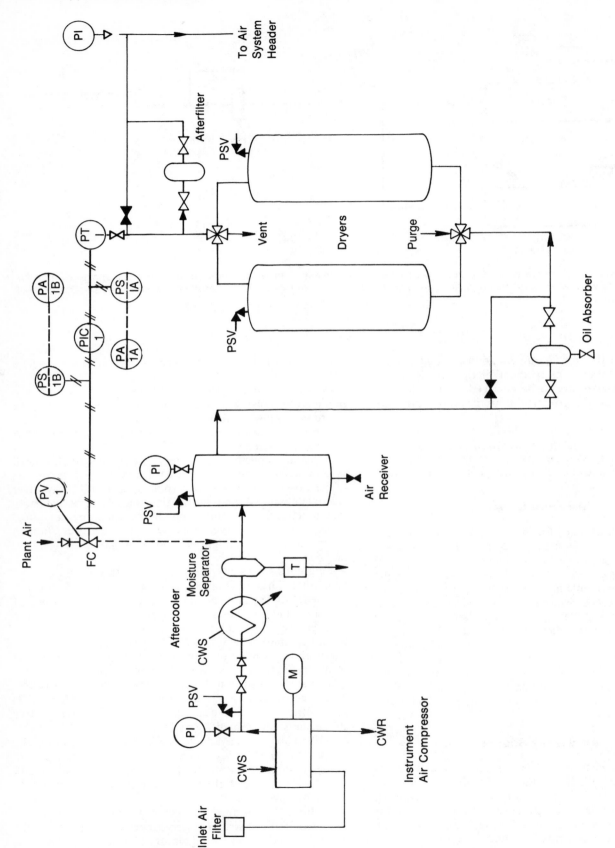

Figure 9.2. This system is typical for large, intermediate or small instrument air requirements. A tie-in from the plant air source using a standard pressure control station is shown as a backup in case the I.A. compressor fails. (Courtesy of Hydrocarbon Processing)

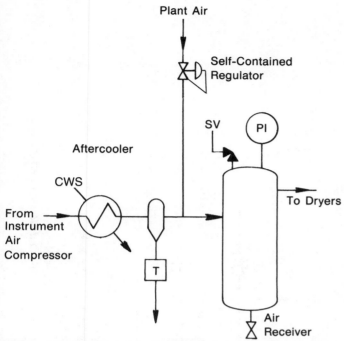

Figure 9.3. A self-contained regulator may be used for plant air makeup to replace the standard pressure control station for automatic air makeup in case of I.A. compressor failure.

use when its back pressure falls to a prescribed minimum (see Figure 9.5). This is not a normal practice, but plant air variations might warrant its consideration.

The purpose of any of the systems is to provide a dependable source of clean, dry air.

Compressor Systems

After the system capacity has been determined, the type and size of compressor must be selected, unless sufficient capacity already exists from the plant or process air systems. When separate compressor systems are used, the compressor and its accessories must be sufficiently large to meet the demands. An undersized compressor, operating almost constantly in an overloaded condition, wears excessively and, in the case of a lubricated model, pumps oil.

Two general types of compressors are used to furnish air for instrument use—the positive displacement and the dynamic.

Positive Displacement Compressors

Positive displacement compressors are machines in which successive volumes of air or gas are confined within a closed space. A reduction in the volume of this space results in an increase in the gas pressure. Both reciprocating and rotary types are included in this classification.

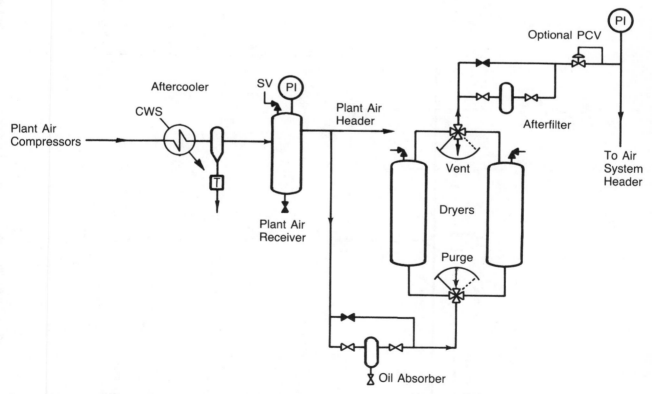

Figure 9.4. Plant air is sometimes used as the only I.A. source when a separate system is deemed unjustifiable. (Courtesy of Hydrocarbon Processing)

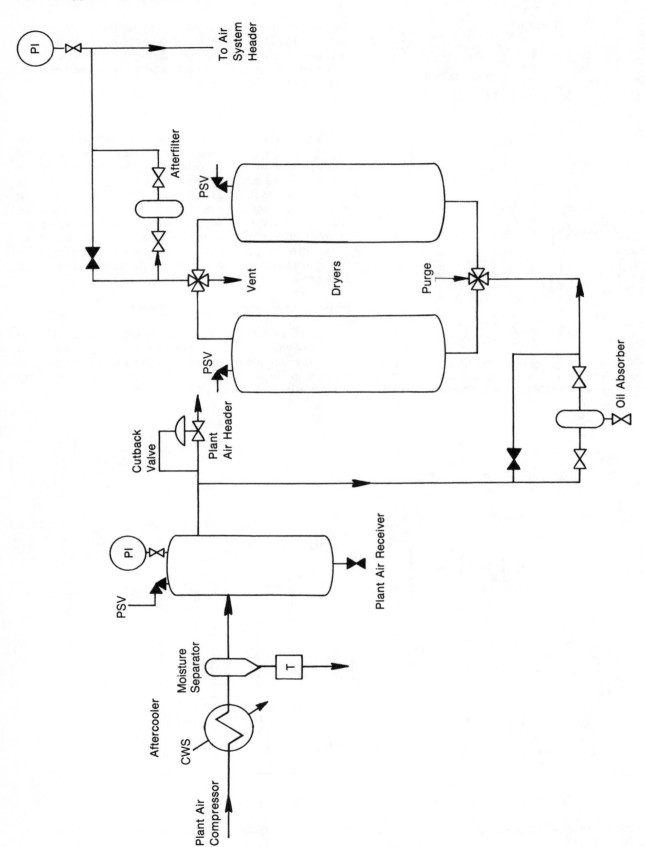

Figure 9.5. When plant air is used as the I.A. source, a cutback valve may be used to shut off plant air use when the pressure falls to a prescribed minimum level.

Figure 9.6. *This single-stage, double-acting, water-cooled compressor is one of many sizes and configurations of reciprocating compressors used for air service. (Courtesy of Ingersoll-Rand Co.)*

Reciprocating

The reciprocating compressor is probably the most common of all types in use today (Figure 9.6). It comes in many sizes and configurations such as single or double-acting, single or multiple cylinder, with cylinders arranged in a number of different ways.

The trunk type reciprocating air compressor (Figure 9.7) operates similarly to the two-cycle engine with the exception that the crankshaft is driven from an external source, no fuel is injected into the cylinder, and the compressed gas is discharged at the peak of the compression stroke into a storage receptacle. Another drive version uses a crosshead to transmit the rotary-crankshaft motion to the pistons.

Pressure levels and flow quantities are unlimited as far as instrument air systems are concerned.

Sliding Vane Rotary Compressor

In the sliding vane compressor, an eccentrically mounted, cylindrical, slotted rotor turns within a large diameter housing (see Figure 9.8).

Rectangular vanes slide in and out of the rotor slots and are held against the casing wall by centrifugal force—as the rotor turns, air is compressed between the vanes and discharged through a port on the compression side. Oil is injected into the compression chamber to provide lubrication and a compression seal. Most of this oil is removed in an air/oil separator for reuse in the compression process. This type compressor produces pressures up to 125 psig in one or two stages at flows of 100 to 600 standard cubic feet per minute.

This compressor is not likely to be used specifically for instrument air but might be used in a system from which instrument air is taken.

Screw Compressor

In screw compressors (Figure 9.9) two helical rotors turn freely with a controlled clearance between both rotors and the housing. Oil is injected into the compression chamber to

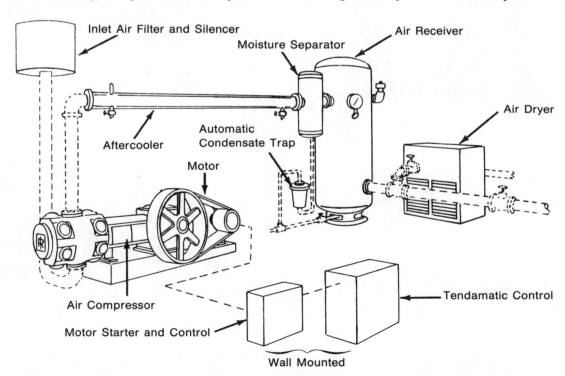

Figure 9.7. *This trunk reciprocating compressor operates similarly to a two-cycle engine, except it is driven by an external power source. (Courtesy of Ingersoll-Rand Co.)*

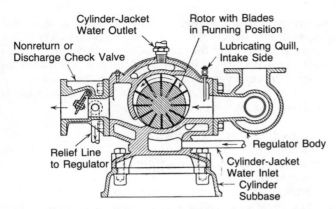

Figure 9.8. This sliding vane rotary compressor is not likely to be used for I.A. unless it is taken from the plant air source. It produces pressures to 125 psig at flows of 100 to 600 scfm. (Reprinted from the Compressed Air and Gas Handbook, 3rd ed., New York: The Compressed Air and Gas Institute, 1961)

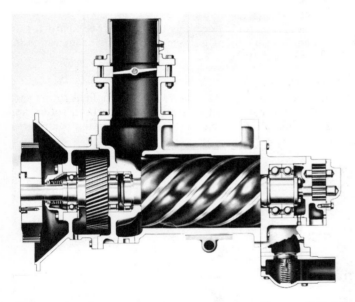

Figure 9.9. This PAC-Air screw compressor of Ingersoll-Rand is available at pressures to 100 psig in flow ranges from 400 to 1,300 scfm. (Courtesy of Ingersoll-Rand Co.)

provide rotor lubrication and a compression seal. Most of this oil is removed in some type of air receiver/oil-separator for reuse in the compression process. At a pressure of 100 psig, the flow range is 400 to 1,300 standard cubic feet per minute.

Liquid Piston Compressor

The compressing medium in this unit is a liquid—usually water. The blades of the rotor form a series of buckets carrying the liquid around the inside of the approximately elliptical casing. As the liquid, following the contour of the inside of the casing, surges back into the buckets at the narrow section, the air in the bucket is compressed and discharged through properly located ports. With continued rotation, the ring of liquid surges out of the buckets into the broad section of the casing (major axis), a new load of air is drawn in through the inlet ports, and the cycle begins again. Two compression cycles are completed per revolution. These units are available in 50 to 100 psig pressure ranges at flows from 20 to 450 standard cubic feet per minute.

Dynamic Compressors

Dynamic compressors are machines in which air or gas is compressed by the dynamic action of rotating blades or vanes which accelerate and compress the gas. Some pressure increase results directly as the gas velocity is increased due to its acceleration by the blades. As the gas passes into the diffuser section of the compressor, a large part of the velocity head is recovered as a pressure increase accompanied by a corresponding velocity decrease. Centrifugal and axial-flow compressors are included in this classification.

Dynamic compressors are characterized by large volumetric capacity and relatively low pressures, although higher pressures are produced by multistage machines. Both centrifugal and axial-flow compressors are inherently high speed machines.

Centrifugal

In a multistage centrifugal compressor with intercoolers (Figure 9.10), air enters the first stage compressor casing and passes into the eye or center of the first stage impeller. The rotation of the impeller imparts a high velocity to the air as it moves radially outward into the diffuser/intercooler area where its temperature is reduced. The air is then directed into the eye of the second stage impeller where

Figure 9.10. Centrifugal compressors are often used for instrument air service. Multistage units can handle large air volumes at pressures to 150 psig and higher. (Courtesy of Elliott, Div. of Carrier Corp.)

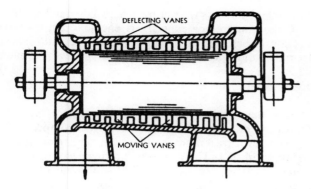

Figure 9.11. Axial-flow compressors are rarely used for I.A. service. They are characterized by essentially constant volume delivery at variable pressures. (Reprinted from the Compressed Air and Gas Handbook, 3rd ed., New York: The Compressed Air and Gas Institute, 1961)

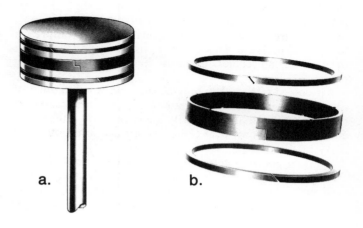

Figure 9.12. This nonlubricated Teflon ring allows reciprocating compressors to be operated without the lubrication that presents filtering problems to I.A. systems. (Courtesy of Ingersoll-Rand Co.)

more compression takes place and similarly through each stage of the machine. Multistage centrifugal air compressors are generally used for handling large air volumes at pressures up to 150 psig and higher.

Axial-Flow

The axial-flow compressor (Figure 9.11) is a machine with aerodynamically shaped, radial blades in both rotor and casing. Air passes into the machine through the first row of rotor blades and is then redirected into the next and successive rows by alternate rows of fixed diffusing blades in the casing. Flow ratings as high as 1,000,000 cubic feet per minute for pressures similar to those of centrifugal machines are available.

Axial-flow machines are characterized by essentially constant volume delivery at variable pressures, whereas centrifugal machines deliver practically constant pressure over a variable capacity range. These machines are rarely used for instrument air systems.

Nonlubricated Compressors

Where high air purity is required, as in instrument air systems, nonlubricated compressors (Figure 9.12) should be considered. The reciprocating compressor is available as a nonlubricated model with carbon or Teflon rings. New filtering problems may arise due to the type of oilless design used, such as carbon dust from worn carbon rings in the compressor. Also, multistaging may be required to attain a desired pressure due to temperature limitation of Teflon materials.

Rotary liquid piston and centrifugal compressors eliminate the need for lubricating oil or oilless designs since no internal lubrication is required on either type. The initial cost of the liquid piston compressor (Figure 9.13) is higher but is offset by reduced maintenance costs. With water as a sealing medium, no aftercooler is needed to remove the heat of compression or foreign matter which may be present in the incoming air. Where the seal water temperature is suf-

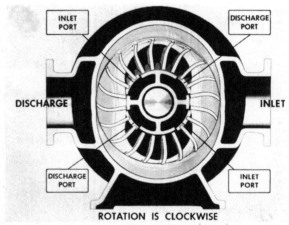

Figure 9.13. The rotary liquid piston is another oilless compressor. Initial cost is relatively high, but it is characterized by low maintenance costs. (Reprinted from the Compressed Air and Gas Handbook, 3rd ed., New York: The Compressed Air and Gas Institute, 1961)

ficiently low, the air delivered to the dryers or distribution lines can be maintained above the dew point, thus eliminating moisture deposits in these lines.

Compressor Cooling

The following discussion makes specific reference to cooling methods used in the reciprocating compressor, but the same methods are applicable to all other types.

Air or gas compression causes temperature increases which become high enough to cause trouble with the lubrication of the cylinder, piston and packings. Cooling of the air and of the compressor cylinder reduces this problem and also the required work of compression, approaching the minimum of an isothermal process. This may be done by circulating atmospheric air or water over the cylinders. Another method is to compress the air only part way to its final pressure, extract some of the heat and then compress it to the final pressure. This is accomplished by passing the air through an intercooler, using air or water as the cooling medium.

Air-cooled intercoolers may consist of a small number of finned tubes through which air under pressure passes. It may be of the radiator type with many tubes. In either type, cooling air is blown over the outer surface of the tubes to carry off the heat.

A water-cooled intercooler consists of a nest of tubes through which air or water passes. The nest is enclosed in a shell or housing. The compressed air in some intercoolers passes over the outside of the tubes through a series of baffles to assure maximum contact with the cooling surfaces of the tubes. In others, the air passes inside the tubes and the coolant is outside.

Compressor Controls

Since compressed air is generally required in varying quantities at a relatively constant pressure, compressor regulators or controls are used to vary the delivery of the compressor. If the pressure rises too high, the pilot (a pressure responsive device) operates to reduce the air delivery; if the pressure falls too low, the pilot operates to increase the air delivery.

Several mechanisms are used as unloading devices to vary the delivery of compressed air.

1. The inlet valves may be held open during both the suction and compression strokes—thus no air will be compressed in the cylinder.
2. Clearance pockets may be used which are automatically controlled by the pilot, reducing the volume of compressed air delivery.
3. On rotary compressors, a valve, which completely closes the intake line, prevents the intake of air.
4. On some compressors, the speed is varied in response to pressure changes on the pilot. As the volume of output air varies directly with the compressor speed, the output follows the demand. This is a common method with reciprocating steam driven engines or gasoline and diesel engines.
5. On small compressors, a type of control known as "automatic start and stop," which responds to a pressure actuated switch, opens or closes the motor circuit as demand varies. Large compressors are usually kept running when unloaded.

The method of control used for a compressor depends on the type of service encountered. Systems with nearly constant demand are usually supplied by a compressor which is operated continuously at nearly full capacity, while on systems with variable demand, either the compressor (start-stop) or its pumping action (unloader-control) may be stopped part of the time to minimize power consumption.

Unloader control is well suited where a demand of 75% or less of the compressor capacity is encountered; start-stop operation is best suited to applications where the normal system demand is less than 40% of the compressor capacity.

Oil Removal

During the compression cycle of the typical lubricated compressor, oil vapor is generated in the compressor by high temperatures and mechanical shearing of the lubricating oil film. The quantity and fineness of size of the oil vapor particles are determined by mechanical factors in the compression chamber—piston speed, clearance, oil viscosity, etc. For example, rotary compressors, which have much higher shear rates in the lubricant film, produce much larger quantities of oil vapor and in a more finely dispersed state.

If the oil vapor is not removed from the compressed air, it slowly coalesces into droplets large enough to settle out in the piping system as liquid oil. This occurs in orifices, pressure reducers, nozzles, etc. Coalesence is a function of air flow turbulence and is not influenced by a temperature reduction in the lines, as in the case of water vapor condensation. A natural application of this characteristic is the creation of a turbulent flow in the presence of oil absorbent material to accomplish the oil removal.

The oil particles vary in size from 5 microns and under and will pass through conventional air line filters designed to remove much larger water droplets and dirt particles. This automatically eliminates many types of mechanical filters or separators since gravitational inertial impingement and other similar separators do not effectively remove droplets smaller than 3 microns. However, certain coalescing filters designed for submicronic aerosols will also remove the larger sized droplets without clogging or even causing an appreciable pressure drop. These filters will also remove the submicron and larger suspended solid contaminants such as pipe scale and piping compounds. This type of liquid oil removal equipment must precede vapor absorbers if low oil effluents and long service are to be achieved.

For vapors under 0.01 micron in size, sufficient Brownian motion exists for them to be effectively adsorbed by activated carbon or alumina, with carbon exhibiting the greater oil capacity (which may offset its greater expense). Filters of this type should be followed by an afterfilter to avoid any dusting problems. Venting to atmosphere prior to adsorber startup also helps eliminate dusting problems.

General Considerations

If an instrument air system is not tied into another air source, two compressors, each with a separate power source, should be used for the system to avoid instrument shutdowns. One unit can function as an automatic emergency substitute for the other, or the two units may be connected so that when one unit's output falls below a specified amount, the second compressor is activated.

The initial condition of atmospheric air introduced into an instrument air system is important. Since the density of air increases with decreasing temperature, the compressor intake location should be chosen to provide the coolest air possible in order to effect a savings in power input per volume of air processed. In addition to locating the intake so as to avoid contaminants (gases, dusts, pigments, steam vents, etc.), an intake filter is necessary to remove grit and dust. This is important not only to the final air condition but also for compressor protection (wear on pistons, cylinders, valves and all other moving parts in contact with the air).

Air should be cooled immediately after compression to remove water and oil vapors. The delivery of hot air to piping systems causes the pipes to undergo alternately expansion and contraction, loosening pipe joints, scale, etc. An aftercooler (heat exchanger), placed in the discharge line between the compressor and air receiver, condenses and removes about 70% of the water vapor contained in the compressed air, thus decreasing the dryer load. The air discharge from the aftercooler may go directly into the system dryer or into a receiver. The air receiver acts as a surge tank to dampen pulsations from the compressor discharge and serves as a reservoir to furnish a supply of air for a predetermined duration (say 5 or 10 minutes) in the event of compressor failure.

Air filters should be used for dust and particle removal and must be sized sufficiently large to prevent clogging and consequent ineffectiveness.

In comparing the installed costs of compressors for instrument air systems, the following observations are made.

1. Reciprocating and centrifugal compressors are the usual choice for instrument air systems. Reciprocating units should be considered for rates below 1,500 standard cubic feet per minute. Nonlubricated models are relatively high, both in installed cost and maintenance cost, but they provide clean, oil-free instrument air systems.
2. For rates above 1,500 standard cubic feet per minute centrifugal compressors should be considered. Installation and maintenance costs are lower, and they are available in nonlubricated models.
3. The liquid piston compressor may be considered for oil-free air quantities up to 500 standard cubic feet per minute—usually more economical than nonlubricated reciprocating compressors.
4. In the 100 to 500 standard cubic feet per minute range, the rotary vane compressor has a low installation cost, but nonlubricated models are not available.
5. In the 500 to 1,500 standard cubic feet per minute range, the screw type may be installed at a lower cost than centrifugal or reciprocating compressors. However, it is not oil-free.

Dryers

Water, the most serious contaminant in compressed air systems, causes rust and corrosion of piping and fittings and carries dust, dirt and other solids to instrument orifices and other restrictions resulting in the plugging of these devices. Pipe capacity is reduced by water, and when ambient temperatures drop, ice formations may plug or burst the lines.

All atmospheric air contains some water vapor. During the compression process, as the air pressure increases, a corresponding increase in air temperature occurs due to the work of compression. Even though the decrease in volume of a given quantity of air could result in the precipitation of water, this does not occur due to the high air temperature. Therefore, the compressor discharge air contains the same quantity of moisture as the compressor inlet air.

When an air-vapor mixture is cooled at constant pressure, the temperature at which the vapor becomes saturated and moisture begins to condense is called the "dew point" temperature of the air.

As the air passes through the aftercooler, its temperature rapidly drops to a point below the dew point, resulting in condensation of from 70 to 90% of its water content. The use of aftercoolers is essential to reduce the load on dryers.

If the air temperature is subsequently increased, the dew point will remain at the lowest temperature to which the air was previously subjected, provided that the condensed water has been trapped out. Otherwise, the condensate will vaporize again and raise the dew point.

A dew point of $-40°F$ at line pressure is commonly accepted as a standard for instrument air systems since temperatures generally do not drop below this value. In areas of extreme cold the dew point may be as low as $-100°F$. The dew point should be several degrees below the lowest temperature the air will encounter in the system.

For indoor installations or where air lines are never exposed to low temperatures, a $-40°F$ dew point is not necessary. In determining the minimum temperature, the cooling effect of adiabatic expansion of the compressed gas to lower pressures must be considered. The greater the gas expands in volume, the lower the temperature drops. Expansion cooling effects should be considered in specifying the dew point requirement.

The two basic types of dryers used for instrument air systems are the refrigeration and the dessicant. Mechanical separators are available (expansion and cyclone types), but they do not produce the quality of air required for instrument air use.

The selection of a particular type depends on required dew point, quantity of air flow, air pressure, desired relative humidity and system operating costs (electricity, steam, water, etc.).

Desiccant Type

Desiccant or adsorption dryers are the most commonly used for outdoor instrument air systems. Desiccants are hygroscopic materials; i.e., they are chemicals that readily take up and retain moisture on their surfaces. One typical desiccant material will adsorb approximately 45% of its weight in water. When moist air is in contact with a desiccant, an equilibrium condition is established in which the

relative humidity of the air determines the degree of saturation of the desiccant particle; the greater the relative
humidity, the more moisture is adsorbed up to the 45% limit
mentioned. Conversely, a saturated particle, when in the
presence of relatively dry air, will desorb moisture until,
once again, an equilibrium condition is reached.

Desiccant dryers are filled with a solid desiccant, such as
activated alumina or silica gel, which removes water vapor
as the air passes through the desiccant bed. The desiccant
becomes wet in the process, and the moisture must be
removed—the desiccant must be regenerated. Dryers are
subclassified according to the method of regeneration used
to remove the water. Some are heat-regenerated (Figure
9.14), and some are pressure-swing or nonheat
regenerated—heatless drying (Figure 9.15).

Heated Dryers

Heated dryers normally consist of two desiccant filled
chambers which are connected in parallel. The desiccant in
one of the two chambers is used to dry the air stream while
the desiccant in the other chamber is being regenerated by
the application of heat. The water vapor driven off is allowed to pass into the atmosphere. The systems usually are
designed with a chamber-switching time of 4 to 8 hours.

Figure 9.15. This heatless desiccant dryer uses a portion of
dry air from the dry chamber to reactivate the desiccant of the
drying chamber. This method is less efficient, but it eliminates
the need for heaters. (Courtesy of Pall Trinity Micro Corp., a
subsidiary of The Pall Corp.)

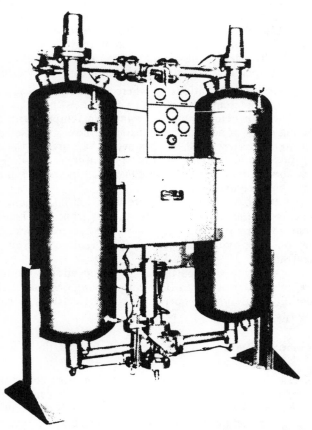

Figure 9.14. Desiccant dryers are classified according to the
method of desiccant regeneration. This one is regenerated by
heat, either steam or electrical. (Courtesy of Pall Trinity Micro
Corp., a subsidiary of The Pall Corp.)

Either steam or electrical heating is usually chosen for the
regeneration heat source.

The dew point of air leaving heat-reactivated dryers
ranges from 0° to −50°F at line pressure, and flows range
up to 30,000 standard cubic feet per minute.

A typical flow circuit is shown in Figure 9.16. Wet air
enters the four-way inlet switching valve (A), passes through
the left desiccant bed (B), through the stainless steel
cleanable outlet filter (C) and outlet check valve (D) to the
dry air outlet.

A small portion of dried air passes through the purge
throttling valve (E), purge indicator (F) and purge flow
orifice (G) which controls the flow rate. Metered purge then
passes through the right purge flow check valve (H) through
the heater tube (I) and over the electric heater (J). Heated
purge then enters the plenum area of the chamber (K) where
it is dispersed downward through the wet desiccant bed in a
direction countercurrent to the drying flow. The purge air,
now carrying previously adsorbed moisture, exits to atmosphere through the four-way and purge exhaust valve
(L). Shortly before the end of the cooling period, the purge
exhaust valve closes to repressurize the reactivated
chamber. Switchover takes place with both chambers at line

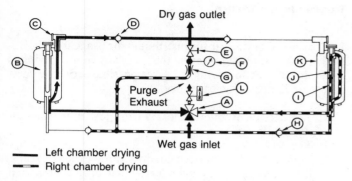

Dry gas outlet

Purge
Exhaust

Wet gas inlet

—— Left chamber drying
═══ Right chamber drying

Figure 9.16. A typical flow circuit for a heat reactivated dryer. (Courtesy of Pall Trinity Micro Corp., a subsidiary of The Pall Corp.)

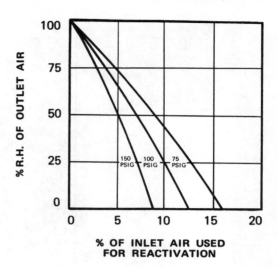

Figure 9.17. The amount of air used for reactivation in heatless dryers depends on operating pressure, air temperature, degree of drying and air flow rate. Typical curves are shown where "relative humidity" is plotted against "air used" for a family of pressure curves. (Courtesy of Deltech Engineering, Inc.)

pressure, eliminating desiccant bed attrition, displacement of switching valve lubricant and line surges. The left chamber is then exhausted to atmosphere and its regeneration cycle initiated, while the fully regenerated right chamber dries air at line pressure.

Heatless Dryers

Heatless dryers are similar to the heated type, except that no external heat is used for regenerating. Basic equipment consists of two chambers, each filled with a like quantity of desiccant and interconnecting piping and valves to allow switching the air stream from one chamber to the other. The chamber-switching time is from 1 to 5 minutes, according to the effluent dew point required. To regenerate the desiccant in one chamber, a portion of the dry air leaving the operating chamber is expanded to a reduced pressure (atmospheric) and is passed over the saturated desiccant bed. The portion of air used to reactivate the desiccant is called the *purge*.

As the purge stream flows over the desiccant bed, it desorbs the water and is vented to atmosphere. The amount of air purged is higher than that purged in the heat-reactivated units.

The exact amount used to reactivate depends on the operating pressure, air temperature, and degree of drying required; it also depends slightly on the air flow rate. It varies from about 9 to 10% of the dryer output for mild climates to about 11 to 12% for cold climates. In extremely cold climates such as −20°F, as high as 18% of the dryer output might be required. Figure 9.17 shows some typical curves for percent of reactivation air versus percent relative humidity at different operating pressures.

Heatless dryers are capable of developing effluent dew points, ranging from −40°F to as low as −100°F at line pressure. Air flows range up to 20,000 standard cubic feet per minute.

The air output of all desiccant dryers should pass through an afterfilter to avoid any dust carryover into the instrument system. All solid desiccants in dehydrators tend to fragment or pulverize in use, particularly on heating and cooling such as in regeneration. It is imperative that an afterfilter be installed downstream of the dryer to avoid dust problems.

Another type of drying system that has some similarity to adsorption drying is the absorption type. It makes use of a deliquescent or absorption-type drying agent such as sodium chloride, calcium chloride, calcium phosphate or urea. These salt particles (contained in a chamber) absorb a portion of the water vapor from the air and, in turn, go into solution. The salt water solution leaves the chamber through the drain, requiring regular replacement of the salts. This method has very limited application for instrument air drying since the dew point can be only a few degrees below the operating temperature.

It may be used where the system requirements allow a 30°F and above effluent pressure dew point and a system air flow of 5,000 standard cubic feet per minute or less.

Refrigeration Type

Cooling is a simple, positive way for removing water vapor from compressed air. The amount of water vapor which it can hold is dramatically affected by temperature. A well-known example of refrigerated dehumidification is freon-type air conditioning. Dehumidified air in a living space is as comfort-producing as a drop in temperature.

The phenomenon of cooling is used regularly in compressed air systems to remove a large part of the moisture from the system. Aftercoolers, either water or air cooled, almost always form a part of the instrument air system. Usually aftercoolers bring the air temperature near the temperature of the lines in the distribution system. They reduce but do not eliminate the moisture problem.

Figure 9.18. Refrigerated dryers are used where dew points need not be below 35°F. An integrally mounted system including compressor, evaporator, expansion device and heat exchanger is shown. (Courtesy of Pall Trinity Micro Corp., a subsidiary of The Pall Corp.)

Refrigerated air dryers carry the cooling process one step further. They use a refrigerant to cool the air to a lower temperature than could be obtained with air or water coolers. Figure 9.18 shows an integrally mounted refrigerated system. The components of the system are a refrigerant compressor, evaporator, expansion device and heat exchanger. They form a closed loop system which removes heat from the instrument air before it enters a dryer. Generally, the heat exchanger is a shell-and-tube type with the refrigerant in the tubes and the instrument air in the shell. The air, being warmer than the refrigerant, gives up some of its heat to the refrigerant.

Refrigeration systems have limited application for instrument air systems, however, for it is impractical to lower the air below a temperature of 35°F. If the temperature drops below 32°F, the condensed water vapor freezes and clogs the heat exchanger with ice.

The use of refrigerated systems have increased considerably since the early 1960s. Though most units are designed with an outlet temperature of 35°F at line pressure, the dew point is considerably lower at atmospheric pressure—from −10° to −20°F. This is sufficiently low for some areas of the country.

When refrigerated systems are used, they should be located inside buildings so that the ambient temperature condition never falls to the freezing point.

Necessity for Dryers

The cost of drying air for instrument air usage is high. Experience has shown, however, that the cost of not drying is even higher. The following list gives possible excessive costs when air is not sufficiently dry.

1. The cost of production losses from lost time when instruments fail
2. The cost of product spoilage
3. The cost of compressed air lost through leaky traps, leaky valves, leaky cylinders, open drains or rusted pipe joints
4. The cost of maintaining pneumatic instruments due to malfunction
5. The cost of trap maintenance and repairs, of draining water pockets and drip legs
6. On new installations, the cost of pitching the air lines, installing drip legs, traps and drain lines
7. The cost of manpower to do work that would be unnecessary if a dryer were used

The first item alone is the primary reason for providing clean, dry instrument air systems.

Design Guideline Criteria

Some design criteria to keep in mind for dryer specifications (desiccant type) are given below.

1. Give air flow rate, inlet temperature and pressure.
2. Specify inlet moisture condition—worst condition is saturated air at inlet pressure and temperature (probably around 100°F).
3. Specify outlet dew point desired at line pressure desired; −40°F is usually acceptable; pressure should be 100 psig.
4. Heat (or heatless) type is preferred.
5. Adsorbent type—usually manufacturer's standard—activated alumina, silica gel, etc.
6. Operation—automatic.
7. Pressure drop through dryer—5 psig is nominal.
8. Relief valves required for dryer chambers.
9. Utilities: (a) for steam—pressure, steam condition; (b) for electric—NEMA classification, voltage and frequency desired.
10. Accessory equipment—prefilters and afterfilters.

Distribution System

The final step in planning a properly balanced instrument air system is the design of the distribution system. It should provide delivery to all air users with a minimum of supply variation in pressure and flow. It should be made of the proper materials. Some conventional rules should be followed to provide flexibility and trouble-free service.

General Layout

There are two basic configurations used for instrument air distribution systems—the loop and the radial.

As the name implies, the loop system provides a header that describes a loop in the plant. Subheaders from the main may be supplied with air from either direction, depending on the pressure caused by air users along each path. Since two-way flow is provided, minimum pressure losses to each air user is assured.

The second system, the radial type, is more commonly used and costs less than the loop. Flow is from one direction only, and users farthest from the supply source are subject to lower pressure levels.

In the radial system, auxiliary receivers could be placed at remote locations to provide capacity for unusual flow requirements. This is rarely needed, however, since header sizing and pressure levels are usually adequate to meet unusual air supply demands.

Header and Branch Sizing

The main header and subheaders or branches in a distribution system should be sized so that the pressure drop to the end point of the branches does not exceed 5 psi from the air dryer outlet. The incremental cost of increasing a header size is small in comparison to the capacity added. For example, about three times as much air will flow through a 1½-inch pipe than through a 1-inch pipe for the same pressure drop; the nominal increase in cost for this material is only about 50%.

The minimum pipe size suggested for running to a user is ½ inch. Smaller sizes can supply a sufficient flow at acceptable pressure drops, but the larger size can be supported more rigidly and as economically, using fewer varieties of materials and with less difficulty. A maximum of four users from a single ½-inch branch is a commonly accepted practice.

Table 9.2 may be used as a guide in sizing main and branch headers.

Materials

Instrument air headers and branches are normally made of galvanized steel pipe and malleable iron galvanized fittings. Galvanizing prevents rust and scale that might otherwise pass through filters and clog instrument nozzles and relays.

The pressure ratings of valves and fittings are usually 125, 150 or 300 psig. The 125 and 150 psig ratings are sufficiently high. In selecting particular valves or fittings, however, 300 psig ratings are often used.

Other materials that are used occasionally include brass, copper and aluminum. Copper pipe is frequently used for air headers in back of panels for instrument air supply to control panels. Copper is too expensive, however, to use for field headers or branches unless some unusual circumstances apply.

Table 9.2. Sizing Main and Branch Headers		
Pipe Headers	**No. Air Pilots**	**Pipe, Size, in.**
Main	25	1
	80	1½
	150	2
	300	3
	600	4
Branch	4	½
	10	¾
	25	1
	80	1½

Takeoffs and Valving

Takeoffs from instrument air headers and branches should be similar to that shown in Figure 9.19 for an individual air user. As many as four users may be supplied from the ½-inch takeoff shown.

The takeoff always should come off the top of the header so that moisture, dirt or scales in the pipe are less likely to get into the instruments served. Valves should be mounted next to the header so that the header is protected in case the branch is broken or needs to be put out of service for some reason such as adding users, replacing valves, etc.

At a location near the individual user, the pipe size is usually reduced to ¼-inch size; sometimes upstream of valve 5 and sometimes downstream of valve 5 (see Figure 9.19).

Filter-regulators are normally used for each instrument device requiring air and their connection size is ¼ inch.

Small lines such as those depicted need to be well supported to prevent piping or tubing breakage which would shut down the instrument. The ¼-inch piping shown between valve 5 and the pressure gauge is sometimes replaced with ¼-inch OD copper tubing if the regulator and gauge are well supported. Otherwise, the piping helps support the regulator and gauge.

Takeoffs from main headers should also come off the top of the header, and valves should be provided for branches so that extensions and changes can be made.

At the ends of headers and main branches, line size valves should be installed so that additions to the system can be made without shutting down the entire air system.

Air headers and subheaders should normally be sized well over the present expected capacity so that additional air users can be added without system shutdowns.

Even when dryers are supplied to keep the air dry, a drain or drains should be installed at low point(s) in the system to provide drainage.

A summary of items to bear in mind in designing instrument air systems follows.

1. The main header should be sized to accommodate no less than 150% of the instrument air requirements for the initial job design.

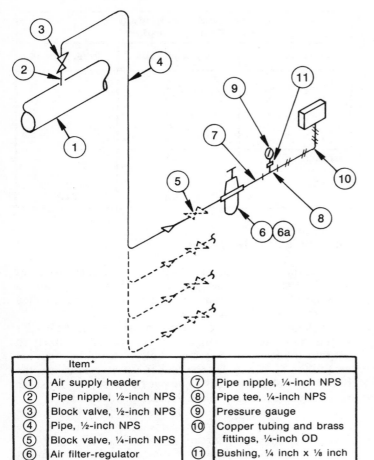

	Item*			Item*
①	Air supply header		⑦	Pipe nipple, ¼-inch NPS
②	Pipe nipple, ½-inch NPS		⑧	Pipe tee, ¼-inch NPS
③	Block valve, ½-inch NPS		⑨	Pressure gauge
④	Pipe, ½-inch NPS		⑩	Copper tubing and brass fittings, ¼-inch OD
⑤	Block valve, ¼-inch NPS		⑪	Bushing, ¼ inch x ⅛ inch
⑥	Air filter-regulator			
⑥a	Air filter (if regulator not required)			

*Ferrous piping materials per Piping Material Class 125A-2.

Figure 9.19. A typical takeoff from an air header to as many as four individual users shows the proper method of header connection.

2. Line sized valves should be provided at the end(s) of the main header for easy addition(s) to the header.
3. Half-inch drain valves should be provided at the low point(s) of the main header system.
4. Valves should be installed on all subheaders from the main header of the same size as the subheader line.
5. Subheaders should be sized to accommodate no less than 150% of the air requirements for the area served.
6. Minimum size of branch line should be ½ inch.
7. Branch lines may serve one or several air users.
8. Takeoffs should be from the top of the main or sub-header.
9. Each header takeoff should supply air to a maximum of four instruments. The filter-regulator should be located as near to the instrument as practicable. On integral valve positioners, the regulator is normally not required. Individual filters should still be install-ed.

10. Supports or braces should be provided for the air supply system (from header to instrument) in order (a) to prevent excessive vibration of the system and (b) to support the weight of the system.
11. A separate block valve, air filter-regulator and a pressure gauge should be provided for each instru-ment. If a pressure gauge and/or air filter-regulator is built into the instrument, separate units are not required.
12. A block valve should be provided near the filter-regulator on the upstream side when the branch valve is inaccessible or when the header takeoff is supplying air to more than one instrument.
13. Direction changes in the piping normally should be made using pipe bends.
14. Copper tubing may be used between the filter-regulator and the instrument air user.

Control Room Air Supply

The control room air supply should come from the main instrument air header, sized to meet the requirements of all instruments expected to be installed in the control room.

It is standard practice to reduce and filter the air for all control room users as shown in Figure 8.16. It provides a parallel set of reducing regulators and filters so that either set may be taken out of service for repair, filter replacement, etc.

Downstream pressure for control board use is normally 20 psig; a relief valve should be provided on the downstream side set at 25 psig. A pressure gauge should be provided to read header pressure.

The panel board air header is normally made of brass or copper. Aluminum headers have been generally avoided because of the softness of the metal, which gives rise to gall-ing of threads, and stripping out of threads. The James C. White Company, however, manufactures an extruded "Unit-Header" having a thick cross-sectional area for drill-ing and pre-drilled and tapped holes along its length (Figure 8.17). Available in 1″ and 2″ diameters, schedule 40, and in 20 ft. lengths, "Unit-Header" reduces installation time.

The header should slope about 1 inch per 10 feet and a ½-inch drain valve placed at the low end of the header, tapped off the bottom, in the event of moisture collection in the header. A line size valve should be placed at the header's end so that header extension can be made without shutting down the air system.

Takeoffs from the air header for individual users should be made from the top of the header. Each user should have a separate block valve. Fifteen to 25% spare takeoffs should be provided for the various panel areas for future additions.

Signal tubing from transmitter to local junction box is usually copper, rather than plastic, for reasons of durability. Obviously a copper tube would endure a small fire whereas a plastic tube would melt more quickly. Most plants are plastic coated copper tubing for corrosion resistance. A few plants use "bright" (bare) copper, although the tubes darken with tarnish.

Figure 9.20. Bundles of numbered copper tubes are. available with the tubes spiraled for ease of making turns in the tubing run. (Courtesy of Thermo Electric Co., Inc.)

Bundles of copper tubes are available with tubes individually identified, and with tubes spiralled within the jacket for ease of making turns in the tubing run (Figure 9.20). Some bundles are armored and then jacketed with plastic for burial. Others are specially jacketed with asbestos for added fire resistance.

Economics favor plastic tubing, particularly in multitube bundles. Bundles are available having special fire resistant jackets, which can provide up to four minutes of bundle integrity while engulfed in a flame of 2400°F.

The susceptability to fire of an exposed plastic tube has been used to advantage. Using a buried steel tube in the process area with a plastic loop at each strategic location to be monitored, a pneumatic pressure is applied through a small orifice. A pressure switch senses loop pressure and trips on low pressure if a loop burns away, depressuring the system.

Other Users of Instrument Air

Occasionally there is a need to use instrument air for purposes other than for instruments. Generally, this should be discouraged in order to maintain the integrity of the instrument air system.

The philosophy of restricting the system to instruments is based on the sound reasoning that the instrument air system is vital to safe plant operation. In many processes orderly shutdown of equipment and portions of the process depend on control actions of various pneumatic devices.

Case Purging for Electrical Area Classification

It is possible in most instances to safely install instruments which are capable of producing ignition temperatures within Hazardous (N.E.C. Division 1 rated) or Normally Non-Hazardous (N.E.C. Division 2 rated) gas-vapor areas by providing the instrument with a continuous purge. Effective safeguards must be taken against ventilation failure.

I.S.A. Standard Practice ISA-S12.4 (1970) covers three purging classifications:

1. Type Z Purging. Covers purging requirements adequate to reduce the classification of the area within an enclosure from Division 2 (normally non-hazardous) to non-hazardous.
2. Type Y Purging. Covers purging requirements adequate to reduce the classification of the area within an enclosure from Division 1 (hazardous) to Division 2 (normally non-hazardous).
3. Type X Purging. Covers purging requirements adequate to reduce the classification of the area within an enclosure from Division 1 (hazardous) to non-hazardous.

Purging is defined as "the addition of air or inert gas (such as nitrogen) into the enclosure around the electrical equipment at sufficient flow to remove any hazardous vapors present and [with] sufficient pressure to prevent their re-entry".

The Standard is based on tests showing that a static pressure within the enclosure of 0.1 inch of water is equal to pressure generated by a 15 miles per hour wind, and that four enclosure volumes of the purging gas through the enclosure is adequate to dilute even a 100% hydrogen atmosphere to below the lower explosive limit. It also limits enclosure volume to 10 cubic feet, and the ratio of the maximum internal dimension to the minimum is limited to 10 to 1. The Standard notes that, while air of instrument quality is acceptable as a purge, ordinary plant compressed air is usually not suitable.

When a purged instrument enclosure is anticipated or required, a thorough reading of the I.S.A. Standard ISA-S12.4 should occur to address all of the requirements for reduction of the area classification. Acceptable installation sketches are given in the Standard.

The danger of indiscriminate use for other purposes is that the air system is bled down so low that the supply jeopardizes control action. Pneumatically operated tools, air cleaning, etc., could easily exhaust a branch header and disturb many control systems.

The best philosophy, then, is to restrict the use of the instrument air system to instruments even though immediate economics often favor using the system for other uses.

10 Slurry Service Applications

William G. Andrew

Slurries are defined as mixtures of solids and liquids. They occur in many processes because of the inherent nature of the process. The presence of solids in flowing streams, storage vessels or other equipment complicates the measurement of process variables which already may be difficult because of other characteristics such as viscosity, density, turbulence, etc.

Slurries are generally described as being homogeneous or heterogeneous. In homogeneous slurries, solid particles are evenly distributed throughout the liquid media. They are characterized by high solids concentrations and fine particle sizes. They often exhibit non-Newtonian characteristics—the effective viscosity is not constant but varies with the applied shearing strain. Polymer slurries are examples of non-Newtonian fluids.

Heterogeneous slurries exhibit concentration gradients. They may have a tendency to settle in the flowing stream, the vessel or other pieces of equipment. They tend to have lower concentrations of solids, and the particle sizes vary and are larger.

Many slurries are encountered with a mixed character. Small particle sizes join with the liquid to form homogeneous mixtures while coarser sizes act heterogeneously.

In any case the measurement of such process variables as flow, level, pressure, temperature, density and viscosity are often difficult to make. Standard instrument devices in these services often present operating problems and require a lot of maintenance. Special types of control valves must often be specified to avoid malfunction or poor operation in slurry services.

The bad effects that must be overcome, reduced or minimized include:

1. The tendency to plug
2. The tendency to coat
3. Wear due to abrasiveness
4. A changed characteristic that makes the device incompatible with a particular measurement method

The purpose of this chapter is to enumerate and describe some measurement methods available for these difficult-to-measure applications.

Flow Measurement

Some of the more acceptable methods for measuring slurry flows include magnetic meters, target meters, thermal meters, sonic meters, magnetic resonance meters, and differential meters using special orifice plates, Venturis or flow nozzles. When differential measurements are used, however, lead lines must often be purged or sealed.

When solids concentrations are relatively low, other type measurements such as rotameters, vortex meters or even some turbine designs may be applicable. Each application should be analyzed individually and thoroughly.

Magnetic Meters

If fluid conductivity is sufficiently high, magnetic meters nearly always merit consideration for slurry flow measurments. Figure 10.1 reveals the unobstructed opening that makes magnetic meters particularly suitable for this type service. Although slurries are susceptible to viscosity and density variations, magnetic meters are unaffected by them. (The mass flow rate would be affected by density variations but not the volumetric rate.)

Fouling of the electrodes is troublesome for some services. Polymer slurries, in particular, exhibit this tendency, either reducing or eliminating the signal. However, cleaning

Figure 10.1. Full pipe size openings, unobstructed flow, make magnetic meters particularly suitable for slurry flow measurements. (Courtesy of Brooks Instr., Div. of Emerson Electric)

methods (both electrical and mechanical) have been introduced to alleviate the fouling problem.

Piping bypasses may also be installed so that meters can be removed and cleaned. This solution is sometimes questionable, though, since dead ends caused by bypasses also present potential plugging problems.

Magnetic meters are available in sizes from 0.1 to over 100 inches in diameter with a wide choice of tube materials.

The rangeability of magnetic meters is theoretically about 30:1. However, the normally accepted practical range accommodation is 20:1.

Accuracy ranges from ±½ to ±2%.

Target Meters

Unlike magnetic meters, target meters do present an obstruction to flow as can be seen in Figure 10.2. However, solid materials can flow relatively unimpeded around the targets and along the bottom of the pipe. There are no lead lines to clog or freeze, and the force bar is suitably sealed where it passes through the pipe.

The Foxboro target meter is available in 2-, 3- and 4-inch sizes and can be flanged or butt-welded into the line. A ¾-inch, 150-pound flanged meter is also available. The ratio of target to pipe size is from 0.5 to 0.8. Accuracy is ±½% of full span with proper flow calibration.

Pressures to 1,500 psig and temperatures to 750°F can be accommodated.

The Ramapo target meter is available in in-line units from ½- to 2-inch sizes, and flange mounted units are available for line sizes from 4 to 60 inches. Larger sizes can be made. Pressures to 5,000 psig and temperatures to 600°F can be handled. A retractable design is available which permits installation and removal of the probe into pressurized lines without system shutdown.

Accuracy is from ±½ to ±3% of span depending on whether actual or theoretical calibration is used.

Where the ratio of target to pipe size is relatively large, solids must be fairly small to flow freely. In the flange mounted units of the Ramapo line, flow is relatively unimpeded but guaranteed accuracies and repeatabilities in the low ranges are dependent on calibration data.

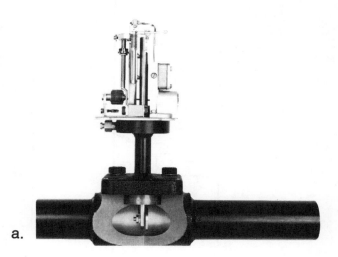

a.

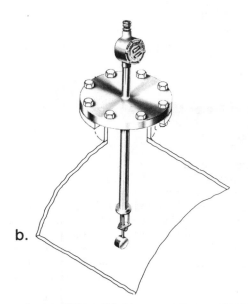

b.

Figure 10.2. Target meters allow solids particles to flow through relatively unimpeded, particularly the flange mounted type which offers little obstruction to the flowing fluid. (Courtesy of The Foxboro Co. and Ramapo Instrument Co., Inc.)

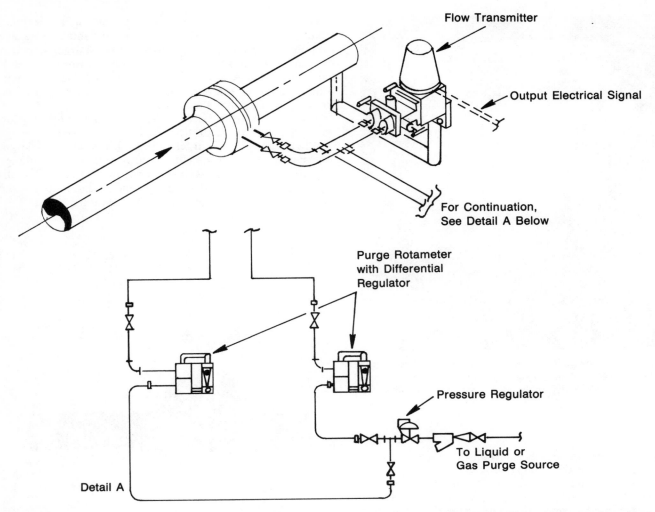

Figure 10.3. Head devices such as D/P cells may be used for slurry flow measurement if lead lines are purged to prevent solids buildup.

Differential Meter Methods

There are several methods of using differential or head meters in slurry services. They usually involve the use of suitable liquid or gas purges on the differential unit lead lines, as shown in Figure 10.3, or use flush mounting diaphragms that eliminate the need for open lines.

Many plants avoid purged lines because they present so many operating and maintenance problems. The purges must be shut off during shutdowns, and process operators often forget to put them back in service. Consequently there is still the possibility that lines may become plugged. The loss of purge sources and accidental closing of purge lines are also possible.

The primary elements used for slurry services include eccentric and segmental orifices (Figure 10.4) and Venturi tubes and flow nozzles (Figure 10.5).

The bore of eccentric orifices is located toward the bottom of the pipe to prevent damming of solid materials or

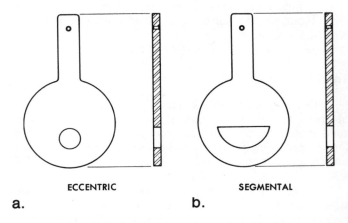

Figure 10.4. Eccentric orifices have offset bores to minimize solids buildup in slurry services. Segmental orifices with the segment near the bottom of the line accomplish the same purpose. (Courtesy of Daniel Industries, Inc.)

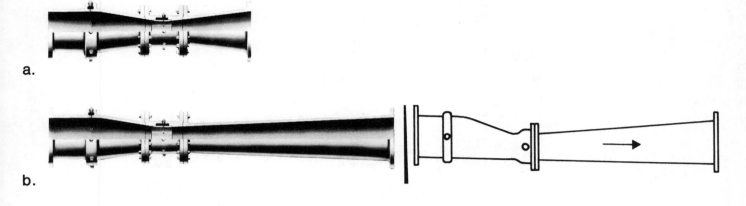

a.

b.

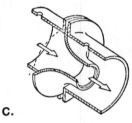

Figure 10.5. Venturi tubes and flow nozzles are designed such that solids accumulation is not very likely. Eccentric Venturis provide additional insurance against buildup. (Courtesy of BIF, a unit of General Signal Corp. and The Foxboro Co.)

c.

foreign particles, thus reducing the error that otherwise would result.

Segmental orifices are placed with the open segment near the bottom of the pipe to reduce solids buildup. Both methods although not ideal, are better than using concentric orifices.

Venturi tubes and flow nozzles, although they need purging to keep lead lines clean, are made so that damming of solids materials is avoided; therefore, they are useful for slurries.

Elbow taps can be used with the Taylor Volumetric DP Unit (shown in Figure 10.6) which allows the use of differential measurement without purging. The flush mounting diaphragms of this unit eliminates dead spots and reduces operating and maintenance troubles. It has the drawback of requiring a shutdown to remove it from the process for maintenance purposes.

The Taylor units are available in ranges from 20 to 800 inches H_2O with an accuracy of $\pm 1\%$ of span. The wafer elements can withstand pressures to 1,500 psig and temperatures to 800°F. The combined accuracy of the system, however, is poor because elbow taps are accurate only to ± 5 to $\pm 10\%$ of span.

Ambient temperature changes also affect the filled system, resulting in additional inaccuracy.

Rotameters

Some rotameter designs work very well in light slurry services. Figure 10.7 shows a straight through design that allows passage of slurries that have only small solid particles in the stream. Low flow rates on this type design are more likely to present operating difficulty because of the tendency to settle out. This design is available in sizes up to 8 inches at relatively high temperatures and pressures.

The accuracy of this type unit is normally $\pm 2\%$ of span. Rotameters provide rangeabilities close to 10:1.

Vortex Meters

The Vortex meter shown in Figure 10.8 is suggested for "dirty" as well as clean fluids. It has no moving parts to be affected by solids movements, but coating of the temperature sensing element would cause difficulty.

This meter is currently available in sizes from 2 to 6 inches at pressure ratings to 600 psig and temperatures from 0° to 600°F. Readout is available in digital form but can be converted to standard analog signals for rate, totalized flow or control applications. Linearity is within $\pm \frac{1}{2}\%$ and rangeability is 100:1. Viscosity effects are negligible.

Piping configurations require that upstream and downstream lengths conform to AGA standards for orifice meters.

Level Measurement

Level measurement in slurry services is somewhat like flow measurement in that the most popular methods for clean services present some formidable difficulties. External chambers for float and displacer measurements provide dead spaces where settling occurs or where sticky substances may solidify. Purging of float chambers may work on some applications, but other methods are usually sought in preference to purging. Fortunately, several other methods work very satisfactorily.

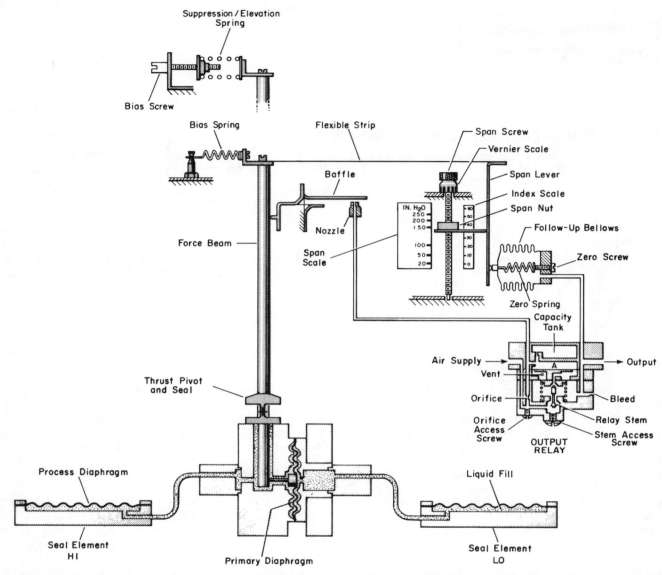

Figure 10.6. *The volumetric D/P unit with the sealed systems can be used with elbow taps for flow measurement in rough slurry services. However, accuracy of this type system is relatively poor. (Courtesy of Taylor Instrument Process Cont. Div., Sybron Corp.)*

Differential Types

There are several different approaches to slurry level measurement using hydrostatic head measurements. One of the obvious drawbacks is the effect of density changes if solids percentages are subject to change. Usually, however, solids contents are stable enough so that density changes do not become objectionable.

Air Bubblers

An old but simple method of level measurement is using air bubblers as shown in Figure 10.9. The dip tube and the purge fluid are the only items in contact with the fluid. Dip tubes may be made from many suitable materials. The prime requirement is to establish sufficient flow to keep the dip tube filled with purge fluid. Once the purge pressure overcomes head pressure, a very small purge flow keeps the measured fluid from entering the pipe. Dip pipes are usually ½ to 1 inch in size, determined primarily by the rigidity needed for the particular application. The lower tip of the bubbler tube should be vee-notched to assure that no large bubbles will form, giving erratic readings. The installation of a clean-out rod is possible through a compression fitting at the tank connection, if chemical buildup at the tip tends to clog the bubbler.

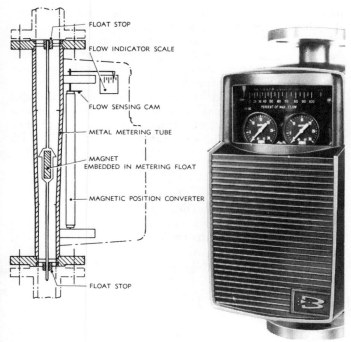

FLOAT STOP

FLOW INDICATOR SCALE

FLOW SENSING CAM

METAL METERING TUBE

MAGNET
EMBEDDED IN METERING FLOAT

MAGNETIC POSITION CONVERTER

FLOAT STOP

Figure 10.7. The straight through armored rotameter is adaptable to minor slurry services. There are no crevices in which solids may settle. (Courtesy of Brooks Instruments)

Figure 10.8. The Vortex meter (with no moving parts) may be used for homogeneous slurries if the fluid does not coat the temperature sensing element. (Courtesy of Eastech, Inc.)

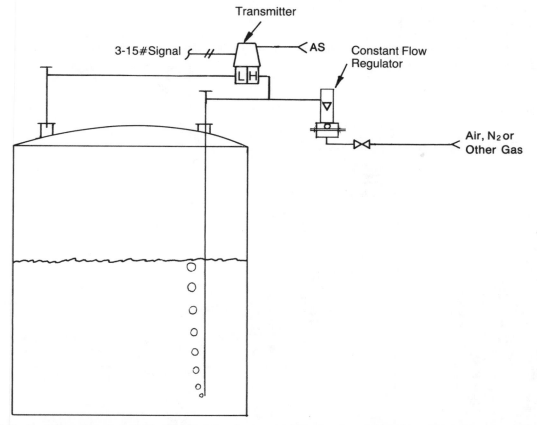

Transmitter

3-15#Signal

AS

Constant Flow
Regulator

L H

Air, N₂ or
Other Gas

Figure 10.9. A dip tube with air bubbler for purging is an old but simple method for measuring level slurries.

Direct readout by a manometer can be used as well as the transmission system shown in Figure 10.9.

A drawback of the air bubbler system is the introduction of the purge gas into the process which is undesirable in many instances.

Diaphragm Types

Figure 10.10 shows two d/p cell designs that were developed primarily as a protection against plugging in slurry services. The flush diaphragm type can be used where purging is not desired but where small dead end cavities are not objectionable. The extended diaphragm type can be used where even small cavities have a tendency to plug or freeze.

Wetted portions of these units can be furnished in a wide variety of materials and can also be coated with plastic materials to inhibit corrosion and prevent process contamination.

Connection flanges are available in sizes of 3, 4, and 6 inches at pressure ratings of 150 or 300 psig. Extension lengths on the extended diaphragm model can be furnished in standard lengths between 2 and 6 inches and in odd lengths between those values.

Accuracy of these units is from ±½ to 2% of span in ranges from 20 to 800 inches H_2O.

The low pressure connection for these units normally comes from the gas space in the vessel. If necessary it can be purged or filled with a suitable liquid to prevent condensation or other similar problems. If this solution is not desirable, diaphragm 1:1 repeaters can be used as shown in Figure 10.11. The 1:1 repeater is a device whose transmitted output is the same pressure as the process pressure exerted on the diaphragm. Its output is transmitted to the low pressure side of the primary measuring device and serves only to cancel out the static pressure of the process, allowing the primary devices to measure the liquid head.

Figure 10.12 shows the use of a volumetric sealed system that uses the same device shown in Figure 10.6 (in the flow section) for measuring slurry levels. Ranges and accuracies are given in the flow section.

The filling liquids in the sealed systems are carefully selected for low thermal expansion coefficients to reduce ambient temperature effects. The error is about 1 inch H_2O for each 50°F temperature change from the calibrated standard of 80°F.

Pressure Bulbs

Pressure bulbs shown in Figure 10.13 are applicable to open tank measurements. The sensing elements are not particularly affected by solids materials, but the requirements for atmospheric pressure on the low side limits their application appreciably.

Ranges for these elements vary from 0 to 6 inches H_2O to 0 to 100 inches H_2O.

Capacitance Type

Capacitance probes can be used very effectively for slurry measurements, either for continuous level or point level applications. They operate on a simple principle and possess an advantage of having no moving parts. Figure 10.14 shows an electronic system for continuous level application in which the active probe length is selected for the level span needed. The primary problem encountered is coating of the probe.

Figure 10.15 shows a pneumatic system on a continuous level application in which the probe never comes in contact

Figure 10.10. Level measurements are made with flat or extended diaphragm units in services where cavities or dead ends tend to plug or polymerize. (Courtesy of The Foxboro Co.)

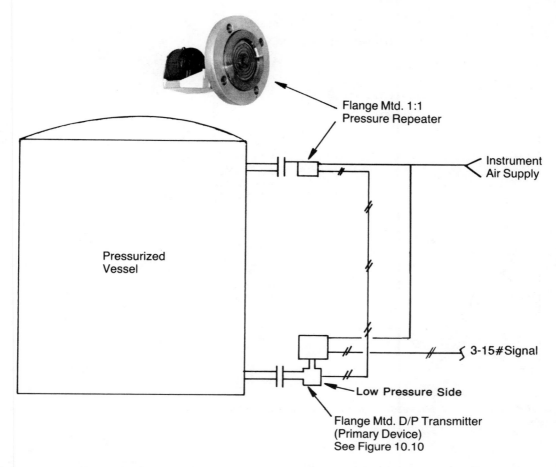

Figure 10.11. In measuring slurries in pressurized vessels, it is sometimes desirable to use 1:1 pressure repeaters on the low pressure side of the primary level devices to eliminate plugged lead lines or lines that might fill with condensate. (Courtesy of The Foxboro Co.)

with the measured fluid, thus avoiding potential coating problems.

Capacitance units can be made to work very satisfactorily. They have been used for several years, and manufacturers and users have accumulated much information relative to their use—such as the dielectric constants of various process materials and mixtures. It is necessary, however, in many cases to follow through with some rather extensive calibration procedures to assure proper calibration and span adjustments, especially for continuous level applications. For on-off control where bridging from probe to vessel is a problem, the Drexelbrook "Cote-Shield" probe (Figure 10.16) employs an intervening electrode. When this probe is kept at the same potential as that of the primary measuring probe, only capacitive reactance current—no conductance current—flows to ground. Level change is therefore indicated when capacitive reactance current changes.

For continuous measurement, the effects of conductive coatings on the probe and the vessel wall are nullified electronically. Drexelbrook's "True Level" circuitry uses a triple electrode probe to ignore changes in material composition. A signal is developed between the measuring electrodes due to variations in fluid density, temperature, and composition.

The accuracy of capacitance units may range from $\pm 1\frac{1}{2}$ to $\pm 3\%$ (or greater) of span. Applications for slurry services are likely to be less accurate than on clean services.

The following conditions should be considered in applying capacitance devices.

1. Dielectric constants change with temperature, about 0.1% per degree centigrade.
2. Chemical changes affect the dielectric constant; moisture has a pronounced effect.
3. Variations in particle size affect dielectric constants.
4. Product coating could ground the probe.

Point level measurement with capacitance probes presents far fewer difficulties than continuous level applications. They are usually mounted horizontally so that

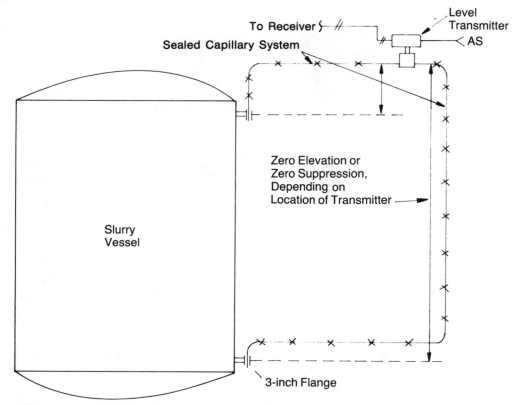

Figure 10.12. *Sealed systems from the vessel to the differential element eliminate the need for purging and can be used for difficult slurry levels.*

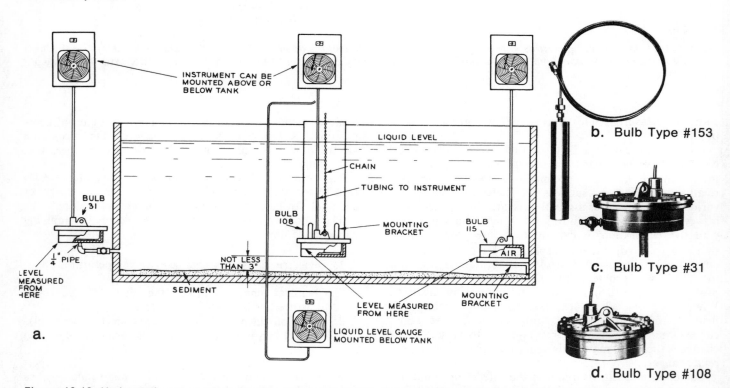

Figure 10.13. *Hydrostatic measurement using pressure bulbs may be used for slurry levels, but applications are limited to open vessels or atmospheric conditions. (Courtesy of The Bristol Division of Acco)*

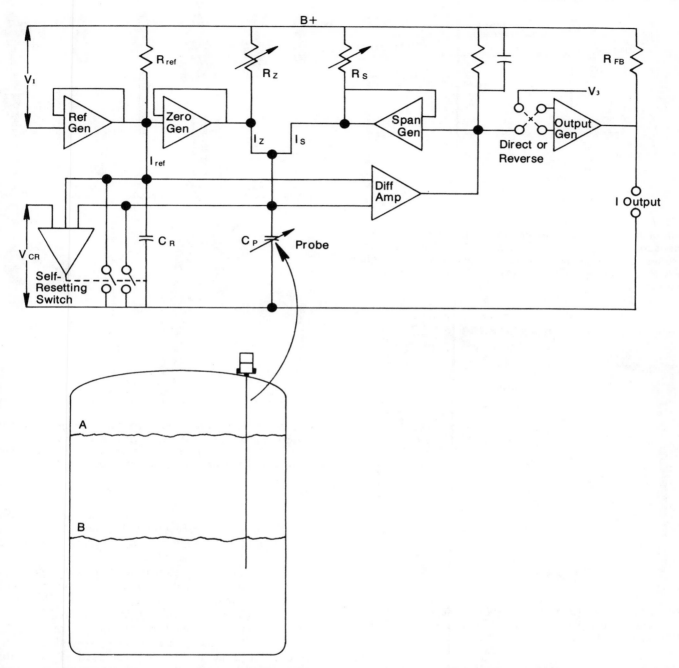

Figure 10.14. Capacitance probes are used effectively on many polymer and slurry services where other type level measurements are ineffective. (Courtesy of Roberstshaw Controls Co.)

the large surface areas of the capacitor plates (probes) are affected by very small level changes. The change in capacitance is rather pronounced and abrupt, and switch action occurs easily.

Capacitance probes for point level measurement are available in a wide variety of shapes and sizes. Figure 10.17 shows several shapes furnished by one manufacturer. Some of these are designed specifically for slurry services.

Ultrasonic Type

Ultrasonic devices have not been marketed very long and have been used on a limited basis to date. The principle of operation, however, lends itself to measurement of clean liquids, solids or slurries. Figure 10.18 reveals the operating principle which keeps the transmitter and receiver both away from the measured liquid.

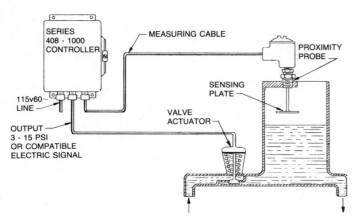

Figure 10.15. Where "coating" problems arise in some level measurements, problems are avoided by using capacitance probes (sensing plates) that do not have to come in contact with the measured fluid. (Courtesy of Drexelbrook Engineering Co.)

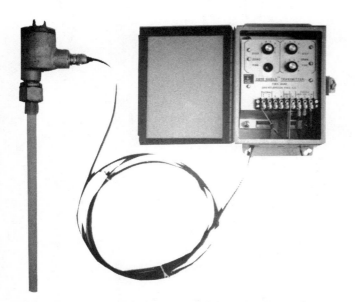

Figure 10.16. The "Cote-Shield" probe employs an intervening electrode to permit only capacitive reactance current flow. (Courtesy of Drexelbrook Engineering Co.)

Because ultrasonic waves travel at different velocities in different media, some calibration and adjustment work is necessary in interpreting and correlating data from such a system unless the manufacturer or user has had previous experience with the fluid and atmospheres involved.

Point measurements can be made with ultrasonic devices in dirty or slurry services as well as clean fluids. Both single sensor and dual sensor systems are used (Figure 10.19).

Nuclear Radiation Type

Nuclear radiation gauges can nearly always be depended on as candidates for level measurement when other methods

Figure 10.17. Many sizes and shapes of capacitance probes are used for the wide variation of applications covered. A few of more than 50 standard probes of one manufacturer are shown. (Courtesy of Drexelbrook Engineering Co.)

fail to qualify. Because of their high cost, the requirement for licensing and the precautions that must be followed in operating and maintaining them, they are usually considered only for the most difficult applications.

Radiation sources are sometimes mounted internally to reduce the source size that would otherwise be required by external mounting. Measuring or detecting cells may be located internally also as shown in Figure 10.20, or they may be externally mounted. In the majority of cases both source and cell will likely be externally mounted as shown in Figure 10.21 if at all feasible. The method of locating sources and measuring cells depends largely on the accuracy and linearity desired for the measurement. Accuracy suffers as the ratio of wall absorption to material absorption increases. It is usually in the range of ±1%.

Other Level Switches

Other level switches which are applicable to slurry measurements include the vibrating reed (Figure 10.22), the conductivity (Figure 10.23) and internally mounted floats or displacers. The latter group works well only if the measured fluid does not coat the float or displacer. Point level switches are operated such that the sensor is always sub-

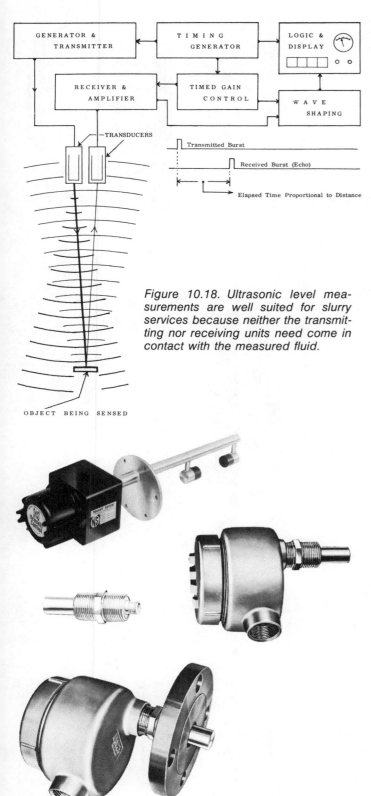

Figure 10.18. Ultrasonic level measurements are well suited for slurry services because neither the transmitting nor receiving units need come in contact with the measured fluid.

Figure 10.19. Both single element and two-element sensors may be used for point level measurements. (Courtesy of National Sonics Corp. and Delavan Mfg. Co.)

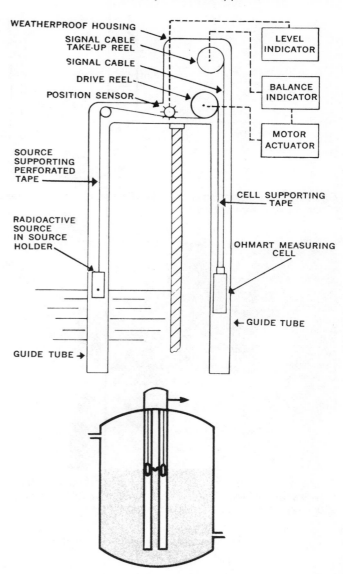

Figure 10.20. Both the radioactive source and the measuring cell may be mounted inside a vessel in a radiation level measuring system when warranted by the application. (Courtesy of Ohmart Corp.)

merged in the product, or is not covered, which ordinarily does not give rise to coating problems.

Level Gauges

In slurry services the normally used externally mounted reflex and transparent gauges cannot be used effectively. Two solutions that may be considered include the welding pad (Figure 10.24) and the magnet (Figure 10.25). The latter type still has a tendency to clog and probably needs purging. The float must be free to travel inside the mounting chamber. The pad type also has a tendency to allow coating

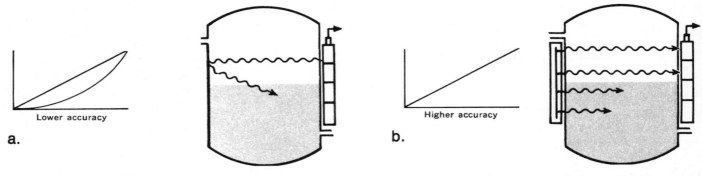

Figure 10.21 External mounting of radiation source and measuring cell is preferable for slurry services if the application permits. (Courtesy of Ohmart Corp.)

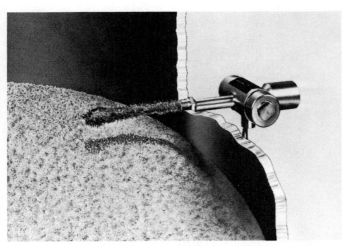

Figure 10.22. Vibrating reed switches work well in slurry services, and probe coating presents no particular difficulties. (Courtesy of Automation Products, Inc.)

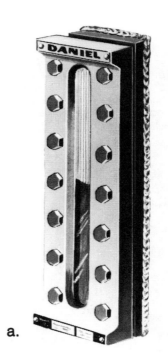

a.   b.

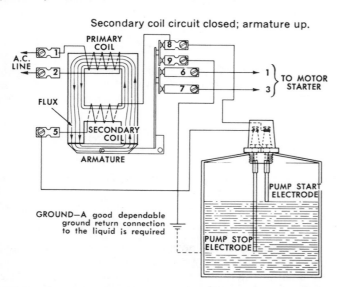

Figure 10.23. Conductivity switches can be used for slurry applications if "coating" is minimal and fluid is sufficiently conductive. (Courtesy of B/W Controller Corp.)

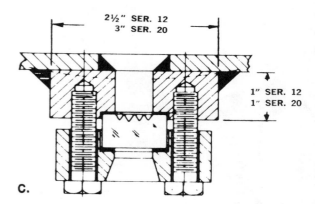

c.

Figure 10.24. Welding pad level gauges may be useful for slurry applications if the glass walls do not "coat" with the measured material. (Courtesy of Daniel Industries, Inc. and Jerguson Gage and Valve Co.)

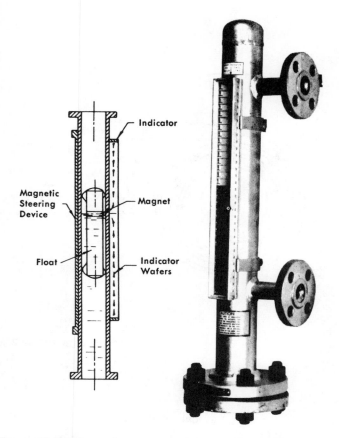

Figure 10.25. Magnetic gauges can be used in slurry services if purging prevents the float from clogging or sticking. (Courtesy of Jerguson Gage and Valve Co.)

Labels in figure: Indicator, Magnetic Steering Device, Magnet, Float, Indicator Wafers

on the glass surface, thus obscuring the level. Both methods have drawbacks but should receive consideration if direct indication is badly needed.

Pressure Measurement

Pressure measurement in slurry services presents few of the problems encountered in flow and level measurements. While it would be undesirable to use bourdon elements (Figure 10.26) and bellows elements (Figure 10.27) in heavy slurry services, using diaphragms for the primary sensing elements or for sealed systems makes pressure measurement relatively easy.

In many instances even bourdon and bellows elements are satisfactory if the solids are suspended in the fluid with little tendency to settle out. Unless the measured fluid hardens, pressure is reflected throughout the element with no decrease in accuracy.

Diaphragm elements as pictured in Figure 10.28 are ideal for pressure measurement in slurries. In most instances, the elements can be mounted flush with the container so that there is very little dead space to present problems.

Figure 10.29 shows a highly effective, diaphragm pressure transmitter used for polymer, highly viscous and solids-

containing services. It is available in ranges as low as 100 inches H_2O to 120 psig for pneumatic applications only. Another diaphragm type shown in Figure 10.30 is available in ranges to 10,000 psig.

Diaphragm elements are also used in strain gauges (Figure 10.31) so that a wide variety of pneumatic and electronic devices are available in all ranges for effective pressure measurement.

The use of chemical seals with filled systems between the diaphragm element and the measuring element of an indicator, recorder or transmitter is also common for slurry pressure measurement. It applies not only to slurry measurements but also to corrosive services where isolation is needed to reduce or eliminate the cost of expensive elements and/or high maintenance costs.

Temperature Measurement

Temperature measurement in slurry services is somewhat like pressure measurement in that it presents far fewer problems than flow or level measurements. There are probably fewer corrective measures which can be taken for temperature measurement than for any of the other primary measurements.

The same type devices are likely to be used as for clean services—thermocouples, resistance elements, filled systems, etc.

The primary difficulty encountered is the "coating" of wells or elements which slows response times. The best solutions in these situations are (a) proper location in the stream or vessel so that turbulence provides a cleaning action or (b) provision of a smooth finish so there is less tendency for solids or sludge to stick or coat.

Two methods of temperature measurement, radiation pyrometry and optical pyrometry, do not require contact with the measured fluid and are as conveniently used for slurry measurements as for clean services. However, their primary use is for high temperature measurements such as those required for furnaces, boilers and other high temperature applications. They are seldom applied to process measurements required by the chemical, petrochemical and refining industries where slurries present problems.

Control Valves

The previous discussions relating to slurries and viscous services have concerned measurement methods primarily. The final control elements used for these services are also of major concern.

Several types of valves are used in slurry applications. The standard globe valve may work quite well if solid particle sizes are small and are suspended in the carrying liquids with little tendency to settle out. However, when the solids tend to fall out and particle sizes vary, other valve styles should be used.

Among the valves particularly suited for slurry applications are ball, diaphragm, split body, pinch, butterfly and angle valves. Brief descriptions of these valves follow.

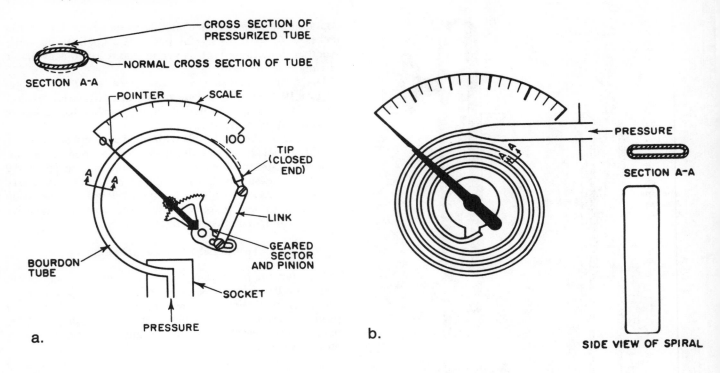

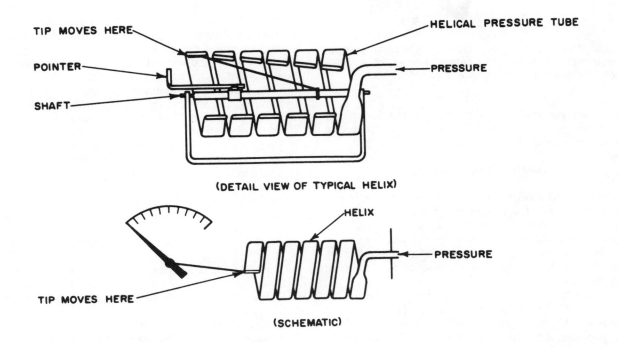

Figure 10.26. In many slurry services, bourdon elements are undesirable, particularly if the solids have a tendency to settle out and/or harden. (Courtesy of Ametek/U.S. Gauge)

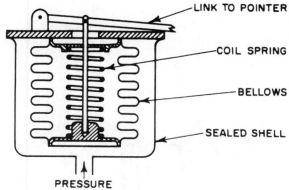

LINK TO POINTER

COIL SPRING

BELLOWS

SEALED SHELL

PRESSURE

a.

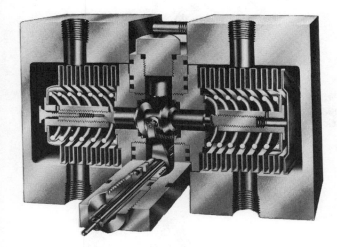

b.

Figure 10.27. Bellows elements are not conducive to fluid measurements where solids might settle and interfere with the free movement of the bellows. (a. Courtesy of Ametek/U.S. Gauge. b. Courtesy of ITT Barton)

TYPICAL APPLICATIONS

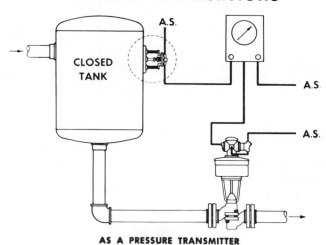

CLOSED TANK

A.S.

A.S.

A.S.

AS A PRESSURE TRANSMITTER

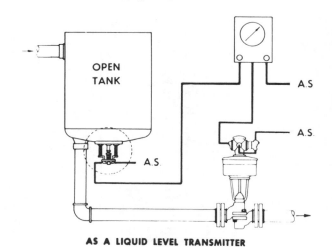

OPEN TANK

A.S

A.S.

A.S.

AS A LIQUID LEVEL TRANSMITTER

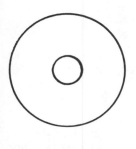

(a) Flat Type

(b) Corrugated Type

Figure 10.28. Diaphragm elements are ideal for slurry pressure measurements.

Figure 10.29. This flat diaphragm pressure transmitter is specifically designed to mount flush with vessel walls and is ideal for polymer and other slurry applications where pressures do not exceed 250 psig. (Courtesy of ITT Hammel Dahl/Conoflow)

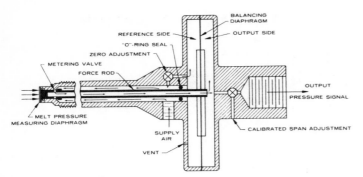

Figure 10.30. This button diaphragm pressure transmitter is available for polymer and slurry services for pressures to 10,000 psig. (Courtesy of Rosemount, Inc.)

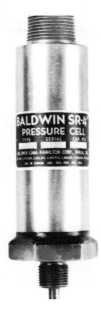

Figure 10.31. This strain cell uses a diaphragm element to sense pressure and produce movement that results in measurement by the strain gauge measuring circuit. It can be used on polymer and slurry services and is available for pressures to 50,000 psig. (Courtesy of BLH Electronics, Inc.)

Figure 10.32. Fisher Controls' Vee-Ball design is well-suited for hard-to-handle fluids such as paper stock and polymer slurries. (Courtesy of Fisher Controls Co.)

Figure 10.33. The Masoneilan control ball design can also handle slurry applications where globe valves are questionable. (Courtesy of Masoneilan)

Ball Valves

Ball valves, vee-balls or partial balls are used consistently in hard-to-handle fluids such as paper stock, polymer slurries and other solids-containing fluids. Some of these were developed primarily for this type service.

Figure 10.32 shows the Fisher Controls Vee-Ball design which is currently available in sizes from 2 to 16 inches. Body pressure ratings are 150 and 300 pounds, and temperatures as high as 730°F can be accommodated.

The Masoneilan Control Ball (Figure 10.33) is quite similar to the Fisher valve. It is available in sizes from 2 to 12 inches at 150- and 300-pound ratings. Temperature ratings are from −250°F to 500°F.

Both the Fisher and Masoneilan valves are available with plastic or stainless steel seal rings to minimize leakage rates of the valves.

Figure 10.34 shows a full ball valve made by DeVar-Kinetics that employs a cage to carry a solid ball into the mouth of the body opening. Its design ensures a self-cleaning action, and tight shutoff is provided.

Figure 10.34. This full ball design was designed for paper stock and other difficult slurry services. (Courtesy of Bell and Howell, Control Products Division)

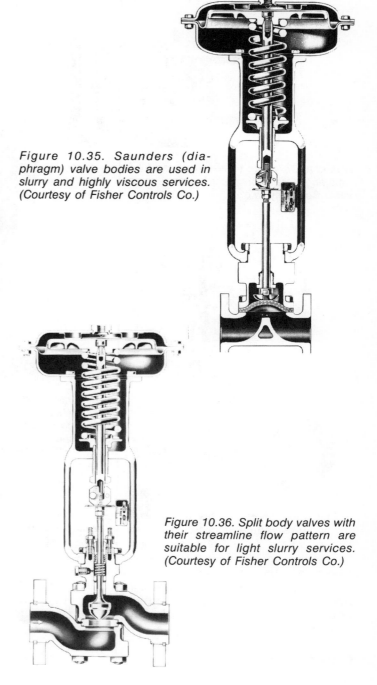

Figure 10.35. Saunders (diaphragm) valve bodies are used in slurry and highly viscous services. (Courtesy of Fisher Controls Co.)

Figure 10.36. Split body valves with their streamline flow pattern are suitable for light slurry services. (Courtesy of Fisher Controls Co.)

These ball and partial ball designs have good control characteristics and high rangeabilities; they are high recovery valves (low permanent pressure loss).

Diaphragm or Saunders Type

Diaphragm or Saunders control valves are constructed in a weir pattern (Figure 10.35) primarily, but straightway patterns are also available. Either pattern is suitable for slurries and viscous fluids.

These valves provide near-streamline flow with self-cleaning action. They have a fairly high capacity at a relatively low cost. The diaphragm isolates the flowing fluid from working parts of the valve.

The valves have poor control characteristics and relatively short diaphragm operating life and are limited to relatively low pressure drops. Operating temperatures are limited by the plastic material used for the diaphragm—around 350° to 400°F. Operating pressure normally should not exceed 100 psig for standard valve topworks.

Split Body Valves

Split body valves (Figure 10.36), because of their streamline flow patterns, are suitable for light slurry services. There are no pockets or shoulders for solids accumulation. They are available in a wide variety of sizes and materials and operate over a wide range of temperatures and pressures—from −400° to 1,000°F and up to several thousand psig.

Soft seats may be used for tight shutoff, and the valves are easily maintained. Costs are less than for standard type globe valves. Control characteristics are good.

Pinch Valves

Pinch valves are not widely used in the refining, chemical and petrochemical processes but find greater use in mining, cement plants, paper mills, etc. Figure 10.37 shows a valve operated by air or other hydraulic means where throttling is accomplished by the variation of pressure on the flexible sleeves. Sleeves are available in gum rubber, neoprene, Buna-N, Buna-S, butyl, viton and hypolon.

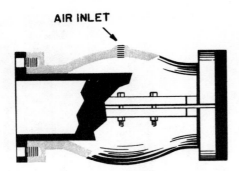

AIR INLET

Figure 10.37. Pinch valves are used for heavy slurry services including metallic ores, coal and paper stock. (Courtesy of Red Valve Company, Inc.)

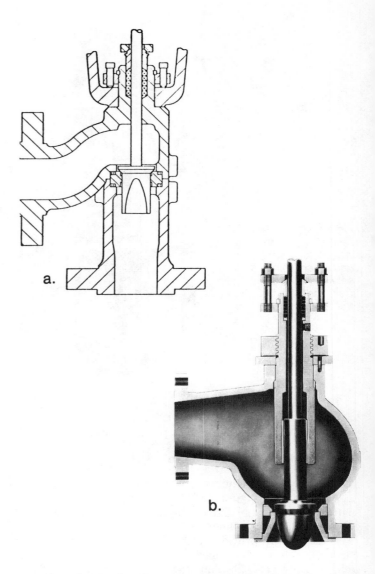

a.

b.

Figure 10.39. Angle valves are useful in slurry services where high pressure drops are required. (Courtesy of ITT Hammel Dahl/Conoflow)

Figure 10.38. Butterfly valves work well in slurry services particularly if they do not operate too near the shutoff position. (Courtesy of ITT Hammel Dahl/Conoflow)

There are several types of construction. Pressure and temperature ratings depend on the valve design and the materials used. In addition to the pneumatic or hydraulic type shown, another design uses a mechanical pinch-clamp to compress the sides together for throttling action.

Pinch valves have poor control characteristics, cannot tolerate high pressure drops, and are slow to respond. They provide unobstructed flow, have high capacities, are low in cost and are easy to maintain.

Butterfly Valves

Butterfly valves may be used effectively for some slurry services, particularly if they operate at relatively high flow rates. Unless they are operated in a near-closed position, they tend to be self-cleaning. Figure 10.38 reveals that as the valve opens, the flow tends to flush along the bottom of the pipe.

Butterfly valves have high capacities, low pressure drops and high recovery characteristics; they require minimum installation space, are economical and are readily available in large sizes.

Operating torques are high; tight shutoff depends on the use of resilient seats which are temperature limited; and the valves have relatively low rangeabilities.

Angle Valves

Several designs of angle valves are available for use in slurry services. Figure 10.39 shows two such designs. These valves come in a wide variety of sizes and materials and can handle high pressure drops, even in the larger 6- and 8-inch sizes.

Angle valves are more likely to be used in the small pipe sizes where some of the other designs are not too applicable. They may also be used in highly errosive and corrosive services where other designs may be lacking.

Miscellaneous Measurements

Measurement and control of almost any variable in a slurry or highly viscous stream presents some unique problems. The previous topics cover the variables that are most often encountered in instrumenting processes, but other measurements and controls also need coverage. The next few paragraphs discuss some of the more infrequently used control variables.

Density

Some density devices recommended for use in light and heavy slurries are described below.

Figure 10.40 is a typical pneumatic U-tube densitometer that handles light slurries. This one is made with a 1¼-inch tube; other manufacturers use U-tubes that are 2 inches in diameter. The range of measurement is from 0.6 to 1.8 specific gravity with an adjustable span from 0.2 to 0.5 specific gravity.

Accuracy is ±1% of span.

The flow rate is 50 gpm maximum. If greater flows are involved, a bypass stream is used for measurement.

Pressure is limited to 50 psig, and the temperature range is from 20° to 180°F standard. Special devices can be built, if needed, outside these ranges.

Figure 10.41 shows a straight flow design that can accommodate larger volumes and heavier slurries (such as ores).

Its range is 0.7 to 1.5 specific gravity; its span is 1.0; and its accuracy is ±1% of span. It uses a 2-inch diameter tube and has a recommended maximum flow rate of 125 gpm. The temperature and pressure ranges are the same as for the previous model.

Figure 10.42 uses a U-tube arrangement for process flow but works on an entirely different principle. The tube is ½ inch in diameter and is welded at the nodes.

A drive coil electrically excites the tube, producing a mechanical vibration. The vibration becomes a function of the mass of the media contained in the tube.

The vibration is sensed by a pickup end consisting of a coil and armature arrangement which produces an AC voltage that is a function of the density or specific gravity of the fluid.

This unit is built for use in Class 1, Group D, Division 1 areas, and has a standard pressure rating of 1,000 psig at 100°F. Its standard basic range is any 0.5 specific gravity units, and its span can be as low as 0.05, specific gravity units.

Nuclear density gauges are suitable for any slurry that flows through pipes if the measurement range falls within its measureable limits. Figure 10.43 shows an in-line type that comes in 1- through 4-inch sizes with full scale ranges as narrow as 0.025 specific gravity units. They are characterized by low maintenance, high accuracy and long life. They are easy to install and are available for Class I, Group D, Division 1 areas, but they are expensive.

Figure 10.44 shows clamp-on density devices that are used for pipe sizes from 2 to 30 inches. Operating characteristics are similar to those listed for the in-line type.

Figure 10.40. This U-Tube densometer can be used to measure fluid densities of light slurries. The U-Tube is 1¼" in diameter. (Courtesy of Halliburton Services)

Figure 10.41. A straight through design densometer is more suitable for larger volumes of suspended solids where higher flow rates and larger particle sizes are encountered. (Courtesy of Halliburton Services)

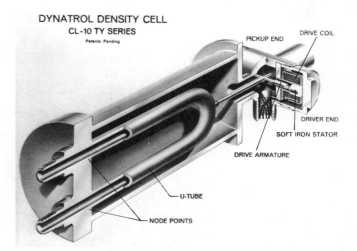

Figure 10.42. Densities can be measured in slurries with this U-Tube device which, when excited at the drive coil, vibrates at amplitudes proportional to the density of the fluid. (Courtesy of Automation Products, Inc.)

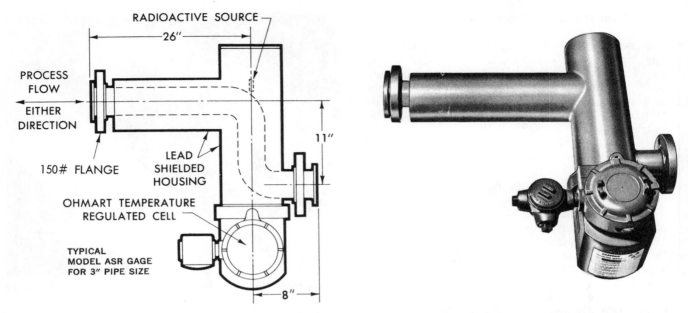

Figure 10.43. Nuclear radiation density gauges are suitable for measuring any type slurry that flows through pipes. (Courtesy of Ohmart Corp.)

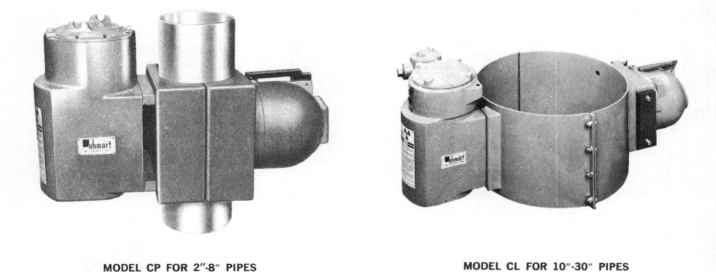

MODEL CP FOR 2"-8" PIPES   MODEL CL FOR 10"-30" PIPES

Figure 10.44. Nuclear density gauges are available in clamp-on types as well as in-line mounting types. (Courtesy of Ohmart Corp.)

Viscosity

A wide variety of viscosity measuring techniques are used for laboratory and process applications in the processing industries. Among them are capillary tube, falling ball, bubble time, sliding plate, ultrasonic, float, falling piston, vibrating reed and rotational types. Many of the techniques are questionable, however, in slurry services.

Several of the types may be made to work in homogeneous slurries if the particle sizes are small. Dif-

ficulties would be encountered in heterogeneous mixtures where particle sizes are varied and generally larger.

The techniques more applicable to slurry services include the ultrasonic, the vibrating reed, and some rotational designs.

Figure 10.45 shows the vibrating reed device which may be vessel or line mounted. It imparts a mechanical vibration to the sensing probe at a frequency of 120 hz. The amplitude of vibration depends on the resistance to the shearing action of the vibrating reed, therefore becoming a

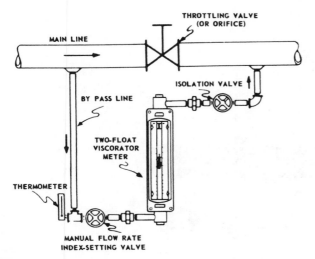

Figure 10.45. The viscosities of slurries can be measured with this vibrating reed device, but temperature compensation is necessary if process temperatures are expected to vary. (Courtesy of Automation Products, Inc.)

measure of viscosity. Vibration amplitude is measured in the pickup end of the unit which consists of an armature and coil arrangement which produces an output voltage proportional to vibration amplitude.

Since viscosity is normally affected by temperature variations, temperature compensation is usually necessary

to cancel the effects of process temperature variations. It is provided in this device by using a temperature sensitive resistance probe.

The unit is made for use in Class I, Group D, Division 1 areas. It can be used at pressures to 3,000 psig at 100°F. Design temperature is good to 300°F as standard.

Ranges are available from 1 to 100 centipoise to 1,000 to 100,000 centipoise. Accuracy is about ±1% of measured viscosity.

The ultrasonic density device measures viscosity by the damping effect the fluid exerts on a single sensing element consisting of a thin strip of metal. The sensing element is caused to vibrate at its natural frequency, and the vibration produces a voltage output proportional to vibration amplitude.

Automatic temperature compensation is available to correct temperature variations to a designated base temperature.

Available ranges vary from 0-5 to 0-50,000 (cps) (g)/cc. Units will operate at temperatures to 650°F and pressures to 1,000 psig. Accuracy is ±2% of span.

Figure 10.46 shows one of several rotational viscometer designs. This one measures the drag force on the rotating probe and translates this force to voltage, current or pneumatic signals proportional to viscosity.

Available ranges are from 0 to 10 cp to 0 to 50,000 cp. Operating temperatures are from −20° to 600°F standard and up to 3,000°F for special applications. Accuracy is ±1% of full scale range.

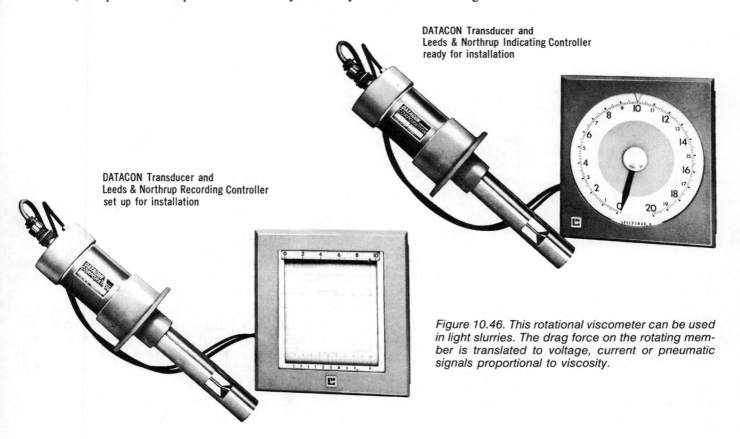

DATACON Transducer and
Leeds & Northrup Indicating Controller
ready for installation

DATACON Transducer and
Leeds & Northrup Recording Controller
set up for installation

Figure 10.46. This rotational viscometer can be used in light slurries. The drag force on the rotating member is translated to voltage, current or pneumatic signals proportional to viscosity.

11 Accuracies and Errors

H. B. Williams, William G. Andrew

A common practice in industry when discussing instruments or instrument systems is to speak of their "accuracy" in terms of their "inaccuracy." For example when a flowmeter is described as having an accuracy of ±1%, it means that its deviation or inaccuracy will be no greater than ±1% of its measurement or control span. What is meant is that the accuracy is within ±1% of span. In either case, the reader of this type information is normally aware of the general use of the term.

There are several other terms besides "accuracy" used to describe the performance of an instrument or of an instrument system. *Accuracy,* for example, may encompass the meanings of several of the other terms. Each term, however, is useful in describing the instrument's capability and performance. Some of the common terms found in instrument specifications are defined below. The definitions given agree in principle with SAMA Standard RC2-011-1964 and other leading sources of specifying terms for instrument devices.

Definitions of Specifying Terms

Accuracy: the capacity of a device or a system to render a true value of a measured variable under reference conditions. This term should not be confused with "repeatability" which is defined below. *See also* Total Error.

Bias: See Systematic Error.

Deadband: the minimum input reversal required to change the output.

Drift: a gradual variation with respect to time from a beginning value when the measured variable has not changed. This term most often applies to changes that occur after a specified warm-up period. A long-term calibration drift usually occurs because of the aging of component parts.

Illegitimate error: an obvious mistake unrelated to process conditions.

Precision: the degree of exactness for which an instrument is designed or intended to perform. *See* Random Error, Repeatability, Resolution.

Random error: the maximum magnitude of the dispersion of measured values about the mean. This value is equal to half the magnitude of repeatability.

Repeatability (or nonrepeatability): the maximum output deviation from the mean of consecutive measurements, having used identical input values arrived at from the same direction and having made full range traverses.

Resolution: the smallest interval between two adjacent distinguishable values.

Systematic error (or bias): a constant uniform deviation of the operating point of an instrument.

Total error (or inaccuracy): the maximum error limit of a device or system, equal to the maximum deviation value plus systematic error.

True value: the errorless value of the measured variable. It must be understood that all measurements are made with reference to some standard, the measurement of which in itself is subject to some error.

Figure 11.1 graphically portrays the relationships between the terms listed above. There are several other terms which are meaningful in explaining the required and actual behavior of process instruments. The list below, while perhaps not all-inclusive, provides brief explanations of some of the more useful terms.

Backlash: mechanical deadband associated with wear or manufacturing tolerance in gears.

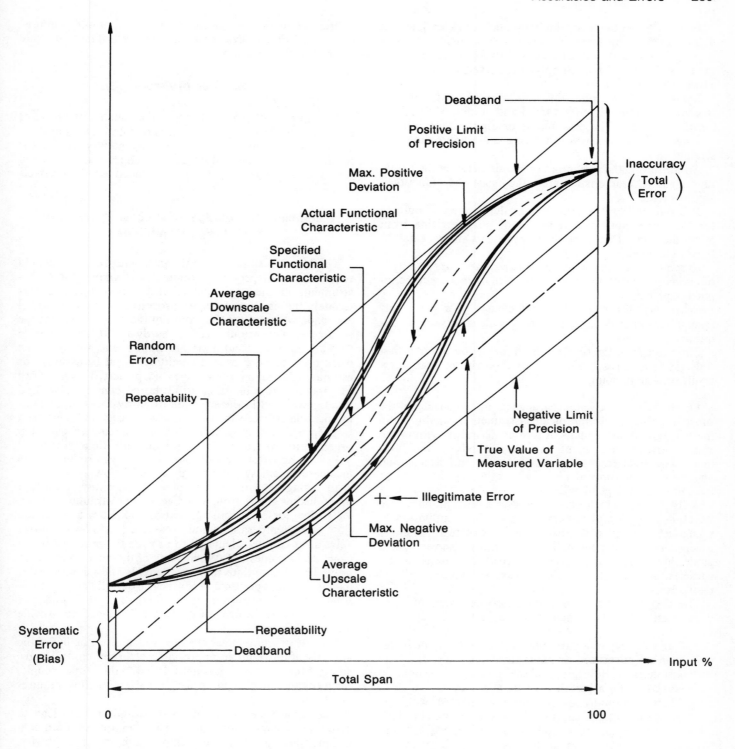

Figure 11.1 Accuracy terms relationships.

Inertia: the property of a body that tends to keep it at rest unless acted on by some net force or that tends to keep it moving in a straight line at a constant velocity unless acted on by some net force. Simply, the larger the mass, the slower the response.

Input/output isolation: the separation of the biasing influences that mutually affect input and output, while allowing the transfer of process analog or digital information.

Linearity: the closeness to which a curve approximates a straight line.

Nonlinearity: the maximum deviation between the actual functional characteristic and the specified functional characteristic.

Reproducibility: the exactness of duplication of outputs for the same input over a period of time, arriving from either direction. Deadband, drift, hysteresis and repeatability are included.

Response time: the length of time required for the output of an instrument to reach a steady state when a specified input is applied.

Sensitivity: the ratio of output change to a given corresponding input change, beginning with steady-state conditions.

Time constant: the time required for the output of a first order system to reach 63.2% of its total rise or decay as a result of a step change at the input.

Of all the terms given, accuracy is usually considered the most important criterion of an instrument's capability. In the selection of instruments, it is often desirable to know the comparative accuracies of various methods of measurement. The requirements for accuracy vary with application and must be considered along with other factors such as economics and dependability.

A list follows which may be used as a guide in hardware selection and control systems design. The compilation of such a list may be a little misleading since there are so many exceptions to the general accuracy ranges shown. As a guide, however, it helps orient the reader to the usual accuracies obtained for various measurement devices and methods.

The user of instruments should also be aware of some other pertinent facts related to instrument accuracy.

1. The accuracy of a particular instrument may be better than its advertised or guaranteed accuracy. In many instances a calibration check against a recognized standard can ascertain its true accuracy.
2. There are many other facets associated with the measurement and control of process parameters which contribute to errors. These are discussed in considerable detail later in the chapter.

Accuracies of Instruments and Measurement Methods

Table 11.1 shows accuracies for many commonly used devices and methods. The accuracy column is in reality the total error expected for the device or method, in percent, referenced to the measurement span. Laboratory calibration of individual devices would in many instances provide errors less than those listed.

Sources of Errors

The sources of errors, other than the inability of a piece of hardware to provide a true measurement, are many. In some cases the amount of error may greatly exceed the hardware in accuracy. Many of these sources are listed and discussed so that they may be avoided or their effects minimized.

Insufficient Knowledge of Process Parameters and Design Conditions

The Utopian situation for the instrument engineer would require that all process parameters be known for all instruments at the beginning of each new project. Unfortunately this is never true. As processes develop, decisions are made which affect the nature of the process and change parameters for various pieces of hardware. It is not feasible to wait until all conditions are known and verified. The work must proceed even if conditions must be assumed or estimated during the early stages of a project. This may lead to possible errors in sizing orifice bores, control valves and relief valves and in the selection of hardware. Proper communication between instrument and process designers tend to reduce this kind of difficulty, but it requires good job organization and adherence to communications details.

Process and design conditions which may introduce errors in measurement systems are discussed below.

In sizing control valves, one of the most important prerequisites is a thorough and accurate hydraulic analysis. When line losses, specific gravities, maximum and normal flow rates, temperatures and pressures are known, control valve sizes can be calculated to provide the process control characteristics desired. When some of these parameters are unknown, assumptions must be made. While a control valve having an equal percentage characteristic will "cover a multitude of sins," misinformation sometimes leads to under-or oversized valves which must often be changed to provide adequate control.

Process or mechanical changes which alter the hydraulic analysis often result in appreciable errors in control valve sizing if the information is not relayed to the proper people. Such changes include pump size, line size, fluid flow requirement, etc.

The same type of misinformation and changes lead to errors in flow measurements. Misinformation on fluid conditions such as pressure, temperature, specific gravity, viscosity, conductivity (for magnetic meter applications), corrosiveness, solids content, etc., result in erroneous calculations and applications.

Rangeability requirements that are different from that specified may cause large measurement errors. For example, if an 8 to 1 rangeability is needed, an orifice measurement whose rangeability is 4 to 1 would not be applicable.

Table 11.1. Expected Error Limits of Commercially Available Instruments

Flow

Device	Total Error
Measuring elements	
Concentric orifice	±½-1%
Segmental orifice	±2½%
Eccentric orifice	±2%
Quadrant edged orifice	±1%
Integral orifice	±3%
Segmental wedge	±1%
Elbow taps	±5-10%
Vortex shedding body	±½%
Swirl body	±¾-1%
Pitot tube	±½-1½%
Pitot venturi	±½-5%
Venturi tube	±¼-3%
Dall tube	±1%
Flow nozzle	±1-1½%
Turbine tube	±¼-½%
Propeller	±2%
Rotameter	±½-2% or
Rotameter (Purge)	±5-10%
Magnetic	±½-1%
Magnetic resonance	±½%
Parshall flume	±2-3%
V-notch weir	±2-4%
Trapezoidal weir	±4%
Hot wire (anemometer)	±2%
Thermal (displaced T/C's)	±½-1%
Target (disc)	±½%
Target (with strain elements)	±½-3%
Material accelerator (dry solids)	±½%
Nutating disc meter	±1-2%
Oscillating piston meter	±¼-½%
Lobed rotor meter	±¼%
Encased turbine meter	±3%
Reciprocating piston meter	±1%
Rotating gear meter	±½%
Reciprocating piston pump	±1%
Diaphragm pump	±1%
Peristalic pump (roller)	±1%

Level

Device	Total Error
Gauge glass	±.02 inch
Bubbler	±1%
Differential pressure	±½-2%
Displacer	±.1-.3%
Float and tap	±1-2%
Float and pointer	±1%
Diaphragm transmitter	±¼-2%
Ultrasonic	±3%
Cable and drum	±.2 inch
RF probe	±1½%
Conductivity probe	±½%
Nuclear	±1-2%
Manometer	±1%
Capacitance (continuous)	±1%
Capacitance (switch)	±2-3%
Diaphragm switch	within diaphragm diameter
Vibrating probe or paddle	±.25 inch
Rotating paddle	within paddle diameter

Pressure

Device	Total Error
Bourdon	±.1-2%
Bellows transmitter	±½%

Device	Total Error
Bellows/indicator	±2%
Diaphragm	±½-1¼%
Strain gauge	±.1-2%
Variable impedance	±¼-½%
Capacitance	±.1-.2%
Piezoelectric	±1%
Dead weight	±.01-.1%
Manometer	±.02-1%
McLeod Gauge (vacuum)	±1%
Thermal transfer (vacuum)	±10-20%

Temperature

Device	Total Error
Glass thermometers	±¼-½%
Thermocouples	±¼-5%
Resistance detectors	±.2-.5%
Thermistors	±½%
Filled systems	±½-1%
Bimetallic	±½-2%
Manual optical pyrometer	±½-1%
Radiation pyrometer	±½-1%
Infrared pyrometer	±½-2%
Color pencils, paint, etc.	±1%
Acoustic thermometer	±2%
Quartz crystal	±.05%

Weight

Device	Total Error
Mechanical Lever	±.1-.5%
Load Cells	±.1-.5%
Spring Balance	±.1-.5%

Analyzers

Device	Total Error
Gas chromatograph	±½-1%
Infrared	±1%
pH	±.005-.05 pH units
Moisture	±2-5%
Conductivity	±½-3%
Specific gravity	
Mechanical weighing	±1%
Vibrating tube or paddle	±1%
Sonic	±.01%
Density	
Vibrating tube	±1%
Radiation	±1%
Turbidity	±2-5%
Viscosity	
Ultrasonic	±2%
Vibrating reed	±1%
1-float	±4%
2-Float	±2-4%
Combustion	±2%
Refractometer	±.006-.01%
Oxygen	±3%
Dissolved oxygen	±1-5%
Sulfur dioxide	±5%
Total hydrocarbon	±2-10%
Total organic carbon	±2-10%

Table 11.1 continued

Table 11.1 continued	
Device	Total Error
Carbon Dioxide	±½%
Chlorine	±5%
Sodium	±5%
NO_2, $NO+NO_2$	±10%
Miscellaneous	
Converters	
Current-current	±½-1%
Current-pneumatic	±¼-½%
Pneumatic-current	±¼-½%
Resistance-current	±½%
Voltage-current	±½%
Voltage- pneumatic	±½%
ORP transmitters	±¼-½%
Speed transmitters	±½%
Velocimeter	±.01-½%
Vibration	±5%
Torque transmitters	±½-1%
Accelerometer (strain gauge type)	±¼-1%
Electronic indicator (D'Arsonval type)	±1-3%
Electronic indicator (digital readout)	±.01-.5%

Fluid or service conditions are not as critical to accurate pressure and temperature measurements as they are to flow measurement. Level applications, however, are more subject to similar errors. Hydrostatic head level devices require that specific gravities be known and that they remain stable. Temperature and pressure variations cause errors proportional to their effect on specific gravity. The same factors affect displacement devices.

The application and selection of analytical devices in particular require complete knowledge of process streams. In chromatographs, unknown components may elute at the same time as one of the measured components, resulting in an erroneous readout. The selection of column filtering materials, column lengths and other application techniques require a thorough knowledge of stream conditions.

The attention to process parameters and design conditions is difficult to overemphasize.

Poor Design

Inattention to details related to the process or to the instrument system results in measurement errors. The engineer and designer must be fully aware of the effect of process conditions and ambient conditions surrounding the components of the control systems.

Materials must be chosen which are compatible with the process and with the environment. Galvanic action of dissimilar metals is costly because of corrosion losses and must be considered. High contact resistance and poor connections are a troubleshooter's nightmare. Hermetically sealed components and fungus-proofed circuit boards can pay for the cost differential in reduction of callouts and maintenance. Air regulators with built-in filters are to be desired over those without, unless separate filters are used.

The selection of round pipe is necessary for accurate flow measurement, particularly for differential measurements. Nonlaminar flows and velocity profile irregularities (caused by using a meter run made of pipe whose cross section is oval rather than circular) can account for errors of several percent. Some grades of pipe can be as much as 2% out of round and still be within specification. The prefabricated meter runs available from some manufacturers are made from pipe which has been checked for ±¼% roundness tolerance. Honed sections can be bought whose roughness has been smoothed to 5 to 10 microinches with ±0.002-inch roundness tolerance. These close tolerances eliminate many errors.

Turbulence caused by upstream bends or projections should be minimized by allowing a sufficient straight length of pipe prior to the orifice or by using straightening vanes.

Transmitters for gas and vapor flow measurements should be above the orifice taps with sensing lines sloped down to the pipe to prevent the collection of moisture in them. Transmitters for liquids should be mounted below the taps, with taps horizontal to prevent sludge from entering the sensing lines. Some users will allow taps to be mounted at a 45° angle up for gases and vapors and 45° down for liquids so that space may be conserved between adjacent pipes in the pipe rack. Sensing lines for vertical meter runs for wet gases, vapors and liquids should be brought to a common elevation before running to the transmitters.

Pulsating flows are difficult to measure accurately since readout and control equipment are made for essentially steady state or slow changing conditions. Location of the metering station upstream of reciprocating pumps or other pulsation sources is advisable. Dampeners are also available to reduce pulsation problems.

There is a tendency in some high vapor-content streams for vapor lock to occur in ½-inch sensing lines. Larger tubing may alleviate this problem, or a shorter sensing line may suffice.

Pulsations in line pressure adversely affect pressure measurement and control. Pulsation dampeners and snubbers should be used for such services. Liquid-filled movements provide viscous dampening in some types of gauges.

The Venturi effect, the phenomenon of a reduction of pressure at a constriction, can be responsible for low readings from pressure instruments. High fluid velocities, especially in gas flows, passing the taps on the pipe can drop the pressure by several inches of water. Since pressure should not be a function of line fluid velocity, taps should be located on a vessel or on large diameter, low velocity sections of piping.

Temperature measurement errors are easily introduced through improper installation design or specification. Thermowells, when used, should be completely surrounded by the fluid (liquid, gas or vapor) which is being measured, rather than being partially submerged. Bulbs and elements should press firmly against the bottom of their thermowells to prevent lags since an air gap is a poor thermal conductor. Sometimes thermocouples which touch a vessel surface to measure skin temperature must have that point of contact purged to prevent moisture buildup and subsequent corrosion. Thermal elements should be placed deeply enough into the vessel or pipe stream to minimize thermal gradients. Thermocouple systems should be grounded at only one point to avoid potentials developed by ground currents. In filled systems, elevation differences between bulb and instrument should be minimized by design to reduce hydrostatic head errors.

In analyzer systems response times can be maximized by designing sample systems for fluid velocities to provide current samples. Sample points should be chosen to give a reaction-complete representative sample. Sample valving systems should be designed to purge themselves after each operation so that new samples will not be poisoned by previous samples.

In control loops excess signal tubing lengths should be avoided. Excess tubing capacity could cause undesirable control lags. Conversely, electronic signal wiring should be large enough to keep loop resistance low. Stranded wire is advisable for ease of installation. Signal wiring should be separated from power wiring or should be shielded with the shield grounded at only one point. Noise is often a real problem where signal wiring is too close to higher energy carrying wires.

Purges or flushing connections should be designed into systems where plugging is a problem.

All these precautions should be taken to prevent errors that would or might otherwise exist. The extent of error due to these possible causes is indeterminate and variable, depending on the situation.

Process Changes, Irregularities, Upsets

Changes in process parameters are often sources of measurement errors. Many measurements are based on assumptions that other parameters such as specific gravity, temperature and/or pressures remain constant. Product specific gravity changes, for example, affect level measurements in direct proportion to the assumed and actual values. They affect control valve sizes in proportion to the square root of the specific gravity difference. Viscosity changes in control valve calculations may account for errors of 2 to 5%, depending upon the Reynolds numbers involved.

Inhomogeneities in the process fluids, such as two-phase flows, mixture ratio changes and bubbles in liquid flows, should be minimized by design to avoid possible measurement errors, particularly in flow and level applications.

Flow measurement errors result when temperature and pressure variations occur in volume measurement methods. When high accuracies are needed, therefore, corrections should be made for density variations that may occur. The extent of error varies in proportion to the square root of the specific gravity difference.

In processes that have rapidly changing flow rates, the wave forms of these pulsations and transients may be significant if high accuracy is important. Dampening properties of the mechanisms within the transmitter and/or the receiver will tend to make the readout "see" the RMS (root-mean-square) value of the input signal.

Flow measurements of saturated liquids and vapors (such as steam) can be inaccurate by several percent because of gas bubbles or condensate. A vent or weep hole may be employed to alleviate this problem. The flow through this small hole will generally amount to less than a percent of the total, if its bore is less than 0.1β.

Level measurements are subject to errors introduced by waves and agitations, specific gravity changes and liquid interfaces. Opacity or transparency changes can foul gauge glasses to the point where reading errors are likely to occur.

Poor Maintenance

The operation and overall accuracy of instrument systems can be maximized by systematic calibration and maintenance. Many guidelines are available for the establishment of proper maintenance and calibration intervals. Inspections usually reveal trouble spots so that correction of potential errors can be made.

Buildups at orifice plates, thermowells, pressure ports, diaphragms, probes, gauges, floats, etc., affect readout and control.

Manometer fill fluids should be checked for loss and for contamination. Sloped manometers must be installed level and maintained in that position.

Static legs for level measurement should be checked periodically for evaporation of fill fluids. Condensates collected in an equalizer leg should be blown down. A dry nitrogen purge can be employed to prevent such a liquid buildup.

Vents plugged by insects may be a problem in many plants. Bug screens and porous plugs are available to alleviate this problem.

Pinched tubing and ferule leaks produce measurement errors.

Kinked wires may cause as many problems as pinched tubing. Open and short circuits and extraneous grounds play havoc with all low-level signals. Kinked thermocouple wire produces a cold working of the wire, causing an inhomogeneity which can become a thermocouple itself, introducing a measurement error. Loose or corroded terminals, weak batteries and weak standard cells are error sources and must be checked.

Thermocouples, RTDs and filled system bulbs should be checked occasionally to be sure that they are firmly seated on the bottom of their thermowells. Where rusting and corrosion at the point of contact is a problem, a periodic purge and a blanket of inert gas may be needed.

Plastic deformation of diaphragms, capsules and bourdons caused by overranging the instrument generally brings about a change in zero, span and linearity. While readjusting the span and zero will sometimes bring the instrument to a near-spec condition, the best action is replacement of the abused part; likewise, for the ruptured diaphragm.

Pneumatic flapper-type instruments should have the flapper-nozzles checked periodically. Nozzle internal diameter will sometimes vary due to wear or buildups. Replacement or rodding out with special wires may be required.

Errors of level, pressure and flow measurements are sometimes caused by excessive purge rates. Liquid purges themselves can be responsible for pressure differentials between the transmitter and the sensing line at the process connection. Purge connections can be located near the process tap, and purge rates should be optimized to minimize pressure drops.

Proper maintenance of instrument systems eliminates many error sources and avoids delays caused by complete failures within the systems.

Miscellaneous

Other sources of error include those caused by people who operate instrument equipment. Recording errors due to carelessness frequently occur.

Errors made in reading charts and scales are caused by several factors. Poor visibility due to bad lighting or dirty instrument windows can lead to mistaken values being recorded. Depending upon scale range, graduations, pen width and distance between pointer and scale, errors of up to 1/2% may be made. Parallax, a visual error caused by the spacing between the viewer's eyes and the scale, can make a difference of a scale division or more. Some instruments employ indicators with a mirrored pointer to eliminate such errors. Closing one eye also improves reading accuracy.

Certain design limitations cause errors. Among these are self-heating effects such as those common to resistance temperature detectors and strain gauge elements. Friction and slippage affect mechanical devices to varying degrees. Component wear and aging and fulcrum point play or "freezing" indicate the need for replacement and repair. Pen and pointer width can cause resolution errors.

Many other potential error sources might be listed, but these will serve to make the reader aware that there are many causes which lead to inaccuracies in instrument systems besides the inherent inaccuracies of the instruments themselves. Some of these are potentially high error sources while others tend to be insignificant. An awareness of their existence by designers, operators and technicians will do much to eliminate excess errors of this type and allow measurements within the claimed tolerances of the instruments and the systems they comprise.

12 Construction and Startup

William G. Andrew

An organized plan for the installation, calibration, testing and startup of instrument control systems is necessary for successfully building and starting up a process. This chapter presents a logical method for accomplishing that purpose. The size of the plant affects to some extent the plan that should be followed, but most steps suggested are applicable whether the job is small or large.

Organizing

The person responsible for installing and starting up the instrument systems for a particular job will know the job size and the scheduled completion date before, or at least shortly after, he starts to work on the job. Schedules already will have been made showing dates by which major categories of work are to be started and completed.

Many projects today follow a critical path method (CPM) or other similar scheduling technique for scheduling design activities, equipment and material purchasing and delivery, and construction activities. This type of scheduling technique is particularly useful for large projects and is used for smaller ones also, especially if the time element for completion is critical. In most instances when a decision is made to build and money is allocated for a project, it needs to be done as quickly as possible in order to provide returns on the investment.

The organization for proper job execution includes the receipt and distribution of all drawings and reference materials; the receipt, storage and distribution of all equipment and construction materials; the scheduling of the various installation activities, including coordination of work between crafts; and coordinating calibration and testing functions in preparation for startup at dates scheduled by project management.

Documents Required

Information necessary for the efficient installation of instruments systems includes:

1. Various types of engineering drawings
2. Instrument index list
3. Instrument specifications
4. Project schedule
5. Purchase order file
6. Vendors' drawings
7. Construction specifications
8. Access to correspondence files

These items contribute to a smooth, efficiently run project. Their use is discussed below.

The engineering drawings which are normally prepared for construction and startup include:

1. Mechanical flow sheets
2. Panel drawings
3. Instrument location plans
4. Instrument loop drawings
5. Instrument installation details
6. Schematic control diagrams
7. Electrical wiring details

Reference drawings needed include:

1. Process flow sheets
2. Piping plans and details
3. Electrical drawings
4. Equipment layout plans and perhaps others

A brief description of each is given.

Mechanical Flow Sheets

Mechanical flow sheets are needed to provide information and knowledge of the entire process, an understanding of the relationship between various pieces of equipment and a comprehension of the control schemes used for the process. They provide general information that allows a quick understanding of how all pieces fit together to form the whole. Building a plant utilizing the many engineering drawings required might well be compared to putting together a jigsaw puzzle. The puzzle goes together more easily when the finished picture is visualized. Similarly, mechanical flow sheets provide an overall view or "picture" of the process that helps during the construction phase.

Instrument Index

The instrument index may be known by this or other terms such as instrument list, instrument summary sheets, etc. It provides a complete list of all numbered instrument items (Figure 12.1). It may contain additional information such as reference documents (purchase order numbers, specification sheets, installation details, etc.), service information, line or equipment numbers, instrument makes and models, etc.

The instrument index usually lists all major items that are associated with a control loop and shows reference drawings where each of these may be located—flow sheets, piping drawings, plot plans, etc. These drawings may be coded or marked in many different ways to schedule and chart project progress.

Panel Drawings

Panel drawings include the layout of instrument equipment, piping details (if pneumatics are used) and electrical wiring details. Piping and wiring details are necessary for terminating tubing and cable runs between the panel and the field. Layout drawings are used to help locate panel instruments when installing tubing and cable and are necessary during calibration and check-out, particulary for interconnections between control systems.

Instrument Location Plans

Instrument location plans normally show approximate locations of all field mounted devices. Since two crafts (electrical and piping) are normally involved in installing control systems, it is usually necessary to have both instrument electrical and instrument piping plans available. Usually the electrical and piping plans show only the devices which require work by that craft. Many devices are shown on both sets of plans since both crafts are involved in their installation and hookup.

Instrument Loop Drawings

Instrument loop drawings are nearly always made for electronic instruments. The drawings show all interconnections between elements of each loop (see Figure 12.2). Less frequently perhaps, they are made for pneumatic systems (see Figure 12.3). Interconnections for pneumatic systems are usually simpler and fewer details are needed. In both cases they are useful during hookup and testing and serve as the primary source for showing all functions performed by the loop. They are also used by the maintenance group for troubleshooting and preventive maintenance checks.

Instrument Installation Details

Instrument installation details (Figure 12.4) along with electrical wiring details provide the information necessary for the piping and electrical craftsmen to install the instrument hardware and accessories with little supervision. Engineering design criteria and operating requirements will have already been considered and satisfied, and these drawings are made to be followed with little or no time lost in analyzing installation requirements.

Schematic Control Diagrams

Schematic control diagrams (electrical and pneumatic) are often necessary for the understanding of complicated control systems. Pneumatic systems seldom offer much difficulty, but electrical systems may interlock with other devices and systems and often require clarifying schematics for better and quicker understanding.

Reference Drawings

Reference drawings of other disciplines (piping, electrical, architectural and structural) often speed up installation work and prevent mistakes. Even though cross-checks are made among piping, mechanical and electrical groups during the engineering design period, occasionally interferences occur. Reference to drawings for other craft work can avoid some costly mistakes.

Process flow sheets also are helpful in understanding control systems. They help to verify instrument measurement ranges that may be suspected of error. They show stream components and give a variety of other useful information.

Vendor Drawings

It is not economical or necessary to reproduce all of the information contained on drawings of vendor equipment; yet much of the information is useful and/or necessary for installation, check-out and startup. Electrical connection details, tubing connections and other similar bits of information may be missing from prepared engineering drawings. When construction begins, all vendor drawings (for instrument and other equipment) should be available and filed for easy reference.

Specifications

Specification sheets for instrument items (see Figure 12.5) should be available for several reasons:

1. To determine if the items meet specifications
2. To determine proper loop functions (in some cases)
3. To check service conditions
4. For calibration information
5. To provide detailed information that is not furnished on any other document

INSTRUMENT INDEX

SHT._____ OF_____

S.I.P. INC.
Engineers & Contractors
HOUSTON, TEXAS

CLIENT A.B.C. Co.
PROJECT File "B" Project
LOCATION United States
CLIENT JOB NO. C-100 S.I.P. JOB NO. E-783

REV.	BY	DATE	APP'D	REV.	BY	DATE	APP'D

TAG NO.	SERVICE DESCRIPTION	LINE OR EQUIP. NO.	MFG.	MOD. NO.	SPEC. SHT.	P. O. NO.	FLOW SHT.	PIPING DWG.	INSTRUMENT DWG. PLAN	INSTRUMENT DWG. INST'L.	INSTRUMENT DWG. LOOP	LOC.	SIZE	CALB. OR ACTION	REMARKS
PT-101	WATER TO HOLD. TANK	2"-P-1241	FOX.	11AH	I-2	I-120	P-1231	P-1472	I-1201	I-1203	I-1204	F	1/2"	200 PSIG	
PIC-101	"	----	FOX.	52A	I-17	I-131	P-1231	---	---	I-1202	I-1204	P	-	0-200 PSIG	
PCV-101	"	3"-CWS 4119	FISHER	657 ES	I-32	I-123	P-1231	P-1472	I-1201	I-1205	I-1204	F	2"	A/O	

Figure 12.1. The instrument index contains information such as item numbers, manufacturers, model numbers, reference documents (P.O. numbers, specification sheets, installation details, etc.), service information and line or equipment numbers. (Courtesy of S.I.P., Inc.)

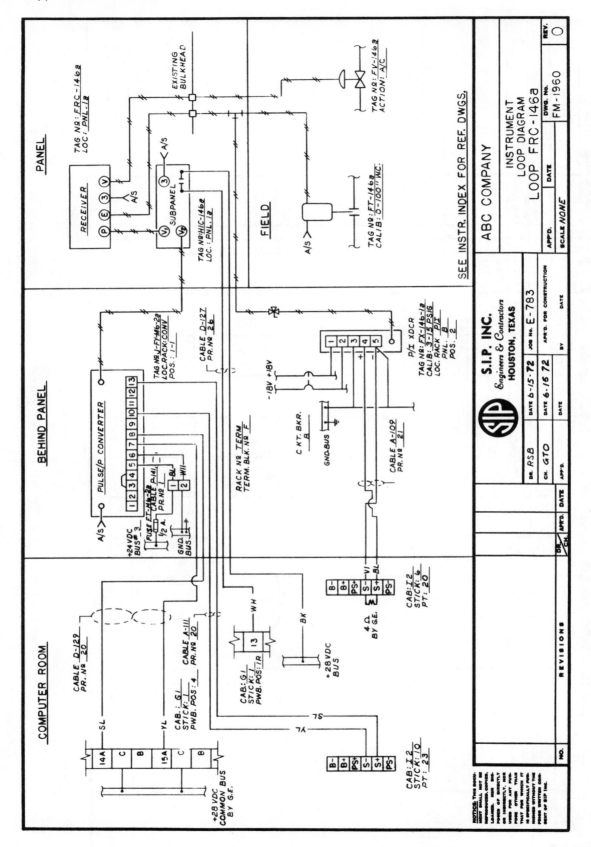

Figure 12.2. Electronic loop diagrams show interconnections among all the elements of the loop, identifying locations and wiring connections. (Courtesy of S.I.P., Inc.)

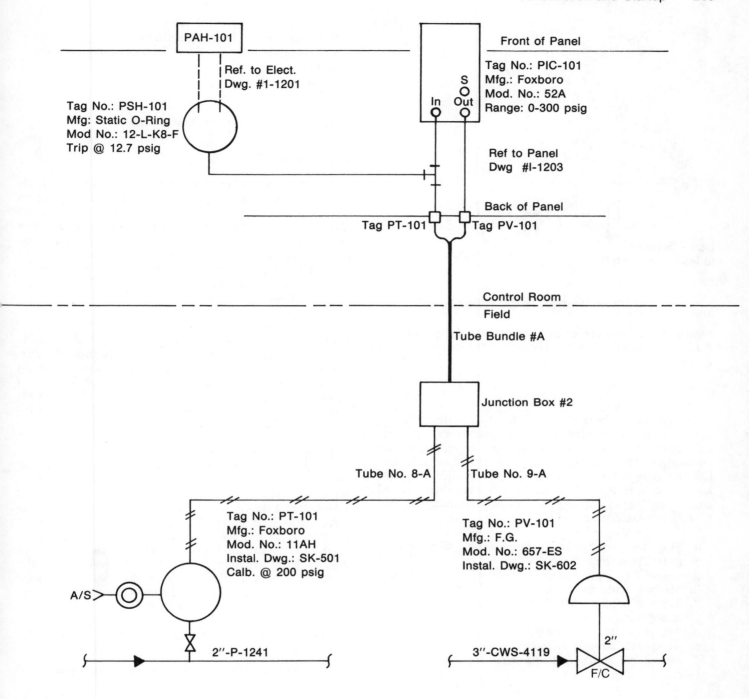

PAH-101

Ref. to Elect.
Dwg. #1-1201

Tag No.: PSH-101
Mfg: Static O-Ring
Mod No.: 12-L-K8-F
Trip @ 12.7 psig

Front of Panel

Tag No.: PIC-101
Mfg.: Foxboro
Mod. No.: 52A
Range: 0-300 psig

S

In   Out

Ref to Panel
Dwg #I-1203

Back of Panel

Tag PT-101   Tag PV-101

Control Room

Field

Tube Bundle #A

Junction Box #2

Tube No. 8-A   Tube No. 9-A

Tag No.: PT-101
Mfg.: Foxboro
Mod. No.: 11AH
Instal. Dwg.: SK-501
Calb. @ 200 psig

Tag No.: PV-101
Mfg.: F.G.
Mod. No.: 657-ES
Instal. Dwg.: SK-602

A/S

2"-P-1241

3"-CWS-4119

2"

F/C

Figure 12.3. Pneumatic loop diagrams are usually simpler than electronic loops and are useful for hookup and testing of the control systems.

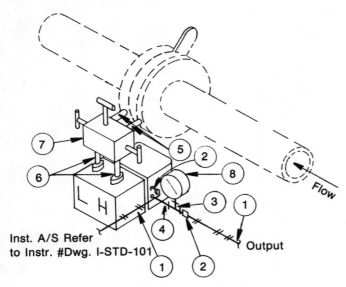

Inst. A/S Refer
to Instr. #Dwg. I-STD-101

Output

Close-Coupled Flow D/P Cell

Figure 12.4. Installation details provide the piping craft sufficient information to install transmitters and control devices properly.

In addition to instrument specification forms, there is usually a set of construction specifications that provide guidelines for installing instrument equipment and accessories. These general specifications include information that may be repeated on some of the detailed drawings, but they also include additional information often omitted from construction drawings and details. Examples of this type of information follow:

1. Size, type and pressure rating of takeoff valves on flow orifice lead lines
2. Size of air supply piping to individual and group users
3. Steam tracing requirements
4. Types of supports for electrical conduit or pneumatic tubing runs
5. Acceptable tubing fittings
6. Acceptable conduit systems
7. Acceptable instrument test and check-out methods, etc.

Planning the Schedule

Although the project completion date has been determined, specific areas of work must be planned to meet that date. Overall schedules probably do not provide details on the various groups of instruments which must be installed to coincide with other work. Major events that the schedule may contain include:

1. Arrival and installation of the main control panels
2. Installation of meter runs

3. Installation of control valves
4. Installation of field transmitters
5. Pneumatic tubing and/or electrical cable runs
6. Testing
7. Calibration
8. Loop check

Dates should be set for these events.

Cost Control

One of the most important functions of the instrument installation supervisor is to control installation and startup costs. The cost of the equipment and, to a great extent, the cost of material for installation are beyond the supervisor's control, but he can influence labor costs by proper planning, scheduling and supervision. To accomplish this objective, he must be familiar with the labor estimate for the project, be able to measure progress against the estimate and execute the job efficiently.

Labor Requirement

From a labor estimate that usually exists, the installation supervisor should become familiar with the total labor requirements for the job. With this knowledge of what must be done and the schedule of other work which must necessarily precede the instrument work, crew assignments can be made efficiently. The estimate and the projected work costs should then be compared.

Progress Status Checks

To chart progress, the responsible instrument engineer may use the instrument index sheets or work up a special report form similar to Figure 12.6 that may suit his needs a little better. Like the index sheet, this report lists all instrument items, but in addition provides columns to show progressive steps in receiving, calibrating, installing, piping, wiring, testing and final check-out of the items. From such a list, total job progress can easily be determined. It also serves as a check-off list and should preclude the oversight of a group or even a single instrument item.

Miscellaneous

Some other items might be considered.

1. Be on the mailing list for all engineering drawings not yet issued or that may be revised.
2. Be familiar with vendor service personnel that may be available for the job—calibration help by primary equipment suppliers or special help from chromatograph or special equipment suppliers.
3. Know whether preliminary or special check-out procedures are required for any equipment items or systems.
4. Contact the owner representative or the contractor representative (whichever is applicable) to work out details of job responsibilities and communication to ensure cooperation and understanding.

S. I. P., INC. Engineers – Contractors Box 34451 Phone 946-9040 HOUSTON, TEXAS 77034	S.I.P. NO._____ PROJECT_____ CLIENT_____ LOCATION_____ CLIENT NO._____	**RECEIVER INSTRUMENTS** SPEC. NO.	

BY	DATE	ITEM	SHT. of

	1	Tag No.	Service

GENERAL

2	Function	Record ☐ Indicate ☐ Control ☐ Blind ☐ Integ ☐
		Deviation ☐ Other _____
3	Case	MFR STD ☐ Nom Size _____ Color: MFR STD ☐ Other _____
4	Mounting	Flush ☐ Surface ☐ Rack ☐ Multi-Case ☐ Other _____
		For Multiple Case, See Spec. Sheet _____
5	Enclosure Class	General Purpose ☐ Weather Proof ☐ Explosion-Proof ☐ Class _____
		For Use in Intrinsically Safe System. ☐ Other_____
6	Power Supply	117 V 60Hz ☐ Other ac ☐_____ dc ☐ _____ Volts
7	Chart	_____ Strip ☐ _____ Roll ☐ _____ Fold ☐ Circular _____ Time Marks ____
		Range _____ Number _____
8	Chart Drive	Speed _____ Power _____
9	Scales	Type _____ _____ _____ _____
		Range 1 _____ 2 _____ 3 _____ 4 _____

CONTROLLER

10	Control Modes	P = Prop (Gain), I = Integral (Auto Reset), D = Derivative (Rate), Sub: s = Slow, f = Fast
		P ☐ PI ☐ PD ☐ PID ☐ If ☐ Df ☐ Is ☐ Ds ☐
		Other _____
11	Action	On Meas. Increase Output: Increases ☐ Decreases ☐
12	Auto-Man Switch	None ☐ MFR STD ☐ Other _____
13	Set Point Adj.	Manual ☐ External ☐ Remote ☐ Other _____
14	Manual Reg	None ☐ MFR STD ☐ Other _____
15	Output	4-20 mA ☐ 10-50 mA ☐ 21-103 kPa (3-15 psig) ☐ Other _____

INPUTS

16	Input Signals	4-20 mA ☐ 10-50 mA ☐ 21-103 kPa (3-15 psig) ☐ Other _____
17	No. of Inputs	1 ☐ 2 ☐ 3 ☐ 4 ☐.
18	Power for XMTRS	External ☐ This Inst ☐ No. of Independent Supplies _____
		For Transmitters. See Spec Sheet. _____

ALARMS

19	Alarm Switches	Quantity_____ Form_____ Rating_____
20	Function	Meas. Var. ☐ Deviation ☐ Contacts To _____ On Meas _____
		Other _____

21	Options	Filter-Reg ☐ Supply Gage ☐ Charts ☐ Int. Illumination ☐
		Other _____
		_____

22	MFR & Model No.	_____

Notes:		Flow Sheet

△					
NO.	REVISIONS	BY	DATE	APPVD.	DATE

REPRINTED AND MODIFIED WITH PERMISSION OF THE COPYRIGHT HOLDER: Ⓒ INSTRUMENT SOCIETY OF AMERICA

Figure 12.5. Instrument specification sheets list purposes and features of individual instrument devices that allow them to meet specific needs.

INSTRUMENT PROGRESS REPORT

PAGE __ OF __

S.I.P. INC.
Engineers & Contractors
HOUSTON, TEXAS

CLIENT __________
PROJECT __________
LOCATION __________
CLIENT JOB NO. __________ S.I.P. JOB NO. __________

REV.	BY	DATE	APP'D	REV.	BY	DATE	APP'D

TAG NO.	SERVICE DESCRIPTION	MFG.	MOD. NO.	P. O. NO.	RECEIVED DATE	CALB. DATE	INSTAL. DATE	PIPING DATE	WIRING DATE	CHECKOUT DATE	REMARKS

Figure 12.6. Instrument progress report sheets list steps for each item such as receiving, calibration, installation, piping, wiring, testing and check-out to chart job progress. (Courtesy of S.I.P., Inc.)

5. Establish a good filing system for drawings, purchase orders, specifications, correspondence and other documents. Establish a system for replacing and voiding out-of-date drawings.

Ordering and Receiving Equipment and Material

Most instrument items, particularly those which have been assigned instrument numbers, will have been purchased prior to the start of field installation. Seldom are field purchases of measurement or control devices necessary.

The field engineer must determine that all items purchased are received and properly stored and that any material not ordered is purchased and received by the time it is needed. In many cases, much of the installation material such as pipe, pipe fittings, conduit, conduit fittings, hangers, tubing, wire, supports and miscellaneous steel shapes will also have been purchased. Field purchases include the following:

1. Miscellaneous pipe fittings and tubing
2. Pneumatic receiver gages for some control valves
3. Air set regulators not previously ordered
4. Replacement instruments for any damaged during shipment, handling, storage, or installation
5. Replacement instruments for any "lost" during storage (Some instruments have a way of "disappearing" into adjacent construction projects.)
6. Replacement devices for instruments whose size is incorrect
7. Tools and test equipment, both special and replacement
8. Replacement recorder charts for any which were used during checkout (Recorder charts are often installed just before startup to minimize this expense.)

Purchase Orders

A purchase order file of all instrument items and all instrument material is essential. Deliveries should be checked against the written purchase orders. A thorough check as items are received often reveals shortages of materials, equipment that does not meet specifications and other discrepancies that cause difficulties and delays unless they are caught early and corrected immediately.

There should be access to process equipment purchase orders also since they frequently include instrument items and materials.

Material Status

Material status sheets are normally kept to provide updated information on the receipt of equipment and material. The responsibility for this function may rest with a purchasing or expediting group, but copies are made available to field personnel so that they may change schedules, if necessary, to avoid labor delays in field work.

Storage of Equipment and Material

Reliable receiving procedures should be established for the receipt of all equipment and materials. As they are received, checks should be made against the written purchase orders to ensure receipt of the correct material. Equipment and materials should be stored properly so that they may be found easily and dispatched quickly and economically.

A warehouse is usually provided for equipment and materials that must be kept dry. Some types of electronic equipment require temperature controlled atmospheres to prevent the formation of condensate and/or corrosion which would adversely affect the equipment. Conduit, pipe, structural supports and other items that can be stored outside are usually placed on racks and pallets to keep them off the ground and prevent undue exposure to dust and mud.

All categories of instruments (transmitters, control valves, relief valves, switches, thermowells, dial thermometers, pressure gauges, etc.) should be stored in groups and arranged by their identification numbers if practical. Material items should also be arranged in a logical manner so their storage and check-out can be made easily and quickly.

Installing Instrument Systems

Normally there are only two crafts involved in the installation of instrument systems—pipe fitters and electricians. It is a fairly common practice for both crafts to have people who specialize in instrument work. It is more common in the piping craft than in the electrical.

One area of work most applicable to specialization is the installation of metal tubing. Installing multiple runs of copper, aluminum, stainless or other metal tubing neatly and compactly in crowded areas is an art that is developed primarily through experience (see Figure 12.7). There are relatively few pipe fitters who are adept at it. The need for the art is diminishing because of the continued growing use of electronic systems and because of the increasing use of plastic tubing for pneumatic systems.

The installation of instruments follows a pattern although the order may vary somewhat. The schedule of work is dependent on pipe installation primarily, although equipment installation, electrical work and other work categories also affect the time and order of installation.

The following paragraphs offer some guidelines to be followed and some suggestions and hints for project execution.

Typical Installation Procedures

The nature of construction work is such that there is a necessary order for doing much of the work and a preferred order for much of the remainder. One of the first things is to

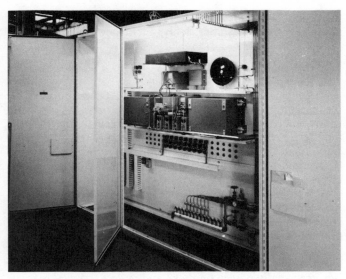

Figure 12.7. The neat, orderly arrangement of metal tubing needed for panel piping requires experienced craftsmen who know the art. (Courtesy of Custom Control Panels, Inc.]

determine whether any underground work is necessary. Very little is normally required for instrumentation. The exception is the use of electrical conduit runs for instrument control cable and thermocouple lead wires. In rare cases, pneumatic tubing is also run underground.

Underground runs are normally used between field junction boxes and a central control building. Even when the entire run is not underground, there may be crossings under roadways within a plant.

Other places where underground runs may occur are

1. From remote storage areas
2. From isolated hazardous areas
3. From loading or unloading areas
4. In large buildings with high ceilings
5. Around large equipment where service equipment must be moved in and out
6. Where monorails are used for access to equipment
7. Between adjacent buildings where roadways are provided for truck or equipment access

When underground runs are used, they must be installed before building floors and equipment foundations are poured and before roadways are completed and paved. Drawings of instrument runs should be checked with regards to interference with existing foundations, sewers, piping, electrical conduits and grounding systems. They should be checked carefully to determine that locations are correct and that the quantity and sizes of conduit are correct.

In aboveground work, very little instrument work can be done until piping installations are well underway. Quite often the first pipe installed is in piperacks where meter runs often occur. Piperacks alongside operating areas may con-

tain control valve manifolds and other instrument items. As piping and equipment are installed, instrument locations can be verified and pipe stands or other type mounting devices can be installed and made ready.

Most in-line devices should be ready when the piping in which they are installed goes into place. This is particularly true where scaffolding must be used for access. All the work requiring its use should be completed while the scaffolding is erected.

As the piping and equipment installation nears completion in an area, field transmitters and other field devices may be mounted and piped. An important consideration is to work efficiently and to avoid interference with other crafts. Instrument items should not be installed so early that they interfere with the installation of heavy pipe or equipment.

When most of the instruments have been installed, instrument air piping may be run to those devices requiring air supply.

Junction boxes for electrical cable and/or pneumatic tubing can be mounted on piperacks or stanchions early in the construction period and still not interfere with other work. Field runs from these boxes should be made only when the majority of instruments are installed and ready for connection or hookup.

The central control panels should be installed several months before project completion and as soon as work in the control center is essentially finished. Painting, welding or any work that might damage or mar the panels and their equipment should be complete before the panels are installed.

Cable and/or tubing runs may be made when the termination points are ready in the field and in the control center.

To accomplish the fabrication and installation work efficiently, many of the work assignments can be made so that efficiency is gained by repetition. Examples that illustrate this approach follow.

1. Assignment of a crew or crews to mounting instrument stands. Have the same crew or crews do all this work for a particular area or plant.
2. Assignment of a crew (or crews) to fabricate meter manifolds and other repetitious work where several or like items may be needed.
3. As in items 1 and 2, organize all facets of work so that crews have work outlined to them and can work long periods of time with little supervision. Other work groupings include: (a) conduit runs, (b) tubing runs, (c) conduit or tubing junction boxes, (d) pulling wire, (e) termination of wire or tubing, (f) instrument air headers and subheaders, (g) instrument air supplies and (h) tubing tray or cable tray.
4. Scheduling the work in piperacks, around towers and in high buildings (and other places that require scaffolding) at the same time, prior to or just after other crafts so as to avoid the disassembly and reassembly of scaffolds.

Coordinating Work Among Crafts

Overall efficiency is achieved on a job when there is full coordination among all the crafts involved. The following list indicates many items that involve more than one craft, and each one should be aware of the needs of the other.

1. Control panels (unless the job is completely electronic, in which case only electricians are involved)
2. Thermowells and thermocouples
3. All process connected electronic transmitters
4. All pneumatic devices that require electrical power supplies
5. All I/P valve transducers
6. All chromatographs and most other analytical instruments
7. Process connected level, pressure, flow and temperature switches that switch electrical power—alarms or interlocks
8. All solenoid valves

Other items could be listed, but these are typical of many that require work by two crafts; and neither can finish without help from the other.

Work done at the appropriate time can do much toward building a plant economically and avoiding costly delays that hinder startup schedules. Typical actions that promote efficiency and decrease costs follow.

1. Have control valves and purchased meter runs available as pipe lines in which they mount are installed.
2. Have relief valves and rupture discs available when flare headers are installed.
3. Make sure that control panels are installed with sufficient lead time for wiring and/or tubing termination, testing and check-out. (Panels should not be installed prior to completion of lighting, ceiling and finishing work in the control room.)
4. In buildings requiring heating and/or air conditioning ductwork, watch for interferences between ductwork and conduit, cable or tubing runs.
5. Check possible interferences between pipe and conduit along pipracks and building walls or ceilings.
6. Watch for interference between instrument locations and equipment removal requirements.
7. Coordinate work between the electrical and piping crafts so that early check-out of equipment can begin early in selected plant areas.

Check List of Good Installation Practices

A successful installation job hinges to a great extent on a common-sense approach to the order of work and adherence to detail in performing work. It also requires that consideration be given to operating and maintenance requirements. Operator actions need to be visualized, and access for maintenance must be kept in mind. The following check list includes items that need to be considered.

1. Make sure all instrument items are accessible. (This should be determined during the engineering phase, but slips do occur, so this should be checked thoroughly.)
2. Do not locate instruments in main traffic lanes.
3. Turn pressure gauges and dial thermometers toward traffic lanes for easy observation by operators.
4. Cover glass fronts on instrument items during construction where heavy traffic or work might cause breakage.
5. Make sure glass and identification plates are covered before painting is done in the vicinity of instruments.
6. Make sure instrument calibration is done prior to installation unless in-place calibration is required.
7. Be certain that door space is sufficient for entry of control panels before completion of the control building and control room.
8. Check instrument air headers and subheaders for moisture and cleanliness before turning on instrument air to the instruments.
9. Disconnect instruments prior to hydrostatic testing of equipment and piping unless the test pressure falls within the range of the device.
10. Do not install orifice plates until lines have been flushed.
11. Do not install positive displacement meters, rotameters or other in-line devices that may be damaged until lines have been flushed.
12. Do not run pneumatic tubing lines too close to hot surfaces.
13. Do not locate capillary lines where they can be stepped on or otherwise damaged. (Use protective covering if necessary.)
14. Keep a progress record using the form shown in Figure 12.6 so that job progress is carefully charted.
15. Keep a record of job labor cost and project the total labor cost by comparing progress and estimated total job labor cost.
16. Where service is required on special equipment, make sure it is installed in plenty of time for check-out by the vendor—alert him early enough so that he can properly schedule the work for which he is responsible.
17. Know the warranties that apply to instrument items or groups of instrument equipment.
18. Know the electrical classifications of all areas of the plant and make sure electrical installations comply with the proper codes.
19. Remove shipping stops from all float and displacement level devices.
20. Check orifice plate sizes for all flow measurement loops.
21. Make sure that sealing fluids are placed in lines to devices that require their use.
22. Determine that all instruments have been tagged with their identification numbers.

Calibration

Instrument calibration normally is accomplished by one or a combination of three methods or practices:

1. Reliance upon the manufacturer's calibration
2. Shop calibration on receipt of instruments or prior to installation
3. Calibration in place (after installation)

Items such as pressure gauges and dial thermometers are seldom recalibrated because factory calibration is normally sufficiently close for the accuracies needed.

Pressure and differential pressure instruments for pressure, flow and level are factory calibrated by most manufacturers at no additional cost, although some manufacturers make extra charges when calibration is requested. Because rough handling during shipment often alters calibration settings, a calibration check or complete recalibration is normal before they are installed in the process. When instruments are recalibrated, it is done more economically in a shop or at a bench set up specifically for that purpose than it is after they are installed. Some companies require in-place calibration, however, on the assumption that calibration may be shifted as a result of rough handling during installation. This precaution is valid for delicate instruments, but most of the devices used are sufficiently rugged to withstand the normal installation handling without calibration shifts.

Factory calibration of filled system temperature devices is often accepted. When calibration is checked, only one point in the temperature range is usually made to verify correct calibration.

Pressure switches used for alarm or interlock functions are normally calibrated at their selected setting(s), but factory calibrations of flow and temperature switches are usually accepted.

Factory calibration of several devices such as positive displacement meters, turbine meters, rotameters, etc., is normally accepted without further checking. Other devices such as magnetic flowmeters and several types of analytical devices have special calibration equipment and/or procedures that need to be followed for field calibration. In most instances these are simple and are made after installation is complete and the system is ready to start up.

When instruments are calibrated prior to installation (acceptable in the majority of cases), the sequence of handling may vary. One of the most economical methods is to route them directly to the calibration shop (a permanent facility or a portable trailer with calibrating equipment) where they are calibrated and then sent to a warehouse or storage place to be held and installed as needed. Reduced cost of handling is incurred if they are handled in this manner. If sufficient manpower is not immediately available or if they are received in small, partial shipments, it may be desirable to wait until larger quantities are ready for calibration.

Until recently it has been a common practice to calibrate instruments at a user's shop when plant additions were made or even at new facilities. At new plant sites (grass roots plants), the shop facility was completed early enough and with sufficient equipment to meet this need. This practice is no longer common. As additions have become larger and more frequent, the inconvenience and overloading of the user's facilities have required that other arrangements be made. Calibration equipment is often set up in mobile trailers so that the only outside requirement is an electrical power source. After preinstallation calibration, each instrument needs only a final check for correct operation.

When instruments are calibrated in-place, the practice is justified on the basis that they may accidentally be dropped or handled roughly while they are being installed. This method is more costly than shop calibration because of the additional labor required to haul equipment to each location and to calibrate under difficult circumstances—at the top of a tower, in a piperack or at other difficult-to-reach locations.

Regardless of the calibration method used, the responsible person should determine that all the necessary equipment is available before calibration begins. Particular attention should be given to chromatographs and other analytical devices to make sure that calibration techniques are known. Sometimes special equipment may have to be purchased or rented to meet special calibration requirements.

Testing

The "testing" of piping, tubing and wiring associated with instruments and instrument systems eliminates a high percentage of difficulties that otherwise could occur when new control systems start up. It includes testing process connections, instrument air supply systems, pneumatic signal lines and all electrical wiring involved in the instrument systems. It assures leak-free piping and electrical systems that have been checked for continuity and shorts.

Testing of the various parts of the system usually falls under the responsibility of three different groups—the process pipefitters, instrument pipefitters and electricians.

Process Connections

Process connections to instruments are usually tested for leaks as the associated pipe lines and equipment are tested. Tests are made with liquids or gases. Hydrostatic tests with water are most common, and air is usually the medium when pneumatic tests are made. The test pressure is usually one and a half times the design pressure of the system.

When leak tests are made, all instrument devices are disconnected to prevent overranging and damage. Although many devices have overrange protection, disconnecting is a safety precaution that is usually taken. When piping and equipment testing begins, the responsible instrument supervisor should correlate his work with the test groups so that no damage results to instruments. A record of completed tests should be kept on the progress status form shown in Figure 12.6.

Pneumatic Lines

Instrument air headers and air supply lines and pneumatic signal lines are cleaned and tested for leaks and continuity before they are placed in service.

Instrument air supply lines are disconnected from instruments, and a source of clean filtered air is used to blow down the air headers and supply lines to remove any deposits in the lines. This is necessary to assure that dirt or small pieces of metal do not pass into the small passages and working parts of the pneumatic devices. After the air system is blown down, it is blanked off and pressure tested.

Transmission and control lines are disconnected and blown through to remove deposits and to establish continuity. Each line is then pressure and leak tested. Tests usually conform to ISA RP 7.1, *Pneumatic Control Circuit Pressure Test,* published by the Instrument Society of America. It stipulates conditions and procedures for the tests. The test pressure for 3-15 psig signal lines is 15 psig. Permissible leak tolerances are based on the signal line length (or volume), and pressure decreases of 1 psi per hundred feet of ¼-inch tubing per 5 seconds is listed as unacceptable. By inference, leaks less than that are permissible. Instrument air lines that terminate in control valve topworks must be corrected to an equivalent length of tubing for the additional volume. A table is given in ISA RP 7.1 for that purpose.

Records of test progress should be kept on the progress status sheet mentioned previously.

Electrical

Electrical testing consists of simple continuity and insulation tests to determine that proper connections have been made and that cables and wiring have not been damaged. Operational tests are made on alarms and interlocking devices to determine that abnormal conditions will function properly when set limits are reached.

Continuity and identification tests are checked by means of a DC test device using a bell or buzzer to "ring out" the conductors.

An insulation test may be made with a 500-volt tester from conductor to ground and with other conductor(s) in the cable grounded.

Other items that may be checked to determime that the job is complete include:

1. Proper loop identification
2. Proper terminal markings
3. Shielded cable requirements
4. Proper grounding method
5. Acceptable loop impedance and loading (resistors may have to be adjusted)

Loop Check

When installation is essentially complete, a final check of all the controls of the unit is made. This is often referred to as the "loop" check because every device in the loop or system is checked out for operation. All elements in the loop should have been calibrated; alarm and shutdown switches should have been set; so the entire loop is checked to make sure that all elements function as intended.

To provide an idea of the kind of detail that should be followed as instruments are installed and checked out, typical check-out procedures are listed below for three devices. They show the type of details that should be checked to determine that specifications have been followed and that proper operation is assured. Some of these steps will have been made prior to the loop check, but the final check-out should verify the completeness of the loop.

Typical Flow Transmitter Check-Out Procedure

The following typical check-out procedure is for a pneumatic d/p cell.

Determine that both orifice holes have been drilled in the orifice flange by pushing a wire rod through each hole.

Check orifice plate size with micrometer. This should be done for all plates when they are received. Make sure each plate is stamped on the upstream side with tag number and size.

Determine that the orifice plate is installed with the proper side facing upstream.

Is there adequate room for maintenance of the d/p cell? Can the cover be removed without disturbing the piping?

Blow down the air supply lines and filter; turn on air supply; set air supply regulator at 20 psig.

Check the instrument for the following items and conditions:

1. 3-15 psi receiver gauge calibrated to read 0-10 square root.
2. Check range of transmitter against required range on the specification sheet.
3. Check identification tag on transmitter.
4. Determine that piping configuration is consistent with the requirements for its service.

Calibrate d/p cell if it is not already calibrated. Proceed as follows:

1. Disconnect meter body from all process piping.
2. Both meter body chambers should be dry; remove drain plugs if necessary to drain liquid; replace drain plugs.
3. Connect a calibrating signal to the high pressure side of the meter body. Use a manometer or other acceptable readout device, calibrated in inches of water (the low pressure side of the meter should be vented to atmosphere). When no pressure is applied to the high side, the transmitted signal should be 3 psig on the output manometer or gauge. If it is not 3 psig, reset the zero adjusting screw until the output reads 3 psig.
4. Increase calibrating signal to full range (as shown by the specification sheet). Transmitter output should

read 15 psig. If not, adjust "span" adjustment screw until 15 psig output is obtained.

5. Recheck zero after "span" adjustment. Repeat until both are correct.
6. Check transmitter output for 25, 50 and 75% of range. Output should be (a) 25%—6 psig; (b) 50%—9 psig; (c) 75%—12 psig. These values should be recorded on a calibration form (Figure 12.8). This form is typical of those used by contractors and/or users to show that all instrument systems have been calibrated, checked and are ready for service. They are dated and signed by the calibrator and an inspector who may be a contractor or user representative.

Reconnect signal lines to the remote recorder or controller.

Check remote recorder at this time.

1. With zero input to the high pressure side, the recorder should show "0" on its chart. Move recorder chart about ¼ inch to mark zero line.
2. Increase input signal to 25, 50, 75 and 100% of range. Move the chart about ¼ to ½ inch each time to ink and mark each setting. Write the signal input value at the corresponding mark on the chart.
3. Decrease the input (transmitter) signal to 75, 50 and 25% of chart range, recording input values as above. Values both on increasing and decreasing signals should be recorded on the calibration form, Figure 12.8, which may become part of a permanent file of the user.
4. If the loop contains alarm or shutdown switches, they should be set while the test apparatus is being used. Their set points and tag numbers should be duly recorded on the calibration form.

Disconnect testing apparatus and reconnect process piping.

Leave all block valves at orifice taps and meter manifold closed and the manifold bypass valve open.

Manufacturers, model numbers and serial numbers of loop devices should be checked and recorded.

Typical Temperature Transmitter Check-Out Procedure

The following procedure is for a "filled system" temperature transmitter.

Check the following against the requirements of the specification sheet:

1. Check instrument tag number and range of instrument. The range should be stamped on the element or inside the instrument case.
2. Is the output gauge installed? The scale should reflect the range of the transmitter.
3. Check the element bulb length and the capillary length.

Blow down air supply lines and filter.

Turn on air supply to the instrument. Set the regulator output to 20 psig.

Make sure the capillary is adequately protected. It should be run in a channel or other protective support to prevent its being stepped on or accidentally hit by passersby. Excess lengths of capillary tubing should be neatly coiled.

Calibrate or check calibration as follows:

1. Immerse the sensing bulb in a known temperature bath. Use a mercury manometer or other accurate readout device on the transmitter output.
2. At an "upper" temperature level (80% or above), check "span" of instrument. Check span adjustment if necessary.
3. If the instrument is indicating or recording, check these devices at the same time.
4. After each change of span adjustment, recheck "zero" at a low temperature reading.
5. When a "zero" change of adjustment is required, recheck the upper temperature point to make certain the span has not changed.
6. Increase input signals to 25, 50 and 75% of span, recording values.
7. Decrease signals in the reverse of (6) above.
8. Set any switches for alarms or shutdown if used.
9. Record all information required on the proper calibration form. The form should be dated and signed by the calibrator and a witness.

Disconnect test equipment and make the system ready for service.

Record manufacturer, model number and serial number of instrument.

Typical Control Valve Check-Out Procedure

The following procedure is typical of that needed for control valves.

Check the control valve specification sheet for agreement on:

1. Control valve size and service
2. Inner valve size
3. Construction materials
4. Valve model number
5. Signal range
6. Diaphragm size if given
7. Valve stroke

Determine that the valve is properly tagged.

Check valve for proper stroke and seating.

Check valve for freedom from friction, correct location and proper flow direction through valve body.

Check valve for proper action—air-to-open or air-to-close.

TESTING and CALIBRATION REPORT

Instrument No. _____ Service _____

TUBING LEAK TEST (Bubbler): Trans. Output: _____ Controller Output: _____

TRANSMITTER: Range: _____ Date: _____

INPUT	% TRANSMITTER RANGE	DESIRED OUTPUT		ACTUAL OUTPUT	REMARKS
	0	3			
	25	6			
	50	9			
	75	12			
	100	15			

RECEIVER: Date: _____

INPUT	% TRANSMITTER RANGE	DESIRED PEN READING		ACTUAL PEN READING	REMARKS
	0	0	0		
	25	25	5		
	50	50	7.07		
	75	75	8.66		
	100	100	10		

CONTROLLER: Action: _____ Check Alignment: _____ Date _____

CONTROL VALVE: Air For Closed Position: _____ Open Position _____ Date: _____

POSITIONER: Posit Input _____ Valve Diaphragm Input _____ Split Range _____

CONTROLLER MANUAL OUTPUT: Stroked Valve: YES — NO

ACCESSORIES: Date _____

Transmitter Output Gauge: Zero Check _____ Full Scale _____

Receiver Ind. Gage Type: _____ Zero Check _____ Full Scale _____

Alarm Settings: Low _____ High _____

Panel Alarm Light Check: _____

Other _____

COMPLETED CALIBRATION & SYNCHRONIZED TRANSMITTER & RECEIVER: YES - NO

Commissioning Certification: *I certify that the instruments listed on this page have been calibrated and are in operating condition in accordance with drawings and specifications.*

_____   _____   _____
signed   firm   date

Figure 12.8. This instrument calibration form is used for pneumatic loops to indicate that all elements of the loop are properly calibrated and ready for operation.

On spring and diaphragm valves, the spring may have to be readjusted when the valve is placed in operation. The diaphragm area and spring are selected to compensate for the unbalance due to pressure drop across the valve. Actual service conditions may be different from conditions specified. The need for checking is particularly applicable to single seated valves.

Valves equipped with a positioner should be adjusted as outlined below. (This assumes a Fisher Type 3560 positioner.)

1. Remove the travel pen from the connector arm.
2. Hook up a suitable pressure source to the "supply" and "instrument" connections. A regulator and test gauge should be provided for varying and controlling the simulated instrument pressure.
3. Provide suitable pressure gauges on the supply instrument and diaphragm output pressures.
4. Turn the nozzle clockwise until it seats in the relay housing. Then turn nozzle counterclockwise two full turns.
5. Set the cam at mid-travel position. In this position, the cam arm will be in a horizontal position. Block the cam so that it will remain in the mid-travel position throughout the beam leveling procedure.
6. Set the instrument pressure at mid-range. For a 3-15 psig range, this is 9 psig.
7. *Beam leveling procedure:* the flapper arm must be on the direct acting quadrant of the beam. Loosen three screws and swing the flapper arm to an almost vertical position (near the cam bracket but not touching it). Adjust the flapper arm pivot so that the diaphragm pressure gauge indicates some output pressure between zero and supply pressure. Tighten the locknut on the pivot pin. Swing flapper arm to the right to a horizontal position (near the bellows pivot pin but not touching it). Adjust the bellows pivot pin so that the diaphragm pressure gauge indicates some output pressure between zero and supply pressure. Tighten the locknut on the pivot pin. Swing flapper arm to the left past the bearing bracket to a horizontal position on the reverse acting quadrant of the beam. To do this, unlock the cam and rotate it by means of the cam arm to move the lower part of the beam toward the back of the case. Block the cam at mid-travel after the flapper arm has been moved to the reverse acting quadrant. Adjust the relay pivot pin so that the diaphragm pressure gauge indicates some output pressure between zero and the supply pressure. Tighten locknut on the pivot pin.

Stroke adjustment:

1. Loosen screws. Move flapper arm along the beam. The beam is labeled to indicate the direction of flapper arm movement required to increase the valve stroke.
2. Vary the instrument pressure over its full range and observe valve travel.

3. Readjust flapper arm position until full valve travel results from full range of instrument pressure.
4. Tighten three screws when proper stroke adjustment is obtained.
5. Adjust starting point—set instrument pressure at value at which valve travel should start (3 psig for a direct valve; 15 psig for a reverse action valve) using a 5/16-inch open end wrench; adjust the relay nozzle until valve travel begins at the set starting pressure.
6. To set a positioner for a split range, loosen three screws. Move flapper arm along beam to increase travel. Vary the instrument pressure, observing the pressure span required to stroke the valve completely. Readjust the flapper arm until full valve travel results from the desired change in instrument pressure. A two-way split would be 6 psi on each valve. Tighten three screws. Adjust the relay nozzle until valve travel begins at the set starting pressure.

Miscellaneous Checks

By observing detailed check lists given for flow and temperature transmitters and control valves, it is obvious that there are many items to be checked on the final loop check to assure that all elements are complete and ready for operation.

Whether the measurement category is flow, level, pressure, temperature or other, the final loop check determines that:

1. All elements are installed, calibrated and functioning properly.
2. All scales and charts are installed with proper ranges.
3. All alarm units and shutdown devices are set properly and accomplish their intended purpose. A list of values for all alarm and shutdown actuators is usually available for proper settings (Figure 12.9).
4. Controller settings for the various modes of operation (proportional band, reset and rate) are at nominal values. Table 12.1 lists some suggested settings.
5. The action (direct or reverse) of the controller is set as prescribed by the specification sheet.
6. The operation of the control valve (where applicable) is as specified (air-to-open or air-to-close).
7. Recorders, indicators, controllers and alarms are properly identified as to tag number, location and service.
8. Process connections are complete. This includes insertion of temperature bulbs or themocouples, orifice plates, filling liquid seal lines or completion of insulation, steam tracing, etc. When items such as these are incomplete, a punch list is made for those that need further attention.
9. On temperature recording or indicating potentiometer instruments, connect a portable potentiometer to each thermocouple in the field and induce a millivolt signal to read one point on the instrument scale or chart. Disconnect one side of the thermocou-

ALARM AND SHUT-DOWN DEVICES

ITEM NO.	REF. SPEC.	SERVICE	TYPE ALARM	H/L	FUNCTION	SET POINT		CONTACTS OPEN WITH MEASURE	REMARKS	REV. NO.
						PROCESS	SIGNAL VALUE			

LEGEND:

P	PRESSURE	-d	DIFFERENTIAL	H	HIGH
L	LEVEL	AN	ANALYZER	L	LOW
T	TEMPERATURE	A	ALARM	INC	INCREASE
F	FLOW	SD	SHUT DOWN	DEC	DECREASE
S	SPEED	SU	START UP		

REV _____ BY _____ DATE _____
JOB _____ BY _____ DATE _____

S.I.P. INC.
Engineers & Contractors
HOUSTON, TEXAS

Figure 12.9. Information on alarm and shutdown switches is kept on this form. The most essential information is the value at which the alarm or shutdown occurs. (Courtesy of S.I.P., Inc.)

Table 12.1. Controller Settings for Various Modes of Operation			
Application	P.B.	Reset	Rate
Flow	150	0.1	0
Press. (liquid)	50	0	0
Press. (vapor)	100	0	0
Diff. press. (liquid)	50	0.1	0
Diff. press. (vapor)	100	0.1	0
Level	100	0	0
Temperature	100	0.3	0.5
Analyzer	100	0	0

ple circuit, and note instrument burnout feature (when applicable).

10. Check the operating action of all pressure switches by applying air pressure or manually opening or closing the switch contacts. Observe the action of the unit alarm sequence, or note its effect on equipment if used for interlock or shutdown.

11. For float switches (cage type), insert a stiff wire rod through the drain valve at the bottom of the float cage. Position rod to operate switch contacts and observe action of the associated equipment.

12. Capacitance or other special type switches should be checked in accordance with the manufacturer's suggested procedures.

13. Check installations for conformity to drawings, accessibilty of instruments and adequate supports and brackets for instruments and piping.

Startup

Startup time may be defined as the span of time between the end of construction and the beginning of *normal* operation. The problems faced during this period fall into five or six categories:

1. Placing instruments in service
2. Tuning the control loops
3. Evaluating process upsets and disturbances that may be attributed to instruments
4. Repairing or replacing defective equipment
5. Recalibrating and starting up analyzers and other special equipment
6. Installing additional controls

Placing Instruments in Service

As instruments are checked out prior to startup, many of them are valved out of service for various reasons—to prevent damage to instruments because they require venting, etc. After process flows are started, these instruments must be *cut in*—placed in service.

Flow instruments certainly fall into this category. Valves to these devices must be opened, legs vented or filled, zeros checked and bypasses closed.

Pressures and other systems (except analytical) are more likely to be open and ready for service when startup occurs. However, in situations where lines may be dirty or in slurry services, these also may be valved off and require attention to place in service. In any event, there are many devices which must be checked and placed in operation as the process starts functioning.

Tuning Control Loops

Prior to startup, controllers are usually set at nominal values. Adjustments must be made to meet operating requirements.

Controller tuning is about as much a function of art as it is a knowledge of process dynamics. Consequently, a person with experience in tuning controls can tune a system as quickly as one with a good grasp of the system dynamics, but with no practical experience.

Much experimentation is often required to tune a control system properly. Some loops require hours of patient work and observation to achieve the desired control. Others can be tuned easily and quickly. For example, loops with only two control modes are not usually difficult to tune for the processes on which two-mode control is used. They have fast time constants and adjustment effects can be noted quickly.

Three-mode control is often used, however, on processes with long time constants, and one must wait for long periods to note adjustment changes. Also, there is a great deal of interaction among the three modes, making it impossible to change one value without affecting the others. Many combinations of proportional band, reset and rate may have to be tried before finding the combination that best suits the process.

Evaluating Process Upsets and Disturbances

As most experienced instrument people know, many process upsets and disturbances that occur during startup time are wrongly attributed to instrument difficulties. This conclusion is drawn when flows, pressures, levels, etc., fall outside expected operating limits. Startup personnel need to know how to distinguish between process peculiarities and disturbances and instrument difficulties. Process upsets and disturbances include equipment difficulties and shortcomings, piping and other mechanical troubles, and similar problems.

Checking out the difficulties mentioned above requires much time. Some typical problems with their answers follow.

1. Failure of a heat exchanger to maintain the proper temperature. *Answer:* Instruments may be working perfectly, but cooling medium may not be at the expected temperature level or the cooling surface may be fouling.

2. Flow records are not in balance. *Answer:* The fluid specific gravity may be different from what was expected, fluid temperature or pressure may be different,

or there may be a sludge buildup at the orifice plate, resulting in an erroneous flow measurement.

3. A chromatograph reveals a high value for a particular component. *Answer:* An unexpected component in the sample elutes at about the same time the desired variable is measured, appearing as the measured variable.

4. A control valve is not delivering sufficient fluid flow. *Answer:* Because of a mistake in calculating pressure losses, the pump is not delivering the expected head at the valve. The pressure drop available to the control valve is less than expected.

5. A capacitance level probe used as an alarm to determine when a vessel is almost empty is acting in an erratic manner. *Answer:* The material which normally should be dry contains enough moisture that it does not flow freely; consequently, the probe sometimes remains covered even when the vessel is essentially empty.

These typical examples illustrate the problem of distinguishing between instrument malfunctions and process difficulties. Some problems are easy to identify; others require much study. These types of difficulties occur more frequently during startup periods when processes are unstable and equipment is being proven. As operational experience is acquired, subnormal operation is less frequent, and process and equipment difficulties are recognized and identified much more quickly.

Repairing or Replacing Defective Equipment

Repair or replacement of defective equipment is usually a minor part of the difficulties that occur at startup. The nature of instrument equipment is such that lengthy run-ins are not practical. At startup time there are early failures of some equipment, and repair or replacement must be made. In some instances, spare parts may be available; in others, equipment must either be sent back to the factory or replacement made from the factory. Delays from such actions are costly and annoying, but unavoidable.

Special Equipment

Starting up special equipment such as chromatographs, analyzers, radioactive measuring devices and new measurement devices may require much attention. When a process is well known and correct information has been given to the manufacturer concerning the application, chromatographs should go onstream easily if they have been properly calibrated and programmed. If processes are new, however, it is easy to overlook (or have little knowledge of) some

physical aspects of the stream. For example, partial condensation of a stream at the sample takeoff point when a gas sample is expected or partial vaporization when a liquid sample is expected sometimes occur. These conditions may be caused by a lack of knowledge of the process or by an oversight by the applications group. In either case, changes must be made in the sample preparation system to overcome the trouble.

In almost any type of analytical measurement, process variations that are unknown when specifications are prepared or which occur during the startup period may result in troubles with the sample handling system or in the anaylzer itself.

Radioactive measuring systems (levels, density, etc.) may be erratic during startup. On a level application, product coating on vessel walls appears as a valid partial level. If the condition is temporary, it must be accepted as such; if it is normal, rezeroing the measuring system is in order.

Initial operation of a density measuring system may be outside the limits of the planned span. This may be a result of operating temperatures different from those initially expected or a different stream mixture.

New measurement techniques often present unexpected difficulties. If the technique is being used on a fluid not previously tested, the effect of the new fluid characteristic may be different from those previously tested. A collection of data may be necessary to substantiate the validity of the application.

Additional Controls

On new processes, it is often necessary to add new controls to provide better operation. Alarms, interlocks, changes from recording or indication to control, or complete new systems may be needed to meet unforeseen operating requirements. Usually spare panel space, spare tubing, spare conduit and wiring, and spare termination points are available so that additions can be made quickly and with a minimum of difficulty.

Conclusion

The installation, calibration and startup of control systems in process plants is challenging. Despite how well-documented the job may be, it needs the services of people who are acquainted with installation and operating requirements. Instruments reveal the process action to the operators. If the action is understood, the operator and the controls have mastery over the process. Instrument supervisors and technicians who understand the functions of the instruments and have some knowledge of the process characteristics can provide much to a smooth, efficient installation and startup of a plant or process.

Digital and Computer Control Systems

13

Michael J. Sandefur

Process computers, distributed control systems, microprocessors, programmable logic controllers and other forms of digital instrumentation represent the most recent steps in the evolution of process control. Forty years ago instrumentation was mostly mechanical. Integral transmitters and controllers were mounted directly on the process line and sensed the process fluid directly with no amplification. Each controller functioned independently, with little or no communication with other controllers. The operator made rounds in the plant, reading instruments, adjusting setpoints, and logging readings. Clearly, increased process complexity and greater demands on plant efficiency and safety required a more centralized mode of operation. As a result, pneumatic amplification and transmission line techniques were developed, allowing the central location of controllers, indicators, and recorders in a control room. Advances were made in packaging, permitting greater instrument densities on the panel boards. Further miniaturization was made possible with the advent of transistorized electronics in the 1950s. However, centralized control rooms and high density panels introduced new problems in signal transmission and necessitated extensive cabling between the process units and the control building. The operator also suffered from visual saturation, due to the proliferation of indicators and control devices on the panel board.

Electronic computers were introduced in the 1950s, and were first applied to process control in the 1960s. Due to the high cost of digital computers, it was necessary to put many of the plant's control loops on the computer to justify its initial cost. All process signals were multiplexed, digitized, and stored in memory. PID control functions were performed by the computer using series approximation techniques. Control outputs were computed and analog signals generated for transmission to the control valves. This type of computer control is normally referred to as "Direct Digital Control," or DDC. Due to hardware complexity and lack of proven software, the reliability and effectiveness of the DDC computer system were questionable.

Despite these limitations, computers offered the possibilities of integrated control, process optimization, enhanced operator interfaces, data reduction, data storage, and a systematic approach to plant management. To offset the poor reliability of the computer system, analog controllers were added as backup equipment. This further added to the cost of installing a process computer control system. It was at this point that the concept of "Supervisory Control" was introduced. With this arrangement, analog controllers performed the primary loop control. The process computer would monitor the process and adjust controller setpoints. This improved overall system reliability considerably over the direct digital approach. The computer was also relieved of some of its computational functions, allowing more CPU time for higher level functions. Costs remained high due to the need for analog instrumentation and panels. Also, wiring complexity doubled to accommodate the computer. For each analog controller, the process variable, setpoint, setpoint feedback, and controller status were hard-wired to the computer's front-end multiplexer.

The advent of the microprocessor in the early 1970s did much to alleviate the remaining problems with computer control. Its most immediate impact was a dramatic reduction in the cost of computer power. Not only did microprocessors make large computer systems more economical, they also made it possible

Figure 13.1. The low cost of microprocessors has allowed the relocation of the controller to the process line or the control valve. Digital transmission of data over a serial communications link provides centralized monitoring and control of the plant with a minimum of field wiring. (Courtesy of Xomox Corporation)

to use many small computers to control the process. By ruggedizing the hardware and utilizing digital transmission techniques, it is now possible to locate the controllers in the process areas (Figure 13.1). A data highway is then routed through the plant to link the microprocessor-based controllers together and provide communications with the control building. To provide the operator interface, a process computer or stand-alone CRT console is connected to the data highway. By using a distributed approach to process control, reliability has been increased, hardware and installation costs reduced, and performance improved to the point where digital control is very competitive with analog instrumentation.

System Classification

Digital and computer control systems have many applications and appear in numerous forms. Classification of a computer system is related to its size, function, and control philosophies. Many of these classifications have evolved as a result of the computer's unique characteristics and limitations at the time it was applied to process control. That is, reliability, cost, and state-of-the-art technology in any given period have dictated the computer's use and, therefore, its system type.

Direct Digital Control (DDC)

The term "Direct Digital Control" has been used to describe large computer systems that provide primary control of the process or plant. All process inputs and control outputs interface to a large multiplexer subsystem which is bused to the control processor. In large systems, the multiplexer equipment is divided among independently functioning units or "nests." This allows a modular approach to system design and expansion. The process variable signals are hard-wired into the nests, where they are digitized, conditioned, filtered, and stored in the computer's bulk memory. This accumulation of process data in memory, along with control algorithms, labeling, and special instructions, is normally referred to as the "data base." The central processor periodically scans the data base, updates the process data, performs the necessary control algorithms, and generates control outputs which are transmitted to the field via the multiplexer unit. All signal processing, indication, and control functions are performed digitally by the computer, hence the name "Direct Digital Control" (Figure 13.2).

Depending on the size of the plant, a DDC computer system can be quite large and complex. Central processor speed and memory requirements are high, due to the many functions that the computer must perform. These functions include multiplexer input and output scanning, input conditioning, data base generation, control algorithm execution, alarming, display generation, report generation, logging, and process optimization. The central processor must also support various background mode software systems, such as FORTRAN compilers, file systems, text editors, data base builders, and numerous utility routines used by system programmers. The computer must also service its peripherals, such as line printers, card readers, programmer consoles, and bulk storage devices. While other computer systems make many of these demands on the central processor, the DDC system has all of them to the maximum extent possible. As a result, system reliability and performance suffer. Since the DDC system requires nearly all of its components to maintain control of the process, any single failure can be catastrophic. The complexity of the system (which is inherent in its design) makes troubleshooting more difficult than with conventional control systems. Means have been devised to circumvent the reliability limitations by adding a second computer system on hot standby, or by backing up the computer system with analog controllers. In either case, considerable cost and complexity are added to the system.

Another aspect of computer control systems which adds a new dimension to system complexity is software. DDC systems, in particular, require considerable effort and attention in the areas of software generation, maintenance, and documentation. Software's impact on system performance has been reduced somewhat by standard fill-in-the-blank software packages provided by the computer vendor. These packages make data base construction, algorithm development, display generation, and other control functions straightforward. Thorough documentation of the computer system's data base configuration is still essential. Also, to maximize the benefits of a direct digital control system, much custom software has to be developed and maintained to provide advanced process control and optimization.

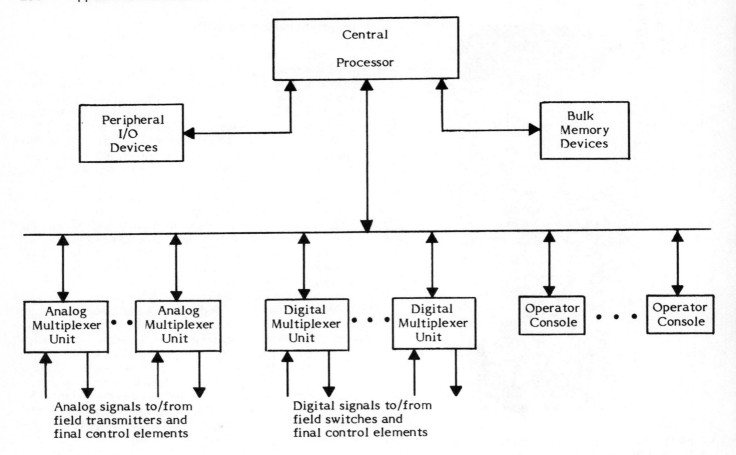

Figure 13.2. The direct digital control (DDC) computer system interfaces directly to the field instrumentation. All process indication, control and alarming are performed digitally by the central processor.

Supervisory Control

As discussed earlier in this chapter, analog controllers were added to a DDC computer system to enhance its reliability. With ''Supervisory Control,'' primary loop control is returned to the analog controllers, while the computer monitors the process and adjusts setpoints (Figure 13.3). Since most control loops involve PID controllers in single or cascade loops, this arrangement works well. The computer is relieved of some computational tasks, and can be utilized for process optimization and plant management. This approach is quite feasible for the user who has existing analog instrumentation and wishes to add a computer system to enhance plant operation. It is also a useful approach for the company building a grassroots plant and desiring the greater performance of a computer control system, while maintaining the familiarity and confidence that a well laid out analog control panel provides. Supervisory computer control is also a viable approach for plants located overseas or in remote domestic areas. In these instances, reliability is essential, and skilled maintenance men and spare parts are usually at a premium.

Total cost for a supervisory control system is higher than for a DDC system due to the need for analog controllers. If a process computer is added to an existing analog control system, then the hardware costs can be less than DDC since a smaller central processor and less peripherals are required. However, system size and cost are a function of what the customer hopes to accomplish with a process computer. The user has more flexibility to choose a system tailored to needs and budgets than with a full blown, direct digital control system.

Installation requirements can vary for a supervisory control computer system. The process variables, setpoints, controller mode, and other pertinent signals must be available to the computer through its front-end multiplexer. This can be accomplished by hard-wiring the controllers and other analog instrumentation to the computer system. To simplify installation, receptacles can be mounted in the control panels and computer multiplexer racks, and the required cable assemblies fabricated in the shop. Installation is a matter of positioning the computer and instrument cabinets and cabling them up. The trade off, however, is increased equipment complexity and cost due to the extensive wiring required between the receptacles and the asso-

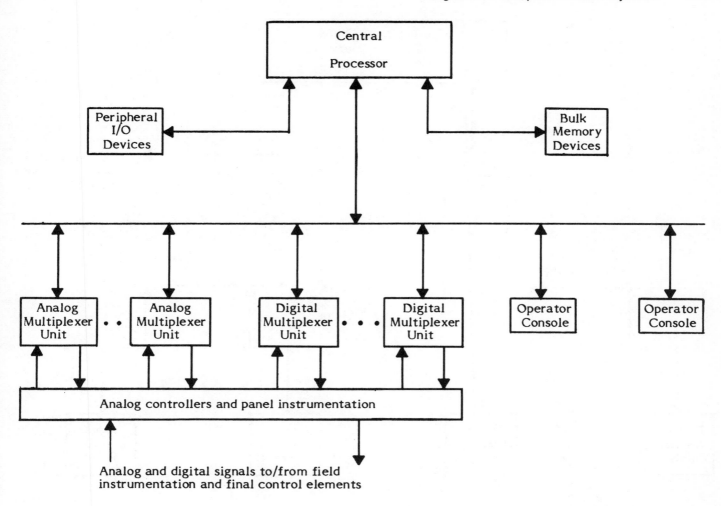

Figure 13.3. The supervisory control computer system interfaces to stand-alone analog controllers and instrumentation, which perform the primary control and indication of the process. The computer system provides coordinated control of the plant by monitoring the analog instrumentation and adjusting setpoints for optimized operation of the process.

ciated electronics in the cabinets. Also, careful design is necessary when interfacing one vendor's computer system to another vendor's analog instrumentation. Problems with fusing, signal commoning, ground loops, and signal power can occur. Coordination between two different vendors can be difficult.

There are several systems on the market which provide an integrated approach to supervisory control. Standard analog instrumentation is configured in nests with the necessary cable and bus interfaces built into the hardware. By adding the vendor's standard interface and cable assemblies, the vendor's process computer can be easily added. This eliminates most of the headaches associated with the hard-wiring approach, and allows the analog system to be upgraded to a digital system at the customer's convenience. Obviously, this type of system has to be bought from a single vendor. However, this may change in the near future in the area of distributed control systems and data highways.

Hierarchical Computer Systems

A hierarchical system is a network of process and/or information management computer systems integrated to serve a common function (Figure 13.4). This function may involve management and control of large refineries, pipeline networks, or the energy production facilities for a whole country. Computer systems are utilized from the primary level of process monitoring and control, up through the various supervisory levels, to the decision-making realms of top management. Information on plant operations is assimilated and processed at the primary level systems and passed up to supervisory level computers. Overall production management philosophies and decisions are made by the higher level systems and transmitted down to the primary level computers. The computer network architecture usually parallels the organizational structure of the company itself. The structuring of computer systems in a hierarchical fash-

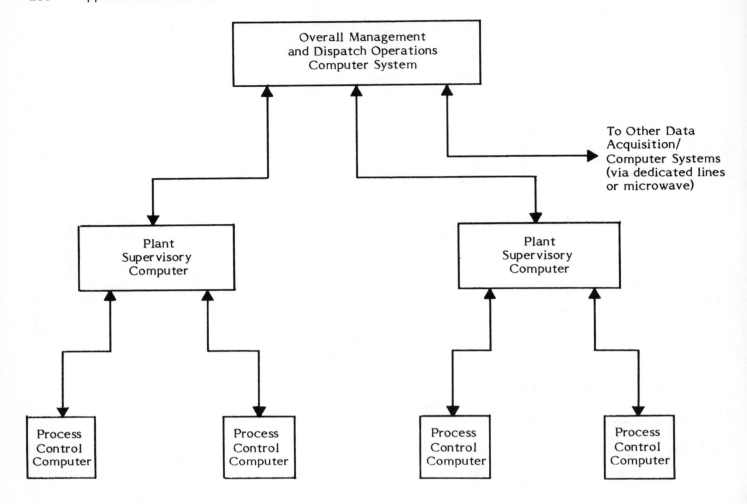

Figure 13.4. The hierarchical computer system provides a logical division of data processing, control and management of complex, industrial operations. Hierarchical systems are especially suitable for operations involving production facilities or transmission networks distributed over large areas.

ion allows clear divisions of responsibility and reduces the complexity of any one computer system. The overall security of plant operations is enhanced by limiting liability, should one or more systems fail. Where facilities are widespread, individual computer systems interfaced to a centrally located, computerized operations center may be the only practical solution for automated control of the network. With hierarchical systems, the primary computers provide direct control of the process and can be DDC, supervisory, or microprocessor based. Higher level computers tend to be larger, general purpose machines dedicated to data processing, information management, and dispatch-type operations.

Distributed Control

With the recent developments in integrated circuitry and the introduction of the microprocessor, it has become feasible to dedicate a digital processing unit to individual control loops. The processing unit is usually based on a microprocessor with the necessary memory, interface devices, and input/output hardware to control a relatively small number of loops. Many independent microprocessor systems are utilized to monitor and control an entire plant. A serial digital transmission line links the controllers and allows a master computer or operator interface device to communicate with them (Figure 13.5). The operator interface is usually a CRT console driven by a microprocessor. No support computer system is required (Figure 13.6). Essentially, the CRT console is a computer in itself, and can perform computer-like functions such as present tabular and graphic displays, drive printers, and store historical data for logging and trending. In many distributed systems, the console is capable of scan and alarm functions, thus eliminating the need for a hard-wired annunciator system. Pushbutton and status indication functions can be performed by the CRT console. A second or third console rounds out the system's reliability and flexibility. Failure of the CRT console does not affect operation of the controller units which provide automatic PID control. Some vendors offer microprocessor-based controllers with indicators

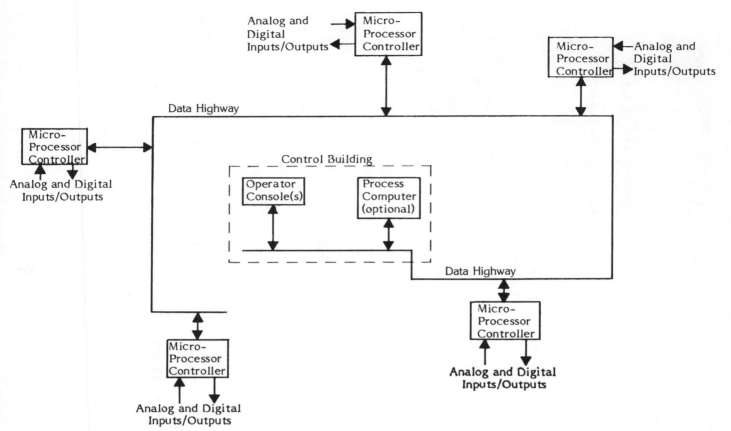

Figure 13.5. Microprocessor-based distributed control systems are a form of direct digital control that divide the data acquisition and control functions among a number of independently operating microprocessor controllers. The controllers are linked together with a data highway that is routed through the plant and connected to the operator's console in the control building.

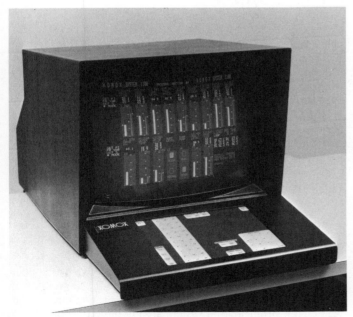

Figure 13.6. A microprocessor-based CRT console provides the interface between microprocessor controllers and the operator (via a data highway) in a typical distributed control system. (Courtesy of Xomox Corporation)

and operator controls mounted on the front of the unit (Figure 13.7). This allows operator control of the process independent of the CRT console.

A number of economic gains can be realized with a microprocessor-based distributed control system. Depending on system complexity, hardware and installation costs can be less than for an equivalent analog system. Floor space requirements in the control building are minimized by distributing the microprocessor controllers in the field. Wiring and cabling costs are significantly reduced with a distributed control system. A process computer can be added simply by connecting it to the data highway using the vendor's computer interface hardware. The computer communicates digitally with the microprocessor controllers and has complete access to the process variables and control outputs. As a result, the computer's front-end multiplexer and wiring costs associated with it are eliminated. The process computer is also relieved of many scan, signal conditioning, alarm, and control functions. The computer can be utilized to perform tasks for which it is best suited, such as process optimization, graphic display generation, and data processing.

Data Acquisition and Multiplexing

Some applications require that process data be accumulated, processed, displayed, and logged. Many times the user also

Figure 13.7. This microprocessor-based controller contains all hardware required for process indication, control and alarm. A keyboard on the side allows convenient changes to the controller's programming. (Courtesy of Process Systems, Inc.)

wants to minimize field wiring requirements and provide a convenient, compact means of displaying the process data. In such cases, various types of data acquisition or multiplexing systems can be used (Figure 13.8). Except for the simplest applications, these are digital systems which perform the same basic functions as a process computer. With large systems, the operator can initiate control outputs, such as starting and stopping motors or operating on-off valves. Data acquisition and multiplexing functions can be implemented with hard-wired, solid-state logic, relays, microprocessors, computers or any combination thereof. In fact, most analog multiplexing is accomplished with reed relays controlled by solid-state logic. A microprocessor performs scanning and supervision at a local level, while a centrally located process computer provides the data display, storage, alarming, and logging functions.

Significant savings in wiring cost can be achieved by concentrating the scanning, A/D conversion, and signal conditioning functions into self-contained packages called remote terminal units (RTUs). RTUs are then distributed throughout the plant, near the field instrumentation. By digitizing the transmitter signals in the field and utilizing digital transmission techniques, the data is much less susceptible to noise. This is especially effective for low level signals, such as thermocouple inputs.

It should be obvious that data acquisition systems overlap in function with hierarchical, distributed control, and supervisory control systems. The distinctions between the various types of digital control systems have mainly been historical in nature. With the introduction of low cost microprocessor systems, these areas have overlapped and merged. The digital system designer can now select the functions and features of any system type and integrate them into a unique package. This kind of flexibility and enhanced performance will make digital instrumentation the clear choice in the years to come.

Logic and Sequential Control

Since computers are digital by nature, they would appear well suited for executing logic and sequential functions normally performed by relays, timers, and drum controllers. The computer can perform these functions, but, because of system complexity, reliability, and software, it usually is not considered a viable alternative. Most logic and sequencing systems perform critical functions, such as process unit regeneration, batch operations, equipment safety, and plant shutdown, which cannot be trusted to a computer. A large percentage of personnel who design and maintain these logic systems have been unsure about computers and their associated problems. Relay logic, however, is tried, proven, and straightforward in function. Most technicians can understand a relay ladder diagram, but few could program a computer in FORTRAN or assembly language.

To circumvent this problem, the automotive industry introduced the programmable logic controller in the late 1960s. The programmable logic controller, or PLC, is a ruggedized, solid-state device capable of simulating relay logic. The PLC is configured by using programming instructions rather than hardwired relays or solid-state logic components. The program consists of relay contact and coil symbols interconnected in a ladder diagram-type structure. To program the PLC, the designer simply lays out a ladder diagram and then enters it into the controller one rung at a time, using a programming box. Special backlighted keys with contact, coil, and branch symbols engraved on them, make program entry straightforward. More sophisticated programming devices have evolved, usually based on a portable CRT console. With these programmers, one or more rungs can be displayed at a time, providing the designer with a picture of the ladder diagram. With the CRT programmers, off-line programming is possible, allowing the designer to make corrections to the program without shutting down the controller itself. Usually, CRT programmers incorporate a microprocessor in their design. As a result, program diagnostics, program labeling, and print-out on a typer is available. This can be very useful for documentation purposes.

Many advances have been made in programmable logic controllers, to the point where their distinction from process computers begins to blur. PLCs can perform arithmetic functions on BCD coded inputs, execute complex sequencing and decision-making operations, structure data in arrays and manipulate them with single line instructions, provide logging functions, and even perform PID control on analog inputs and outputs. In fact, most PLC manufacturers utilize microprocessors to enhance performance and reduce cost. From the beginning, a programmable controller has always been a small, specialized computer packaged in a new box. This, however, should not deter the would-be user from considering PLCs for logic and sequencing applications. There is such a wide range of products available today that users should have no trouble selecting the controller best suited to their needs.

Planning a Computer System Project

It is important to take a systems approach towards the planning, management, and execution of a computer control system project. Techniques used to design, engineer, and build a conventional instrumentation system are not directly applicable to the digital system.

Analog instrumentation involves controllers, indicators, recorders, and other discrete components that are individually specified and bought. The control panel is then laid out and de-

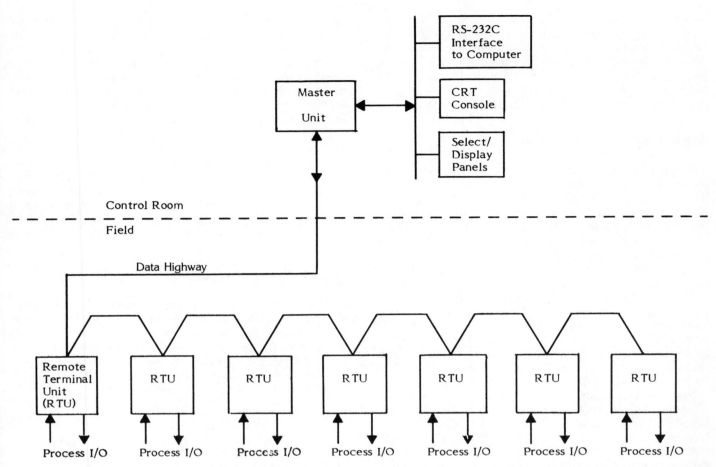

Figure 13.8. The data acquisition or multiplexer system functions as a stand-alone system designed to reduce field wiring costs and present process data to the operator via compact display panels or CRT consoles. The multiplexer system can also interface to a process computer system for expanded operations.

sign specifications established for rack construction, wiring, instrument mounting, and other assembly details. Construction and assembly of the control panel are usually performed by the instrument vendor or by a separate panel shop.

The successful computer system project follows a different set of procedures. The instrument engineer must analyze the system's functional requirements, hardware configuration, software, operator interface, and other needs. The engineer must scope out the size of the system early in the project and provide necessary spare capacity for future expansions that may occur during project execution and later in the field. Basically, the digital or computer system performs the indication and control functions which discrete components provide in an analog system. Changes or additions to a computer system can be more costly and time consuming due to the interactions of hardware, software, and system configuration. Gross error in the point count or desired system performance not only affects the system configuration, but could also require a change in equipment type. Major changes usually result in slipped schedules and substantial cost adders. The vendor can be helpful in anticipating such problems, but only if the computer system has been thoroughly scoped early in the project. A systems approach must be taken at the beginning if the project is to proceed on schedule,

stay within the budget, and meet the expectations and needs of the client.

Manpower Requirements

The first step in meeting manpower requirements is to assign a project engineer who is familiar with the design, specification, engineering, and execution of digital control systems. Depending on the size and complexity of the system, the project engineer probably will be required to work full time on the project. Increased demands will be made of the project engineer during the advanced stages of engineering, design review, software development, inspections, and acceptance testing.

The project engineer will require assistance. For small systems with 500 process points or less, the project engineer will require a designer to handle the details of interfacing with the instrument, electrical, and architectural disciplines. Numerous drawings, specifications, data base sheets, instrument loop drawings, and other documentation must be developed, checked, and issued for construction.

For systems with 1,000 points or more, the project engineer will require an organized team of engineers, designers, and technicians to successfully execute the project. Clear areas of responsibility, goals, and milestones must be established for each member. Typically, engineers are assigned responsibility for the different steps in system engineering and design, while the project engineer establishes system configuration, monitors project schedules and costs, interfaces with client and vendor, and in general, maintains control of the project. In all cases, the project engineer must be willing to delegate authority.

Scoping the Computer System

Scoping the system can be one of the most challenging and creative aspects of a computer system project. It is also one of the most critical steps, since everything that follows is built upon it. It is essential that the system be specified as closely as possible at the beginning of the project. Using the system specification, firm prices can be quoted, manpower assigned, and manufacturing schedules established. However, none of this can be accomplished unless the project engineer and the client have a clear idea of what they need.

In reality, the client usually has some concept of what is required in the way of computers or digital control systems for the plant. The client may carry this to the point of preparing a detailed specification from which the project engineer can obtain quotations. Typically, the client's specification must be rewritten to include additional detail on equipment types, quantities, point counts, and specific operational requirements.

System configuration will be dependent on many factors. Many questions need to be asked and answered: How many analog and digital points will be brought in? How many control loops will the computer system support? Will there be a centralized control building? How much floor space is available? Are there suitable locations in the plant to locate remote equipment such as RTUs or microprocessor controllers? Does the client wish to minimize field wiring by distributing the system, or would the client prefer the computer system in one location where maintenance can be performed conveniently? Should analog instrumentation perform the primary control function and the computer act in a supervisory role? Is the client receptive to microprocessor-based instrumentation for primary control instead of analog controllers? What mix of digital and analog instrumentation will be required? What degree of system security or reliability is desired? Will a backup computer be required? What is the fail-over philosophy if a dual computer system is implemented?

Other factors that the project engineer must consider are the functional requirements of the system: How complex are the control algorithms to be implemented on the system? How much process optimization will be performed by the computer, and how will it affect system loading? Will the computer system provide process alarming, or will a hardware annunciator system be required? The project engineer needs to determine historical data and report generation requirements, which establish bulk memory type and size. With small systems, some of these factors are not as critical, since the vendor's minimum standard configuration usually can handle the client's requirements. On large, customized computer systems, this information is essential if the vendor is to give a realistic quotation. With large systems, some vendors may not be capable of handling the job.

This will not be apparent unless the system has been thoroughly scoped and specified before sending out inquiries.

Attention must be given to the type of operator interface required. In most cases, a CRT-based operator console is the preferred interface. In a few instances, the client may require digital display panels or even dedicated indicators and pushbuttons, but these arrangements are usually confined to data acquisition and multiplexing systems. Sometimes a shared video display system is requested by the client. "Shared Video Display System" can be interpreted several ways. Some vendors offer a microprocessor-based system which can be interfaced to their analog instrumentation to provide this function. If this option is not available, then a supervisory process computer is required, and the resulting interface with existing instrumentation involves extensive wiring, cabling, and other expensive resources. These requirements should be identified early in the project and quoted by qualified vendors.

Steps in Project Execution

Once the project engineer has been assigned, feasibility studies have been performed, and meetings arranged to scope out the system configuration, the project is well underway. To ensure continued progress, the project engineer should map out and schedule the various steps required to successfully execute the project. These steps are listed here and will be discussed in greater detail in the following paragraphs:

1. Prepare a detailed system specification.
2. Establish a list of qualified vendors.
3. Issue Request for Quotations (RFQ).
4. Evaluate bids and select vendor.
5. Issue purchase order.
6. Arrange kickoff meeting with vendor and client representatives.
   a. Establish communication channels with vendor and identify necessary information and drawings required by both parties.
   b. Establish schedules and milestones with the vendor for manufacturing, system assembly, software development, documentation, and testing.
7. Assign personnel and start system engineering.
8. Established design requirements for affected disciplines (electrical, power, grounding, air conditioning, architectural).
9. Review vendor's design documents and release for manufacturing. Conduct design reviews and visual inspections at factory as required.
10. Monitor design changes and costs, and negotiate with vendor as required.
11. Generate computer data base forms and submit to vendor.
12. Define CRT displays and report formats.
13. Develop custom software.
14. Establish spare parts and maintenance requirements.
15. Develop test plan for customer acceptance of the system.
16. Conduct acceptance test at vendor's facilities. Make final arrangements for shipment and storage if required.
17. Commission system on-site.

18. Issue final revision of system specifications and affected engineering documents.

Computer System Selection and Procurement

The selection and procurement stage in the execution of a computer system project can be a major effort and require many months to complete. For small systems, it may only represent the final culmination of the project planning effort. The distinction is mainly a matter of system size and the general practices followed by the engineering firm involved. On multimillion dollar computer projects, much time and effort is spent before a final system configuration is developed and a qualified vendor selected. Even after the purchase order has been released, negotiations may continue on terms and conditions, and on equipment and services to be provided.

System Specification

Before inquiries can be sent to prospective vendors, a detailed definitive specification of the computer system must be prepared. Basically, the system specification is the final documentation of the system scope developed earlier by the project engineer and client. It is a detailed written definition of system hardware, software, functional requirements, and necessary services to be performed by the vendor. While the specification will be in a continuous state of revision during the life of the project, every effort should be made to make it as complete as possible initially. The system specification and the resulting vendor quotation form the basis for all future negotiations between the vendor and the contracting firm.

Detail and functionality of the systems specification will be determined by several factors. System scope and definition as determined by the project engineer and the client are the most obvious ones. Customer preferences for certain hardware configurations, such as dual computers or a microprocessor-based distributed system, will shape the final form of the specification. The availability of process and plant design data will also have a crucial effect on the specification. Educated guesses on system size and configuration may be required so that firm prices can be quoted.

Required vendor services, such as data base generation, graphic display building, custom software development, and extensive acceptance testing, should be identified in the system specification so that these items can be quoted. Hardware interfaces between systems supplied by different vendors should be specified in detail. Typical areas of concern are: Who will supply sensing power for analog and digital inputs/outputs? What type of interface will be required (hard-wiring, connectors, special electronics, etc.)? All requirements for communications with other digital systems should be identified and the necessary protocols established.

Less tangible factors, such as schedule requirements or degree of competitive bidding required, may affect the time and effort spent in developing the specification. If the specification is written around a particular vendor's hardware, or puts stringent requirements on component design, manufacturing, and testing, then few vendors may respond to the inquiry.

Preparing the Request for Quotation

Once the system specification has been finished, a Request for Quotation (RFQ) should be prepared and sent to the vendors whom the project engineer and client have determined capable of supplying the required system. This document identifies what type hardware, software, and services are to be quoted. The system specification is attached to each inquiry and forms the basis for quotation. The RFQ also identifies all necessary terms and conditions, packing and shipping instructions, system design documentation, and spare parts requirements. To ensure proper response to the inquiry, it is important that the project engineer establish communications with qualifying vendors prior to the release of the RFQ.

Bid Evaluation

After the bids have been received, the task of bid evaluation begins. This can be a very difficult task, especially when many vendors have submitted quotations. Often, all of the vendors appear capable of supplying a functional system suitable to the client's needs. The project engineer may be tempted to make a decision based on the lowest bid. However, two considerations should be kept in mind: One, all systems are not equal; and two, all "bottom lines" on comparable systems are not equal in scope of material and services provided. It is the responsibility of the project engineer to determine which systems are acceptable from a technical and performance standpoint. It is also the project engineer's responsibility to establish a "bottom line" for each vendor's bid. This is accomplished by tabulating each bid, breaking out hardware, software, options, and services provided. Each of these categories should be based on the system specification and the RFQ. The project engineer then determines if the vendor has responded to each item in the bid tabulation. This is best accomplished when the vendors are instructed to respond to the RFQ and system specification on a paragraph-by-paragraph basis. In addition, all components, services and options should be listed and priced separately in the quotation. Many times the vendor will ignore these instructions and only quote a bottom line figure. This is usually done to save time and present a "low bid" to the potential client. To circumvent this problem, the project engineer must maintain communications with the vendors and request *written* price clarifications. If the vendor is not cooperative, then the project engineer is wise to eliminate the bid.

As the bid tabulation is completed, a picture will develop as to who the successful vendor will be. Certain vendors will be disqualified, which simplifies the task. Other vendors may have systems which offer more attractive features. Finally, the bottom line of the bid tabulation, if prepared properly, will give a true comparison of system cost for the remaining vendors. Final selection will be a function of system features, costs, services provided, the vendor's past performance, and the client's preferences. If close attention has been paid to the steps of system specification, request for quotations, and bid tabulations, then final vendor selection can be made intelligently, and the best possible system will be purchased.

Pre-Purchase Negotiations

Depending on the size and complexity of the system, the vendor, contracting firm, and the client may enter into negotiations before the purchase order is released. Final details on system configuration, hardware, software, and services provided in the quotation are discussed and additional pricing established, if required. On large jobs, it is important that such meetings are executed prior to issuing a purchase order to ensure that there are no costly misunderstandings between vendor and contractor. If an impasse is reached, the contractor still has the option of selecting another vendor without suffering cancellation charges.

Purchase Order

The purchase order is released next, with a revised copy of the system specification reflecting the system configuration, equipment types, quantities, and services to be provided by the vendor. Again, it is essential that the specification is as detailed as possible and reflects all system requirements. As a result of the pre-purchase order negotiations, there may be changes to the system configuration. This should be reflected in the detailed specification.

Project Kickoff

The project engineer should schedule a kickoff meeting between the successful vendor, contractor, and client. On smaller projects, the negotiations discussed earlier may be conducted during this meeting. However, the main function of the kickoff meeting is to establish how the work is to be done, when it will be done, and who will do it. Schedules will be developed for engineering drawings and specifications, design review and approval, manufacturing, system assembly, testing, software development, delivery, installation, and startup. Communication channels are established, and key personnel identified. Manning requirements are established for the project.

Information requirements for the vendor and contractor need to be identified. For the vendor, this usually consists of the field terminal assignments, selected options, functional requirements yet undefined, and final system configuration. The contractor requires dimensional drawings of the cabinets, power, grounding and heat load requirements, identification of power, grounding and signal terminations, and other particulars related to system support and interface. The contractor needs a firm date as to when final design documents will be made available for approval. In turn, the vendor must know when this approval can be expected so that manufacturing schedules can be planned in advance.

Hopefully, upon completion of the kickoff meeting, all participants will leave with a clear-cut plan to follow. Work will start on the project and proceed according to the agreed schedule. The project engineer will have met the key personnel involved, established rapport with them, and be in a good position to successfully manage the computer system project.

System Engineering and Design

The engineering and design of the computer system are built upon criteria established in the detailed specification and discussed in the kickoff meeting. Questions posed by the vendor and the project engineer on hardware configuration, process interface, and system functions must be answered and documented. Additional drawings and specifications addressing specific design areas of the computer system must be issued and transmitted to the vendor and affected design disciplines. This stage of the project represents a period of information gathering, decision making, and communications with the vendor and client.

Point Count

An exact point count must be developed to determine the total number of analog and digital inputs/outputs serviced by the system. This information is taken from the latest mechanical flowsheets and instrument index. A percentage of spare I/Os, based on live points serviced by the system, should be calculated and a total point count established. As additional points are added to the system, the point count will be revised upward to maintain the required percentage of spares.

Rack Layout

Once the point count has been established, the type and quantity of input/output cards, mux chassis, controllers, terminal panels and other supportive hardware can be identified. From this equipment list, the system racks can be identified and laid out (Figure 13.9). Because the vendor normally performs this function, it is important that the point count, grouped according to signal type, is developed as soon as possible. The project engineer should provide input on overall rack layout, since this affects the electrical and architectural design disciplines.

Field Terminal Assignments

After the rack layout has been finalized and termination panels located, field terminal numbers are assigned. Although the vendor can perform this task, it is recommended that the project engineer make the field terminal assignments. This will ensure that terminal arrangements are structured in a logical and functional manner. The terminal assignments can also be transmitted immediately to the electrical design group without the delays normally encountered when vendor drawings are received.

Control Panel Interface

For supervisory control systems, the computer's multiplexer must interface with an analog control panel. This interface can be accomplished with point-to-point wiring performed in the field. To minimize installation costs, the multiplexer racks and control panels can be wired with receptacles, and multi-conductor cable assemblies can be used to make the final connection. However, coordination between different vendors on matters of signal commoning, fusing, shielding, and pin assignments can be very difficult. Control panel interface with a large computer system must be designed carefully. Errors discovered once the computer rack and control panel fabrication are underway can be very costly to correct.

Clear guidelines for grounding and shielding are not always available, especially in large customized computer systems. Grounding practices are further complicated in supervisory control systems when the computer must interface to an analog control panel. In such cases, an integrated approach to grounding must be followed for both the computer system and the analog instrumentation. A single common ground reference must be established for both the analog and digital systems. All signal interfaces between the two systems must be analyzed to ensure that ground loops are not inadvertently introduced into shields and signal returns. Analog and digital input cards for the computer system should be selected with adequate input filtering, isolation and common mode noise rejection characteristics.

Guidelines for grounding and shielding equipment racks, power supplies, signal lines, and cable shields are listed. They should be used in conjunction with the vendor's recommended practices.

1. Each equipment rack should have an instrument ground bus bar. The bus bar should be electrically isolated from the rack itself. The 115 VAC neutral, green ground wire and other chassis or safety grounds *must not be connected to the instrument ground.*
2. Due to the vendor's hardware design, some computer systems and most computer peripherals have the instrument or digital ground connected to the equipment chassis. Ground connection is made through the green safety wire. Care must be taken to ensure that the green wires are connected to an insulated bus bar in the power distribution panel and then grounded at the power service entrance only. Most peripherals plug into 115 VAC outlets. A green wire should be provided for each outlet and grounded in the same manner.
3. Equipment racks should be grouped together and a master instrument bus bar provided for each group. Each rack bus bar should be solidly bonded to the master bus with a single #4 insulated cable. In turn, each master bus bar is solidly bonded to a single ground connection of one ohm or less to true earth. Separate 2/0 insulated cables should be used for each master bus bar. Daisy-chaining is not recommended.
4. DC power supplies used to power instrumentation, digital electronics and field transmitters should be referenced to the instrument ground in the associated equipment rack.
5. Individually shielded twisted pairs with an overall cable shield should be used for field analog signals, and for cable assemblies between the computer mux racks and the analog control panels. Digital signals need not be individually shielded. However, individual twisted pairs are recommended for field digital signals; they are optional for interconnecting cable assemblies. Analog and digital signals should be assigned to separate cables.
6. Signal returns must not be grounded anywhere along their length except where the final connection is made to the power supply common in the rack.
7. Cable shields must be grounded at one end only. Preferably the shields should be grounded in the rack where signal power is supplied.
8. When available, optical isolation should be specified for all equipment interfacing to the data highway for distrib-

Figure 13.9. Analog and digital multiplexer chassis, microprocessor controllers, analog instrument cards, power supplies and terminal blocks are organized modularly in equipment racks for maximum reliability, ease in installation, and to facilitate troubleshooting and repair. (Courtesy of Honeywell, Process Control Division)

Care should be taken to ensure that all necessary signals are brought from each analog controller to the computer mux racks. As a minimum, the process variable, setpoint feedback, controller mode, and the computer setpoint are required for each loop. Cascade loops also require the master controller output signal, the auto/cascade relay status, and the digital output from the computer mux rack to the auto/cascade relay. These signals are necessary for bumpless transfer of control for all modes of operation. Should the computer fail, the setpoint must remain at its last value until the controller is switched from computer control. A special output board in the computer system or a memory update card in the control panel is required for this purpose.

An integrated analog controller and process computer package simplifies a supervisory control system installation considerably. All necessary interfacing is built into the equipment. Connecting the process computer system to the analog instrumentation is a matter of adding the vendor's standard interface hardware and cable assemblies.

Grounding and Shielding

A single ground reference should be provided for all logic, instrument, and shield grounds. The vendor normally provides a system of bus bars and ground cables which provide the signal return and ground reference for the central processor and its peripherals. All vendor-recommended grounding practices should be followed.

uted control or multiplexer systems. Optical isolation should be specified for all field digital inputs.

Field Wiring Interface

Usually the vendor supplies the necessary terminal blocks to interface the multiplexer hardware with the field wiring. The terminal blocks mount on the edge of the multiplexer I/O boards or in separate termination panels. Standard cable assemblies connect the terminal panels with the mux chassis. In either case, the I/O cards can be removed without disconnecting the field wiring.

Special terminal arrangements are necessary when loop fuses, zenier barriers, or interposing relays are required. In order to provide a clean interface with the field wiring, Wiedmuller-type terminal blocks can be assembled in the rack. Special fused blocks with blown-fuse indicators can be used where required. However, additional wiring is required to connect the terminal blocks with their associated relays, zenier barriers, and multiplexer equipment. The field wiring can be terminated directly on the relay or barrier terminals, but on large systems, this may cause confusion and errors in field wiring design and installation.

Most digital outputs to the field require interposing relays with contacts rated for the applicable load. For 115 VDC service to inductive loads, the relay contacts must be oversized and specifically rated for the required DC current and voltage levels, both for continuous current flow and for surge (due to inductive kick when the contact opens). It is also a good practice to fuse each contact output separately to prevent a short on one output from affecting others. Fusing for analog and digital inputs is usually provided by the vendor on a per-card basis.

Signal Power and Commoning Requirements

If the field wiring is connected directly to the computer system multiplexer, then the vendor normally provides the necessary DC power for energizing transmitters and sensing contact inputs. Digital outputs may be dry contacts or energized. Motor status inputs normally receive sense power from the mux racks, while start/stop outputs are dry contacts. On digital inputs and outputs, it is advisable to avoid commoning. Run two wires for each contact input and output. This provides a clean interface and avoids wiring problems when connecting to other control systems.

With supervisory systems, the multiplexer racks are connected to a control panel, usually provided by another vendor. Problems can occur concerning which vendor supplies sensing power and how signal pairs should be commoned and fused. As previously stated, it is best if signal commoning is avoided. Most digital input cards service a group of 8, 12 or 16 digital inputs with a single common return. In this case, commoning should be done at the input card only. Sense power should be provided in the mux racks for digital inputs, and dry contact outputs provided to the control panels. It is important that the vendor wires sensing power into each signal loop inside the cabinet. Otherwise, additional terminals and wiring are required when making field connections.

In all cases, interface wiring with field instruments and other vendor's control systems must be planned carefully to ensure proper operation. Criteria related to signal power, commoning, and fusing must be established, documented, and transmitted to the affected vendors and design disciplines.

Data Highway Communications

Communication requirements must be established for remotely located equipment, such as remote terminal units (RTUs) or microprocessor controllers. Distributed control and data acquisition systems normally communicate over twisted pair or coaxial cable, which functions as a data highway. Two separate cables are routed through the plant to enhance system reliability. The data highway links the remote units together in a daisy-chain, loop, or radial pattern (Figure 13.10). Total cable length is limited to about 5,000 feet for coax and about 20,000 feet for two twisted pairs. In general, these maximum lengths are for the total length of cable of the data highway, including branches and loops.

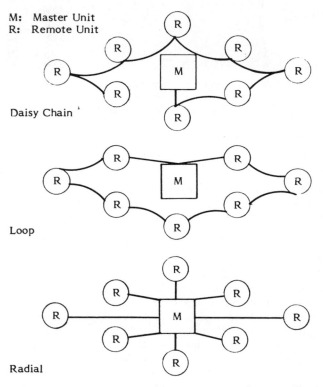

Figure 13.10. The data highway associated with a distributed control or data acquisition system can be routed several ways in the plant. Daisy-chaining (a) is most common and usually results in the least amount of cable required. Closing the data highway on itself to form a loop (b) increases the system's reliability by providing a separate path if the cable is severed. In both cases, a redundant highway is normally routed through the plant separately from the primary highway to enhance the security of communications between the master and the remotes. A radial pattern (c) provides the greatest reliability, and allows greater scanning speeds for large systems.

Figure 13.11. CRT stations with operator keyboards are grouped together to form an operator console. Hard-wired instrumentation, such as pen recorders, pushbutton stations and annunciators can also be incorporated in the console's design. (Courtesy of Honeywell, Process Control Division)

To communicate with remote units over longer distances, modems are required. Communication is generally provided by dedicated telephone circuits in the plant. For pipeline control systems and dispatch operations, microwave channels normally link the central computer with its remotes. In either case, transmission rates must be established and circuit grades specified. It is usually the responsibility of the client's communications department to provide the necessary telephone circuits. Transmission rates should be limited to a maximum of 2,400 baud to minimize demand on the communications channel and improve system performance.

Operator Console Design

The operator console normally consists of one or more interactive CRT stations with operator keyboards (Figure 13.11). Depending on the client's preferences and the vendor's standard options, special features may be required. Pushbutton stations, pen recorders, or hard-wired annunciators are sometimes mounted on the operator console. A separate control console can also be installed next to the CRT station to house the hard-wired instrumentation. In general, dedicated instruments should be avoided. The CRT console provides a much more efficient and cost-effective operator interface with the process.

The computer system vendor may offer light pens, joy-sticks, or track-balls to position the cursor on the CRT screen. With some systems, cursor control provides the main operator input into the system. The choice of cursor control is mostly a matter of personal preference. There are arguments that light pens are too fragile, and that joy-sticks and track-balls are susceptible to coffee spills. In either case, the vendor usually provides a standard keyboard with pushbuttons that facilitate cursor control. On the other extreme, the vendor may supply a bank of dedicated pushbuttons which allows the operator to select a particular display, table or report with a single keystroke. Most systems utilize a combination of keyboard entry, cursor control, and dedicated pushbutton functions.

Printers are normally incorporated in the operator console design for logging and reporting purposes. On larger systems, it is desirable to dedicate one printer to alarm logging. Low speed,

single character typers are adequate for this purpose. The printer can be free standing or located on the console (table top).

Depending on the vendor's limitations, modems may be required for remotely located operator consoles. This does not present a problem for the printers, but could significantly reduce the CRT response time. Distributed control systems have less severe distance limitations, since the operator console can be located anywhere on the data highway.

Backup Systems

Large computer control systems that provide critical process data acquisition, alarming and control functions may require a backup system to ensure adequate system availability. The overall system configuration is usually established in the proposal stage, but the final hardware and software functions must be determined.

The most susceptible components in a computer system are the central processor and its primary peripherals, such as disc units, magnetic tape transports, display generators, and auxiliary control and bus interface devices. These units should be backed-up with a separate computer system. Generally, the multiplexer equipment is shared by both the primary and backup systems, since the mux racks are modular in construction and failures are limited to the particular rack involved.

Communications between the two computers and fail-over philosophies depend on the vendor's standard configuration and the client's preferences. The data base can be shared by the two systems, using a common bulk memory storage device such as a disc unit. Theoretically, this allows the backup system to upload the current process configuration and pick up where the failed system left off. Sometimes the hot standby approach is used, where both systems have access to the process interface devices and maintain their own data bases. However, only one system is allowed access to the operator consoles and to generate control outputs. Should the primary system fail, the backup system is instantly available to take over.

In practice it can be very difficult to implement a satisfactory fail-over system that works. One approach is to program each computer to monitor the other and seize primary control when certain conditions occur. Depending on what failure conditions are selected and what interlocks are provided, the two computers may be locked in an endless power struggle with control flip-flopping from one system to the other. However, if too many interlocks are incorporated in the fail-over logic, then neither system may respond when one system fails.

Generally, the vendor has a standard approach to dual computer control that has been proven to work effectively in past applications. The customer should thoroughly test the fail-over system. All means of failing the system should be attempted to determine the system's response. Any malfunctioning of the fail-over system should be evident as a result of the testing, and the necessary changes can then be made.

In distributed control systems, microprocessor-based controllers provide primary control of the process. Controllers used for critical service should be backed-up with either a manual control station or another controller. These backup systems can be simple patch panels, allowing the operator to switch in a backup device or controller for the failed unit. It is not necessary to provide a backup controller for each primary controller. Some vendors offer automatic backup of their microprocessor controllers. Should a controller fail, a backup unit is switched in place automatically. All process wiring, data highway connections, device addressing, and loop configuration are transferred to the backup controller. The operator sees no interruption to plant operation other than an alarm indicating which highway device failed.

Logic and Sequencing Requirements

The logic and sequencing requirements for programmable controller systems should be finalized, documented and forwarded to the vendor. It is important to firmly establish the logic at the time the system specification is written. The type of programmable logic controller and the amount of memory required are determined by the program's size and complexity. Finalizing the logic allows the vendor to determine if sufficient memory is available to hold the program. Twenty percent spare memory should be reserved for future sequencing requirements.

If the logic is relatively simple, a ladder diagram can be drawn to document the logic. Usually the programmable controller has restrictions as to the number of contacts, branches and coils in each rung. The latter diagram should reflect these restrictions so that the program can be loaded directly.

Complex sequential logic with interlocks and branch operations is very difficult to lay out in ladder diagram form. Instead, tables should be developed depicting interlock criteria and final control element positions for each step of the sequence. The logic can also be written with input conditions, interlocks, timing functions, branch decisions, and output commands identified in narrative form. However, with this approach, it is difficult to completely define the logic. Every combination of input conditions and operator commands must be anticipated and contingency logic developed for them. On large programmable controller systems, this task becomes unwieldly. Usually the system check-out reveals most of the loopholes that were missed in the programming stage.

Logic flowcharts are best suited for defining sequential operations. Because flowcharts are sequential in nature, it may be difficult to represent static logic, such as interlocks or Boolean expressions. However, once flowcharting is mastered, it is possible to completely identify all logic requirements in a complex sequential operation. Logic flowcharts can be readily interpreted by programmers and technical personnel, and can clearly communicate the controller logic requirements to other affected groups, such as the electrical or process design disciplines.

Special Control Systems

Special functions, such as plant shutdown logic, may be provided by a separate control system. In this particular case, hard-wired relays, solid-state logic, or a programmable logic controller can be utilized. If logic requirements are small, the computer vendor can implement the system with hard-wired relays. It is advisable to avoid distributing the logic throughout the equipment racks. All contact inputs and outputs should be brought to one area and all points labeled. Complex logic or sequencing systems should be located in a separate rack. A programmable logic controller should be used to minimize cost and enhance system flexibility.

It may be advantageous to subcontract the logic system to another vendor. However, the logic system should interface to the computer control system so that the operator is aware of the PLC's operations. The interface between the logic and computer systems should be defined and communicated to the two vendors, to be incorporated in their final system designs.

A mimic board will be required for all but the simplest programmable controller systems. A mimic board provides the operator with information on valve positioning, equipment status and the current sequence operation being performed. Pushbuttons, thumbwheel switches, and digital readouts provide additional operator interface. When possible, the mimic board should provide a graphical representation of the process. The mimic board can be mounted in the control panel or on the CRT operator console if associated with a computer control system. It is also possible to hard-wire the PLC's I/O to the computer system and depict its operation using the computer's graphic display capabilities, if available.

Communications with Other Computers

A large refinery or chemical plant usually requires many digital control systems for efficient operation. Separate computer systems can be dedicated to major process areas or units. Many analyzer systems use a computer to provide centralized control, process raw data, and format the results for logging purposes. Multiplexer systems are used for display of thermocouple points and control of pumps, motors, and valves. Tank gauging systems and royalty meter stations often utilize microprocessors or mini-computers in their operation.

To provide coordinated control of the plant, the process computer system may be required to communicate with other computer or microprocessor-based systems. An RS-232C interface is normally specified for digital systems within 50 feet of each other. The RS-232C interface defines the connector type, pin assignments, voltage levels and functions necessary to provide a compatible hardware interface between different makes of digital equipment. For longer distances, standard 20MA serial current loop interfaces or modems are required.

The type of interface defines hardware requirements and signal levels. However, a protocol must be established between the two computer systems. The protocol establishes data word structure, parity, interrogation commands and responses, data base structure, tag numbers, and other criteria necessary for reliable communications. It can be a real problem to establish agreement between two vendors on the protocol. Each vendor has his own way of structuring transmissions, or preferences as to which system shall be active or passive. In general, the larger, more flexible computer system should adapt its protocol to the needs of the more limited multiplexer or data acquisition system. In any case, a detailed protocol must be developed and agreed upon. The project engineer may be required to meet with representatives of both vendors and "lock the door" until agreement is reached.

Once the protocol is established, it is a good idea to test it. This can be accomplished by using a test computer to emulate the other digital system. The two systems can be linked while still in the factory, using modems, acoustic couplers, and a standard telephone set. A long distance connection is made; the telephone receiver on each end is placed in the acoustic coupler; and communication is established between the two systems.

Interface with Other Design Disciplines

Information must be communicated on a timely basis to the affected design disciplines associated with the computer system project. These include the electrical, HVAC (heating, ventilation and air conditioning), and architectural staffs involved in the plant design and construction.

Electrical Design Group

Power requirements for each major component of the computer must be established. The number of 115 VAC and 220 VAC circuits must be identified. Voltage, maximum continuous current, inrush current and duration, and load power factor are required for each circuit. This information should be readily available from the vendor. If a large number of circuits are involved, the power requirements should be documented by the vendor in a written communication. This information is normally used to size the UPS (uninterruptible power supply) that provides power for the computer system and plant instrumentation. The vendor should be held to the stated power requirements so that the UPS can be sized accordingly.

Requirements for power filtering, isolation transformers or motor generator sets should be identified and communicated to the electrical design group. These requirements are determined by the vendor, who usually is in a position to supply the necessary hardware. If a UPS is incorporated in the design of the plant, then a motor generator is not required. However, the vendor may recommend power filtering or isolation transformers to eliminate harmonics generated by the UPS inverter.

Identify and locate all 115 VAC receptacle outlets in the computer room for both UPS and non-UPS loads. Power, ground and field wiring connections to the computer system must be communicated to the electrical design group. Grounding configurations must be established to ensure that system performance is not degraded by noise introduced through faulty grounds. Signal and power cables should be routed in separate trays, with high and low level signals assigned to dedicated cables. Interconnecting cable quantities and sizes should be forwarded to the electrical group, so that cable routing can be determined and wall penetrations located.

HVAC Design Group

Operating ranges for temperature, humidity and suspended particulates must be established for the computer system. Maximum allowable concentrations of H_2S, SO_2 and other corrosive elements should be identified. Heat loads for each component are required so that air conditioning equipment can be sized and bought. This information is normally supplied by the vendor.

Architectural Design Group

A floor plan for the computer equipment, instrument racks, and operator consoles should be laid out. Consider front and rear access, grouping of related equipment, future expansion and aesthetics. Room should be provided for work areas, book racks, and shelving for tape reels and disc paks. For large systems, it is desirable to have a maintenance area next to the computer room. Operator consoles should be conveniently located in the control room, with consideration given to other control panels and instrumentation. Maximum cable lengths must be kept in mind when locating the computer equipment. Generally, the central processor, peripheral controllers and I/O bus hardware are located next to each other. Line printers, mag tape units or disc drives must be located within 50 feet cable length of the central processor. Operator consoles and other CRT devices usually must be within 200 feet cable length of the central processor.

Normally, a raised floor is used in the computer and control rooms. If possible, the floor grid should be laid out with the equipment rack locations in mind. Cutouts in the floor can then be minimized by lifting up a row of tiles and centering the cabinets over it. This provides easy cable entry to the equipment and allows the raised floor to be used as a plenum for directing air conditioning to the hardware.

Suitable locations must be established for remote terminal units and distributed microprocessor controllers. If possible, the equipment should be located in an air-conditioned building. Electrical substations, analyzer houses, and remote control buildings are suitable locations. NEMA 12 enclosures purged with clean, dry instrument air should be used for all outdoor installations and for dusty or corrosive environments indoors. A sun roof should be provided for outdoor installations in hot climates.

Design Review, Changes and Cost Control

The successful computer system project requires close coordination with the vendor, especially during the system design and review stages. Communication is accomplished with drawings and specifications issued by the vendor, which detail the exact hardware design and layout. The drawings are submitted to the project engineer for review and approval. Corrections are made and the drawings submitted to the vendor for re-issue and final approval.

Refinements in system design or customer requested enhancements often result in cost overruns. If the system was thoroughly specified during the proposal stage and engineering finalized prior to hardware assembly, then changes should be minor and cost impacts minimal. However, if system configuration and desired performance were not completely defined during the proposal stage, or if major scope changes have been made after the release of the purchase order, then significant costs over the original purchase order amount may be incurred. Some of the costs will be valid; some will be inflated and unreasonable. In such cases, the issue will ultimately come to a point where negotiations between the contracting firm, vendor, and client are required. Once the negotiations are concluded, system design and cost should be finalized and future changes avoided.

Vendor Drawings

As a minimum, the vendor should submit the following drawings for approval:

1. Dimensional drawings of equipment racks, cabinets, consoles, teleprinters, and display panels.
2. Rack layout drawings identifying electronic components, power supplies, relays, junction boxes, circuit breakers, terminal strips, panels, bus bars, mounting hardware, panduit and cable routing for each equipment rack and console.
3. Wire and cable lists for all point-to-point wiring, internal cable assemblies, and interconnecting cables between equipment racks and consoles.
4. Field terminal drawings, identifying all process and shield drain connections by instrument tag number.
5. Schematic drawings identifying AC and DC power distribution, fusing, circuit breakers, junction boxes and their locations in the rack. All power, safety ground, and instrument ground connections should be identified on these drawings.
6. Manuals describing system operation and diagnostics (usually applicable to custom systems).

These drawings should be thoroughly reviewed (1) to ensure inclusion of all hardware, design features, and options detailed in the system specification and purchase order, and (2) to ensure compliance with all design criteria and hardware assignments established during the system engineering phase. Copies of the marked-up drawings are submitted to the vendor for revision and final approval.

The vendor should provide a schedule and a timetable for the drawings that will be submitted for approval. The project engineer should maintain close control of vendor drawings that have been submitted, reviewed, and approved for construction. It is a good practice to identify changes on each drawing and assign revision numbers. When a large number of drawings are involved, it is important to ensure that only one drawing revision is outstanding at any time. On large customized systems, it can become an impossible task to review the system design when several versions of the drawings are outstanding. This is especially critical when the affected drawings contain information required by other vendors or design disciplines.

Design Reviews

On large, customized systems, it is advisable to meet with the vendor and client on a periodic basis to review system design and determine the status of the project. Schedules, design changes, change orders, test procedures and spare parts requirements are some of the items usually discussed. It is important to maintain close contact with the vendor to ensure that all system specifications are met, and that the project stays on schedule.

Whenever possible, design reviews should be combined with a visit to the vendor's plant. Early in the project, the vendor's staff requires much information on system configuration, equipment layout and terminal assignments. In most cases, it is more advantageous for the project engineer to meet at the vendor's facilities and work directly with the personnel to provide them with critical information. The trip also allows the project engineer opportunity to inspect the facilities and meet the key personnel involved with the project.

Inspection Trips

After system assembly has begun, inspection trips should be made on a periodic basis to see that the hardware meets the system specifications and design criteria. Problem areas unnoticed by the vendor can be identified by on-the-spot inspections, and then corrected. This is especially important for customized systems, where some aspects of the design are not specified in detail. Usually the project engineer establishes functional requirements and attempts to specify the customized hardware or item as completely as possible. However, what the project engineer or client envisions and what the vendor builds are not necessarily the same. Timely inspection trips can ensure proper design and prevent costly changes later in the project.

Design Changes

Design changes and the cost adjustments normally associated with them are inevitable with complex digital and computer control systems. These changes can be minimized by careful planning and detailed specification of the system during the bid evaluation and project kickoff. The extent of the design changes and cost overruns is usually determined by:

1. System size, complexity and the extent of customization required.
2. Degree of detail identified in initial system specification.
3. Thoroughness in bid tabulation and vendor evaluation. In particular, identification of services that are to be provided by the vendor, such as data base generation, loading and testing.

4. Enhancements, changes or restrictions imposed by the client after the project is underway.
5. The vendor's past track record on other projects in the areas of schedules, additional costs, design compliance, willingness to accommodate changes, installed system performance, system support, and services.

Design deficiencies and changes should be identified as early as possible, and cost studies initiated to alert higher level management of impending problem areas and possible cost overruns. Minor changes are normally handled by written correspondence and followed up with a change order. However, on large complex systems, these cost adders are not always resolved through the normal channels, and tend to accumulate. Ultimately, a point is reached where the total cost overrun becomes unacceptable and a meeting must be scheduled to negotiate the changes and associated price increases.

Cost Tabulation

Negotiation with the vendor concerning the cost of the project is best performed from a position of knowledge and strength. A detailed breakdown of hardware, software, services, and prices as originally quoted by the vendor is needed. This is best accomplished by insisting on a paragraph by paragraph response to the detailed specification *during the bid evaluation stage*. Subsequent requests for information after the contract is awarded may not accurately reflect the true price structure. The vendor should be required to provide detailed breakdowns of any additional pricing for design changes occurring after the release of the purchase order.

All correspondence with the vendor should be gathered and organized in chronological order. A detailed tabulation of all system costs and price adders should be prepared. Each major component, option, design change and service to be provided by the vendor should be listed separately. Original purchase order prices and cost increases, both contested and uncontested, are broken out for each item. The difference between the contested and uncontested costs for each item is calculated and a total prepared for the entire system and services provided. From this tabulation, the project engineer can determine exposure (percent increase in costs over original purchase order amount), and decide what action is required.

Estimates should be made for all questionable cost adders. Extrapolations can be made using the detailed price breakdowns for similar items or services provided in the original quotation and any subsequent pricing. These estimates should be incorporated in the cost tabulation to provide a target or goal for negotiations.

Negotiations with the Vendor

Once all correspondence and pricing information have been gathered and correlated, a meeting should be set up with the vendor, contracting firm, and the client. The meeting should be conducted at the contracting firm's facilities. Prior to the meeting, the vendor should be provided with a list of problem areas, an estimate of the cost differential, and be informed of the seriousness of the problem. In general, it is a good practice to pursue every possible concession from the vendor and to present a well-padded cost differential in the initial stages of the meeting. Usually the vendor inflates the pricing for changes once the contract has been awarded, and the only satisfactory approach is to pad the cost differential to be negotiated.

The negotiations normally open with a presentation of the problem by the project manager or purchasing agent. Each cost item is scrutinized and questioned when necessary. The vendor responds to each point with a defense of the quoted prices. It is at this time that the project engineer draws from the calculations and cost tabulation and presents his case. The vendor should not be allowed to view the project engineer's reference material except in an abbreviated and edited format. Also, some of the arguments should be reserved for later sessions.

Initially, both sides present their case, and some resolutions are reached. Cost reductions are achieved by specifically identifying the required material or services in the original quotation or in subsequent written clarifications to the bid. In areas where an impasse is reached, a relaxation of the system specifications or elimination of nonessential items can result in cost reductions. It is also helpful to present a well-padded counter bid to the vendor at the beginning of the meeting. The counter bid should be itemized and include uncontested cost adders to demonstrate its validity to the vendor. However, for this and other tactics to work, representatives from the contracting firm and the client, with the authority to cancel the purchase order in question and influence the awarding of similar contracts in the future, must be present.

These detailed discussions on the various cost adders can go on indefinitely. In fact, it is to the vendor's advantage to stay at this level of negotiations. As long as a resolution is not reached, the original pricing stands. To prevent this situation and to maintain control of the meeting, it is essential that a chairman be appointed at the beginning, preferably a senior buyer or project manager. A useful tactic that the chairman can exercise is to recess the meeting and call for a caucus. The point here is to break up the fruitless discussion and allow each side to regroup. This represents the end of the first stage in negotiations. It is important to leave the detailed discussions and begin negotiating major items of the contract. Ultimately, an agreement on a new bottom line figure for the entire system and all services must be reached.

For small, standard systems, the negotiations may end at this point because the cost differential is usually small and agreement can be readily reached. However, for large, customized systems, costs and disagreements are on a much greater scale. As a result, the negotiations move into a second phase where the bottom line is negotiated. This is mainly a "horse trading" situation where the negotiators dicker, make threats and call bluffs. Again, it is absolutely essential that the individuals present have the necessary stroke to make or break the contract.

Certain tactics can be utilized by the negotiators to strengthen their position. The negotiators can threaten cancellation of the project if a satisfactory agreement is not reached. However, this usually is not feasible if the project is well underway, due to cancellation charges and schedule setbacks. The vendor is aware of this, but cannot be certain how resolved the customer would be if the issue is pressed. A much more effective ploy is to inform the vendor that future contract awards will be pending the outcome of the current computer system project. This should be the central theme of the negotiations, but real emphasis should be reserved for the final sessions. Another effective tactic is to delay any scheduled partial payments. The negotiation session is a delay in itself, as a change order is usually pending the outcome of the meeting. This approach can be effective against the small firm with limited financial resources. However, if the vendor is much larger than the contracting firm or client, the tables can be reversed, since the customer may be the one making high interest payments.

Depending on the amount of money involved, the negotiators must be firm in their demands and should initially strive for every concession possible. Compromises are to be made in the latter stages of negotiations; not in the beginning. A "nice guy" approach or appeals to the vendor's sense of fair play usually have the opposite effect desired. It must be remembered that most companies are out to make money and will usually seize any opportunity to do so. A calm, firm, and logical approach is best. Use of caucusing, when required, is highly recommended. If necessary, recess the meeting until the following day to allow each side time to think over their position and determine what concessions are required.

Change Order

The negotiations should be concluded with a written agreement on the final cost of the computer system and all services required. Every effort should be made to anticipate future changes and incorporate them into the agreement at this time. A change order should be issued immediately to reflect the agreement and authorize payment of additional funds.

The project usually runs smoothly from this point, but any significant changes could jeopardize the results of the negotiations and open the door for more cost overruns. Changes should be avoided as much as possible. Any additional price increases, if not resolved immediately, should be allowed to accumulate. Successive negotiation sessions may be required if the cost overrun becomes significant again. However, this should not occur if system design has been finalized in the initial negotiation meetings.

Software Configuration

The digital or computer system must be programmed with the process information, control algorithms and operator interface instructions necessary for proper operation. Without the configuration data, the control system is incapable of performing its intended functions. System performance and pay off are ultimately determined by the software that is implemented.

Software configuration is normally initiated after the system has been bought, and is developed concurrently with the system engineering and design. For small, standard systems, the bulk of the software specification can be done in the latter stages of the project. Normally the point addresses and hardware assignments need to be defined early, as this information is required for equipment layouts, wiring lists, and field terminal assignments. On larger systems, it is essential to assign the necessary personnel and start the software configuration early, due to the

tremendous amount of data to be gathered and coded. Sufficient time must be allowed for loading and debugging the configuration data prior to acceptance testing.

Data Base Generation

Most microprocessor and computer-based control systems come with standard fill-in-the-blank software for data acquisition, process control, alarming, and operator displays. Data base generation requires identification of the necessary process and control information for each point, and the filling out of appropriate forms. Information for the data base is gathered from many sources, such as piping and instrument drawings, instrument loop drawings, instrument specification sheets, instrument indexes, and field terminal drawings.

The first step in developing the data base is to identify all process inputs and outputs by type (analog and digital). Each process input and any associated outputs are grouped together and called a "point." Each point record is built by coding the configuration data, such as hardware addresses, input conditioning requirements, filter constants, alarm limits, control algorithms, tuning constants and multiple loop configuration on to standard forms. Transmitter ranges, engineering units, deviation limits and tag descriptions are required for each point for display purposes.

Data base generation for computer systems with a thousand points or more requires a major effort in data collecting, compiling and coding. Personnel familiar with computer systems and their software requirements should be dedicated to the task. Process and instrument information must be organized for each point in the system. It is important to ensure that the latest data is available and new points identified. Decisions must be made as to what input processing, control algorithms, alarm types and other software functions are required for each point. Data base forms must be filled out, reviewed, and updated. Depending on the type of control system, the forms are either punched on cards or entered into the system directly through a CRT console or programming device. The data base is then loaded, and errors diagnosed by the system are corrected. Normally, the data base is in a constant state of revision, which continues through acceptance testing, commissioning, start-up and system operation.

It may be advantageous to have the vendor perform some or all of the data base generation, including coding, keypunching, loading and debugging. This requirement should be identified early enough in the project to be priced by the vendor during the competitive bidding stage.

Displays

Distributed control and process computer systems normally incorporate a CRT console in their design. Process indication and control are provided through various types of displays, which are generated by standard software supported by a process computer, or are integral to the CRT console. In general, the displays include plant overview, group and detailed formats (Figure 13.12). The user assigns points to various groups to perform indication, alarm and control functions. Once points are assigned to a particular display, other information

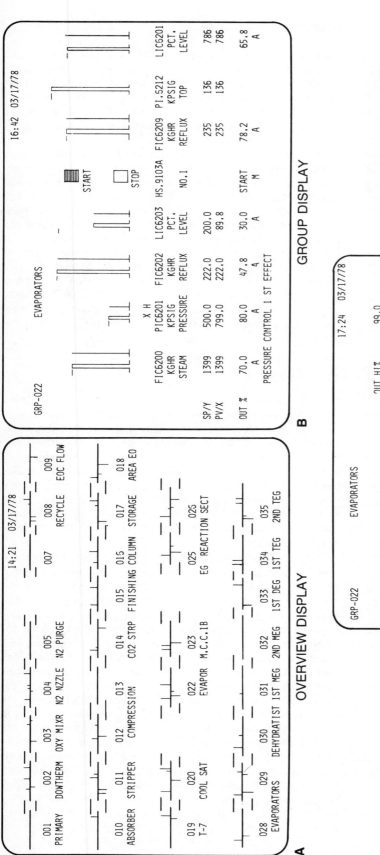

A   OVERVIEW DISPLAY

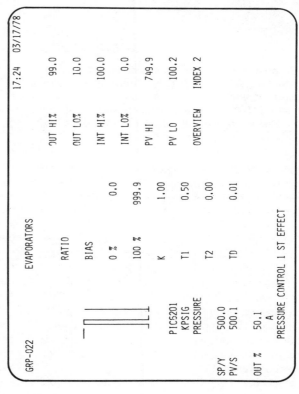

B   GROUP DISPLAY

C   DETAIL DISPLAY

Figure 13.12. Most microprocessor-based distributed control systems provide standard CRT displays for operator interface with the process. For example, Honeywell's TDC-2000 provides an overview display (a), which allows monitoring the operation of the entire plant, group displays (b), which provide the primary indication and control functions for a group of eight loops, and detail displays (c), which provide tuning constants and configuration data for a single point.

Figure 13.13. Graphic displays provide a customized operator interface that can be tailored to suit the needs of the user. Properly designed displays can offer the operator a greater understanding of plant operations and allow more efficient responses to process upsets. (Courtesy of The Foxboro Company)

pertaining to that point can be assigned (such as controller modes, alarm limits, and scaling factors). The vendor normally supplies standard forms for display assignments. The forms are then entered directly into the operator console in an off-line mode of operation.

Some distributed control systems and most process computers offer a graphic display capability (Figure 13.13). Plant flow diagrams, schematics, one-line diagrams and other customized displays can be built. The displays are first laid out on special grid paper which establishes the field size of the CRT screen. Vessels, heaters, pumps, valves and other equipment are drawn with interconnecting process lines. Process data indication areas are defined for the display. Decisions are then made as to color, intensity, blink action and character size for each shape and process data readout.

It is a good practice to keep the displays simple and avoid cluttering the screen with a lot of alphanumeric information. On some systems, tabs can be set on the display, allowing convenient paging to another display. This feature can be used on an overview display to provide a zooming effect. More information can be presented on the screen by positioning the cursor in the area of interest and using the tab function, which addresses a detailed display of the selected area.

Once the displays are laid out on paper, they are entered into the system through the CRT console. With limited graphic systems, shapes and lines are drawn using a special character set provided by the CRT display generator. With a full graphics system, circles, vector lines and other shapes are drawn by defining shapes mathematically. The computer actually draws the display using preprogrammed shape parameters, such as end points, slope, center, and radius.

Reports

Typical plant operations require the capability to format and print out process data, alarms, software configuration and system diagnostics. Most systems can interface to a printer, which provides a hardcopy of alphanumeric information displayed on the CRT console. Process computer systems are capable of storing historical data on selected points and of generating reports on both a periodic and a demand basis.

The vendor normally provides standard report formats, such as shift, daily, monthly, yearly, production summary, alarm history, and operator action reports. Report types, description headings, tag numbers, point order, and report scheduling are specified on standard fill-in-the-blank forms. Historical data points must be identified in the data base, along with sample periods, process averaging, and data storage requirements.

Custom reports are normally generated on computer systems that support a high level language, such as FORTRAN. The computer's operating system must allow FORTRAN programs access to the data base and to historical files stored in bulk memory. Special formatting, point processing and data manipulation can be easily implemented using the high level programming capabilities that a process computer offers.

Custom Software

In many applications, the vendor's standard fill-in-the-blank software provides the necessary data acquisition, display and control functions. Most distributed control systems offer 30 or more control algorithms, multi-loop control, input conditioning and processing, alarming, operator displays, multi-variable trending and logging as standard features. Process computer systems offer an even larger selection of preprogrammed functions. There may be special processing and control requirements, however, that cannot be satisfied with the vendor's standard software.

Optimized operation of major equipment items and process units can require complex control schemes involving simultaneous solution of state variable equations, Z-transform transfer functions, and other forms of process modeling and multi-variable control. Material and steam balances normally are required for efficient operation of the plant. There may be requirements for special processing of plant information and generation of calculated variables to provide more meaningful data to the operator. In these situations, it is necessary to develop custom software.

Normally, the project engineer and client work in a joint effort to develop the required software. In some cases, the client may wish to develop custom software after the computer system is installed and operational. With large, customized systems, where special software is necessary for plant operation, it may be expedient to contract out some of the software development to the vendor. Functional requirements, process models, algorithms, block diagrams, and flow charts should be developed by the user and submitted to the vendor. Again, it is important to identify any software development requirements early in the project and to obtain firm pricing from the vendor.

Maintenance and Spare Parts

Maintenance and spare parts requirements should be addressed during the bid evaluation stage of the project. While the vendor may not quote the final cost for the desired level of support, a base line should be established from which future pricing can be developed. In general, the bid inquiry should instruct the vendors to list and price their recommended spare parts for start-up and one year of operation. If the computer system is to be installed at a remote site or overseas, then the vendor should demonstrate his capability to maintain and support the equipment. Prices, terms and conditions should be quoted for on-site maintenance.

Normally, the vendor can recommend the required spare parts. The type and quantity of spare parts stocked will be determined by the system's complexity, the reliability of the hardware, and the extent of critical functions performed by the system. The vendor may stock parts locally, which reduces spare parts requirements. As a minimum, spare circuit boards should be stocked for all major items of hardware. In remote locations, or where the system operation is critical, it is also advisable to stock whole components, such as the central processor unit, controllers, CRT monitors, disc drives and typers.

The vendor should issue the spare parts list early in the project to allow purchase and delivery of spare parts in time for plant start-up and operation. Each part should be identified with a part number, description, and a reference to a maintenance manual. The manual should have an illustration of the part, identify its function, and describe how it is used to repair the system.

Customer Acceptance Test

System testing is performed at several levels by the vendor and the user. The vendor performs numerous tests on each equipment item and its individual components as part of a quality control program. Additional inspection and testing takes place as the system is assembled. Finally, the vendor tests each input/output and exercises all components of the system. Typically, problems are identified in the system hardware and software during the vendor's factory test. After all the bugs are worked out and the vendor is satisfied that the system is fully operational, the customer is invited to conduct an inspection and acceptance test at the factory.

Equipment Inspection

All equipment racks, operator consoles, peripheral devices, and remote terminal units should be thoroughly inspected. As a minimum, the following items should be checked:

1. General appearance and workmanship.
2. All components specified in the purchase order for proper installation and labeling.
3. Wiring bundling and Panduit runs for neat installation.
4. Tight wire connections.
5. Cable connectors fully inserted and secured.
6. Correct tagging for all wiring and cables.

7. Operation of all cooling fans.
8. Condition of all light bulbs, LED indicators and fuses.
9. Terminal blocks and connectors labeled properly.
10. Defective relays and relay sockets.
11. Operation of all power supplies.
12. Power and instrument grounds for tight connections. Also check for electrical isolation between instrument and power grounds with ohmmeter.
13. Operation of doors and latches.
14. Gaskets and sealing of NEMA-type enclosures.
15. Equipment rack construction and paint.
16. CRT and keyboard operation.
17. CRT color, resolution and convergence.
18. Bad keys in keyboard.
19. Hardwired instrumentation, such as pushbuttons, annunciators, and trend recorders for proper installation.
20. Operation of peripheral devices, such as printers, disc units, tape drives and card readers.
21. Cable assemblies between equipment racks and peripherals. Look for loose connections, bent/broken pins, poor construction or assembly, and improper labeling.

Software Configuration

The data base, loop configuration, and displays should be checked against the data entry forms to ensure that the software configuration has been entered into the system correctly. Point tag numbers, hardware addresses, input conditioning, control algorithms, loop configuration, point descriptions, and display formats should be checked. Reports should be examined for proper descriptions and format. The software configuration can be checked conveniently by dumping a copy of all point displays, configuration data and formatted reports on the line printer. The print-out can then be placed next to the configuration forms to facilitate easy verification.

System Functional Test

Every component in the system should be exercised during the customer acceptance test. Each input and output should be tested to check wiring, point addressing, front-end hardware and shared components. Analog inputs are simulated with a resistor box or potentiometer, and a milliammeter, if transmitter power is supplied from the instrument racks. Otherwise, a 24 VDC power source is also required. Analog outputs are checked with the milliammeter. On large systems, it is often expedient to use ganged potentiometers to simplify the test procedures.

Digital inputs and outputs can be simulated several ways. Contact inputs can be shorted with test leads and alligator clips. An ohmmeter is used to test contact output operation. Programmable controller systems having a large number of digital inputs and outputs require a test panel with toggle switches and lights to simulate field inputs and outputs. For systems performing complex sequencing, the lights and switches can be mounted in a pegboard, with a copy of the process flowsheets pasted over the board. Lights and switches are set near valve and pump symbols to facilitate verification of the controller sequence and interlocks.

All inputs and outputs should be checked. The displays associated with each point should be called up. Point scaling, signal conditioning and controller output action should be checked. A/D calibration should be tested for each mux unit or controller by ramping the input signal and monitoring the display for proper indication of the process variable. Thermocouple linearization and cold junction compensation should be checked for each mux unit, using a precision millivolt source and thermocouple voltage tables. Wiring and point addressing can be checked on thermocouple inputs by shorting the terminals and reading ambient temperature on the corresponding display.

Backup systems should be tested by failing the primary components and observing the system's response. This is accomplished by powering down the device, and in some cases, removing a circuit board to simulate a specific type of failure. System diagnostics are tested by simulating failures and observing what alarms are initiated and what diagnostic information is presented on the CRT console. Redundant data highways should be tested by disconnecting one cable and then the other, and observing the system's response. The data highway card for each controller should be removed to test the system's response to failures in the data highway electronics.

A system availability test period should be scheduled for large computer-based systems. Test duration can vary from a week to several months, and is determined by whether the system provides critical functions for plant operation, system location (remote, overseas or local) and customer preferences. Dual computer systems should be tested thoroughly to ensure that the fail-over system functions properly and that no data is lost. At least 50% of the inputs should be loaded and adjusted on a periodic basis so that input data is available for processing, display, and verification. The availability test period should be revised backwards whenever the system fails a key test. Given a major failure and loss of data, the test period should be set back from several days to several weeks, depending on the severity and duration of the failure. As with other system requirements, lengthy availability tests should be identified during bid evaluation and incorporated in the vendor's pricing.

Correcting Deficiencies

Failed components, defects in workmanship, faulty design, software bugs and errors in configuration should be corrected before signing off the system. Repairs should be made on the spot and added to the inspection checklist for approval. The checklist for complex customized systems may go through several revisions during the acceptance test. All items on the list should be inspected and approved once the repairs or corrections have been made. If the corrections are extensive enough, a return trip may be required to ensure that all work has been completed. Final payment on the system is normally made after the system is approved. This should provide incentive for the vendor to correct deficiencies expeditiously. There may be some changes, however, which the vendor will insist are out of scope. Normally these changes can be negotiated on the spot and a change order issued. Major changes should be avoided, as they could hold up system delivery and incur large cost overruns.

Shipping Arrangements

The project engineer should make the final shipping arrangements with the vendor before concluding the acceptance test. Packing, crating, storage, and method of shipment should be finalized. Also, assistance that the vendor will provide during system installation and start-up should be specified. The project engineer should confirm spare parts schedules to ensure that delivery is in accordance with the final delivery date for the computer system.

Installation and Commissioning

The digital control system is shipped to the plant site after customer acceptance testing has been completed. The control building should be completed to the point where the computer room is enclosed, air conditioned and dust free. If the project schedule requires an early shipment of the system, then arrangements should be made to provide air-conditioned storage facilities. If possible, the equipment should be vacuum packed before shipment to seal out dust and moisture.

Prior to system installation, provisions should be made for moving the equipment into the computer room. Floor plans for large systems should be laid out with a large center aisle and double doors on the end of the room, to facilitate moving and placing the equipment racks. If the computer room has a raised floor, then field wiring, power cables, grounding conductors and cord sets should be placed in position before installing the floor tiles. It is a good practice to have the vendor conduct a pre-installation inspection of the site to ensure that all necessary preparations have been made.

In most cases, the vendor will provide the necessary personnel to supervise and assist in the installation of the system. The contracting firm is responsible for making field wiring, power, and ground connections to the computer system. Each step in the installation should be checked and a final inspection made to ensure that all connections have been made properly. Each component is powered up, and test routines are performed to determine that it is functioning properly. The data base is then loaded into the system via cassette tapes, floppy disc or disc pack. The vendor performs a system availability test to demonstrate to the customer that the computer system is fully operational and running.

Once the vendor has turned the system over to the user, a loop check is performed for each point. Copies of the process flowsheets, instrument loop drawings and the software configuration forms should be readily available. All field instrumentation should be wired to the system's input/output racks, and the field wiring tested for shorts, ground faults or open circuits. Each analog and digital input should be exercised and the corresponding displays checked to verify the A/D calibration and the overall operation of the system.

The computer system is normally commissioned one loop at a time during plant start-up. At this time, the computer system has been checked out, powered up and the data base loaded. As each loop is brought into service, the associated computer points are brought on-line, and the loop is put on manual control. If the system functions properly, the system is placed on PID control, and the loop is tuned. For supervisory control systems, the analog controllers are brought on-line first, and then the computer points are commissioned one loop at a time, in a data acquisition mode only. With distributed control systems, the microprocessor controllers are brought on-line one loop at a time. If the distributed system incorporates a process computer, then the associated computer points are also commissioned simultaneously with each loop.

Normally, data base changes are required as each loop is brought on-line. Changes in tuning constants, alarm limits, control algorithms, or loop configuration are often required. After the necessary corrections have been made, the loop is placed on computer control. If the system is to perform direct digital control of the process, then the computer control point should be turned on. If the computer system is to perform a supervisory role, then the analog controller associated with the loop is switched to computer mode so that setpoints and outputs can be manipulated from the system's operator console.

Once the computer system is on-line and the data base is operating properly, other system functions are commissioned, such as logs, reports, material balances, and custom programs. After the system stabilizes, and operator and plant management interfaces have been established, then process optimization routines are implemented and brought on-line. As new applications for the computer are found and future expansions planned, the system's hardware and software configurations will be in a constant state of change throughout its useful life span.

Conclusion

The successful installation of a digital control system requires careful planning at all stages of project execution. A systems approach to the design, specification, vendor evaluation, engineering, software development, testing, and commissioning of the control system is essential. Communication channels must be established between the vendor, client, project group, and the affected design disciplines. Manpower requirements should be identified, and personnel with the necessary expertise assigned early to the project. Tight control of the system design and expenditures should be maintained to minimize cost overruns and schedule slippages. A coordinated effort in all of these areas will result in the smooth and timely execution of a digital or computer control system project.

Index